Graeme

D1685593

CORROSION CRACKING

*Proceedings of the Corrosion Cracking Program
and Related Papers presented at the
International Conference and Exposition on
Fatigue, Corrosion Cracking, Fracture Mechanics
and Failure Analysis*

2–6 December 1985
Salt Lake City, Utah, USA

Edited by
V. S. Goel

Organized by
 American Society for Metals
in cooperation with:

American Society for Testing & Materials	Japanese Society for Strength and Fracture of Materials
American Society of Mechanical Engineers	Japan Light Metal Association
American Welding Society	Swiss Association for Material Testing
American Society of Naval Engineers	Institution of Engineers of Ireland
The Metallurgical Society of AIME	Danish Aluminum Council
National Association of Corrosion Engineers	Verein Deutscher Eisenheuttenleute
Institute of Corrosion Science and Technology	Society of Automotive Engineers
Society for Experimental Mechanics	Society of Naval Architects and Marine Engineers
National Society of Professional Engineers	American Society of Agricultural Engineers

International Association for
Bridge & Structural
Engineers

Library of Congress Catalog Card Number: 86-071193
ISBN: 0-87170-272-X
SAN: 204-7586

Printed in the United States of America

SESSION CHAIRMEN

Takeo Yokobori
Society for Strength & Fracture of Materials
Japan

E. Neville Pugh
National Bureau of Standards
USA

H. J. Westwood
Ontario Hydro Research Division
Canada

Christina Berger
Kraftwerk Union
West Germany

James A. Smith
Naval Research Laboratory
USA

Rabvindranath Bandy
Brookhaven National Laboratory
USA

Michinori Takano
Tohoku University
Japan

J. S. Snodgrass
Reynolds Metal Company
USA

David Matlock
Colorado School of Mines
USA

N. Henry Holroyd
Alcan International Ltd.
United Kingdom

Masanori Kawahara
Nippon Kokan, K.K.
Japan

W. L. Server
EG&G Idaho, Inc.
USA

Yoshito Fijiwara
Nippon Yakin Kogyo
Japan

D. T. Read
National Bureau of Standards
USA

Shane Findlan
J. A. Jones Research Company
USA

V. Charyulu
University of Idaho
USA

Marcia Domack
NASA Langley Research Center
USA

M. Pevar
General Electric
USA

Sten Johansson
Linkoping Institute of Technology
Sweden

FOREWORD

This volume contains part of the total number of papers presented at the "International Conference on Fatigue, Corrosion Cracking, Fracture Mechanics and Failure Analysis," held in Salt Lake City, Utah, USA, from 2-6 December 1985. Response to this conference was so good that it resulted in a large number of papers. To satisfy the needs of different interest groups and to keep the proceedings of the conference in a manageable form, it was decided to publish it in four separate volumes:

Analyzing Failures: The Problems and The Solutions
Corrosion Cracking
The Mechanism of Fracture
Fatigue Life: Analysis and Prediction

The above paper collection volumes may be obtained from the American Society for Metals. This conference covered a wide range of topics, some of fundamental interest and some of application interest. To facilitate an early publication, the editing has been kept to a minimum. We hope the the technical merits of the papers outweigh any grammatical or minor stylistic deficiencies.

The advances in the concepts of design are pushing the operational limits of engineering materials and so maximum performance is expected out of the materials. Due to the general economic crunch, almost everyone wants the maximum life out of their equipment. The electric utilities want their plants to run more than the designed plant life (mostly 40 years), aircraft companies want their planes to fly longer, the transportation industry wants that its bridges last indefinitely, and the chemical industry wants their plants to keep on producing products. There is also an increased awareness on the part of the public for safety and reliability of components, because failure of components in large aircraft, nuclear plants or other large structures can lead to large-scale disasters like the Bhopal tragedy in India, the Three Mile Island accident in the USA and the string of airline disasters in 1985.

All of this shows that today materials are expected to show maximum performance, provide long life for maximum economy and at the same time ensure safety and reliability of components and systems. For all this, we need to understand the materials better and apply the principles of fracture mechanics, corrosion and fatigue to the solution of practical problems. This conference was planned to provide a forum for the exchange of ideas and allow a better understanding of the theory and applications of the materials science which can ensure safety in combination with the expected life and performance goals for materials.

The theme of this conference was "Technology Transfer" among the various groups who apply theory to the application of practical problems. There are many specialized meetings in this area which permit workers to come together and discuss problems in their specific application areas. However, there is no single meeting or conference which brings together workers in the various application areas such as Aerospace structures, Army-Navy Applications, Bridges and Architectural Structures, Transportation Industry and Nuclear Industry to learn what is being done in other areas which they may be able to utilize to their advantage. This conference was aimed at bringing together workers from different applications areas to give them a wider perspective. Hence, this conference was of interest to engineers, metallurgists and also to the engineering managers who remain concerned about product failure and liability.

The success of this conference was based on the contributions of the speakers, session chairmen and members of the Technical Review Committee and the Organizing Committee who generously supported this Conference. I would like to thank all the participants on behalf of the American Society for Metals and the co-sponsoring societies for their generous contribution of time and effort towards the success of this Conference.

Dr. V. S. Goel
Chairman, Organizing Committee

TABLE OF CONTENTS

Engineering Applications

Radiation Embrittlement

Test Techniques

EFFECTS OF FREQUENCY, STRESS RISING TIME AND STRESS HOLD TIME ON CORROSION FATIGUE CRACK GROWTH BEHAVIOUR OF LOW ALLOY Cr-Mo STEEL

A. Toshimitsu Yokobori, Jr., Takeo Yokobori Tomokazu Kosumi
Department of Mechanical Engineering II
Tohoku University
Sendai, Japan

Nobuo Takasu
Nippon Kokan KK
Kawasaki, Japan

Effects of Frequency, Stress Rising Time
and Stress Hold Time on Corrosion Fatigue
Crack Growth Behaviour of Low Alloy Cr-Mo
Steel

By

A. Toshimitsu Yokobori, Jr*., Takeo Yokobori**,
Tomokazu Kosumi***, and Nobuo Takasu[†]

ABSTRACT

For da/dN under 3.5% NaCl solution in low alloy Cr-Mo steel, the experiment were carried out, stress frequency f being controled by stress rise time t_R and stress hold time t_H as well, in order to clear whether da/dN depends on cycle only, or on time only or on both cycle and time. From the results and the analysis, a new experimental formula was obtained for da/dN in terms of f, t_R or t_H non-linearly. A line of considerations as stochastic process is proposed for the significance of the equation of this type, and as a special case da/dN and da/dt in corrosion fatigue (t_H=0) are characterized in terms of frequency dependence.

INTRODUCTION

There have been many literatures[1)-6)] on the effect of stress rise time t_R, stress hold time t_H and frequency f on crack propagation rate under corrosive condition. It, however, appears to have been no systematic studies at the point of that each of these factors can be chosen independently and thus formulated. The present study is one for that purpose and a continuation of our studies[7)8)] on the problems. Frequency being controled by t_H and t_R as well, the experiments were carried out and based on the analysis of the results, a new experimental formula for da/dN under corrosive condition is proposed in non-linear terms of frequency f, stress rise time t_R and stress hold time t_H. A line of consideration as stochastic process is proposed for the significance of the equation of this type, and as a special case da/dN and da/dt in corrosion fatigue (t_H=0) are characterized in terms of frequency dependence.

MATERIALS AND SPECIMEN USED

The specimen used is low alloy Cr-Mo steel, and the chemical composition is shown in TABLE 1. It was machined as shown in Fig. 1, and then oil tempered after heating for 10min. at 850°C in salt bath, and water quenched after heating for 60min. at 520°C. Static mechanical properties is shown in TABLE 2. After heat treatments, the slit of 0.3mm width and 2mm depth was provided by crystal cut at one side of the specimen. The specimen surface was grounded by #6~2,000 emery paper, and then polished by buffing with cromium oxides.

TESTING APPARATUS AND METHOD

In the present testing machine, stress rising time t_R and stress hold time t_H can be selected independently, that is, the stress frequency f can be controled by t_H and t_R as well. The testing machine of cantilever type was used. Examples of the stress wave realized by this machine are shown in Fig. 2. A sine wave was used for fatigue, and therefore, the stress gradient with respect to time equals to zero at the instant when maximum tensile stress σ_{max} is reached, and, thus, the shock effect can be avoided. 3.5% NaCl water as soluted in distillated one was used. Test specimen temperature and pH was kept as 18.0 ± 3.0°C and 6.3 ±

*   Associate Professor, Department of Mechanical Engineering II, Tohoku University, Aboba, Aramaki, Sendai, Japan.
**   Professor Emeritus of Mechanical Engineering, Aoba, Aramaki, Sendai, Japan.
*** Graduate, Department of Mechanical Engineering II, Tohoku University.
†   Nippon Kokan KK, Kawasaki, Japan.

1

TABLE 1. Chemical Composition (wt%)

C	Si	Mn	P	S	Cr	Mo
0.38	0.34	0.84	0.018	0.004	1.07	0.17

TABLE 2. Static Mechanical Properties

NPa		%	
Yield Stress	Ultimate tensile strength	Elongation	Reduction of area
1140	1187	11	37

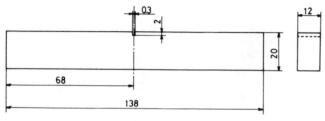

Fig. 1. Specimen shape and dimension for stress corrosion cracking and corrosion fatigue. (in mm)

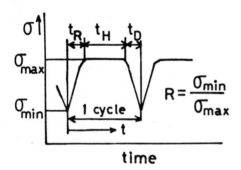

Fig. 2. Stress wave used.

0.2, throughout testing, respectively. The crack length was continuously measured by the microscope with the magnification of twenty times. The stress intensity factor was calculated by the following equation:[9]

$$K = \frac{6M\sqrt{a}}{BW^2} Y \qquad (1)$$

where

$$Y = \left\{ 1.99 - 2.47\left(\frac{a}{W}\right) + 12.97\left(\frac{a}{W}\right)^2 - 23.17\left(\frac{a}{W}\right)^3 \right.$$
$$\left. + 24.8\left(\frac{a}{W}\right)^4 \right\}$$

and $\frac{a}{W} \leq 0.6$. M=bending moment, W=specimen width, B=specimen thickness, a=crack length.

TESTING CONDITIONS

The two series of tests were carried out.

TABLE 3. Testing Conditions

The value of K_{Ii}	Stress Rsyio	Ridr Time (s)	Hold Time (s)	Frequency (HZ)
$K_{Ii} > K_{ISCC}$ $K_{Ii} = 38.6$ K_{ISCC} $= 36.4$ (MPam$^{1/2}$)	0.16	4.5 (8)	0	1.11×10^{-1}
			11.0	5.00×10^{-2}
			40.2	2.00×10^{-2}
			77.4	1.15×10^{-2}
			186.0	5.13×10^{-3}
		9.75	0	5.13×10^{-2}
			11.0	3.28×10^{-2}
			40.2	1.68×10^{-2}
			77.4	1.03×10^{-2}
			186.0	4.87×10^{-3}
		15.0 (8)	0	3.33×10^{-2}
			11.0	2.44×10^{-2}
			40.2	1.42×10^{-2}
			186.0	4.63×10^{-3}
$K_{Ii} < K_{ISCC}$ $K_{Ii} = 19.9$ (MPam$^{1/2}$)	0.31	0.81 (7)	0	6.17×10^{-1}
			1.8	2.92×10^{-1}
			11.0	7.92×10^{-2}
			40.2	2.39×10^{-2}
		4.5	0	1.11×10^{-1}
			1.8	9.26×10^{-2}
			11.0	5.00×10^{-2}
			40.2	2.03×10^{-2}

That is, one series is for the condition that the value of K_{max} of the initial stress intensity factor at starting stage of the crack propagation, K_{Ii} is above K_{ISCC}, that is, $K_{max} > K_{Iscc}$ throughout the crack propagation. Another series is for the condition that the value of K_{Ii} is less than K_{ISCC} and K_{max} increases from below the value of K_{ISCC} and then becomes larger than K_{Iscc} as the crack extends. For each series of the tests, t_R and t_H are changed respectively as shown in TABLE 3. In the present article, stress decreasing time t_D was kept equal to t_R, as it is reported[1)-3)] that the effect of stress decreasing rate is smaller as compared with that of stress increasing rate. The range of $t_H = 0 \sim \infty$, and the range of t_R is from 0.81 sec to 15.0 sec, respectively. The stress ratio $R = \sigma_{min}/\sigma_{max}$ was shown in TABLE 3, where σ_{max}=maximum stress in cycle, and σ_{min}=minimum stress in cycle.

EXPERIMENTAL RESULTS, ANALYSIS AND DISCUSSION

EFFECTS OF STRESS RISE TIME t_R AND STRESS HOLD TIME t_H ON CRACK PROPAGATION RATE UNDER THE CONDITION OF $K_{Ii} > K_{Iscc}$.——Fig.

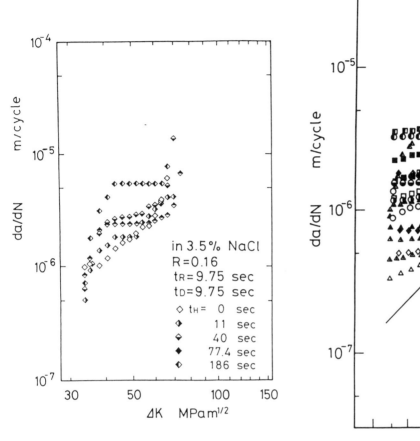

Fig. 3. da/dN versus ΔK with t_H =0~186 sec and t_H=9.75 sec. $K_{Ii} > K_{Iscc}$.

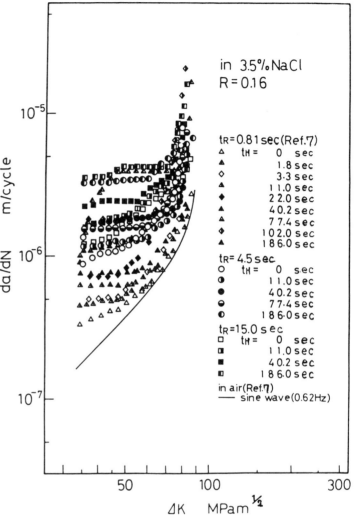

Fig. 4. da/dN versus ΔK with parameters, t_H= 0~186 sec and t_R=0.81~15.0 sec. $K_{Ii} > K_{Iscc}$.

3 shows the effect of stress hold time t_H on the plot of the logarithm of crack propagation rate per cycle, da/dN against the logarithm of stress range intensity factor ΔK for stress rise time t_R=9.75 sec, as an example. With respect to the results of other series of tests, in Fig. 4 log da/dN is plotted against log ΔK with both t_H and t_R as parameters. From Figs. 3 and 4, it can be seen that da/dN and the width of the plateau of da/dN independent of ΔK, increase with increase of t_H. From Figs. 3 and 4, it can be seen that t_H at which the plateau appears becomes shorter with decrease of t_R, and da/dN becomes nearly independent of t_R when t_H is longer. Fig. 4 also shows that the effect of t_R is large when t_H is short.

In Fig. 5 the value of log da/dN at ΔK=46 MPa $m^{\frac{1}{2}}$ in the plateau of log da/dN versus log ΔK

diagram as shown in Figs. 3 and 4 was plotted against stress frequency f as changed by t_R and t_H, where f is denoted as follows:

$$f = \frac{1}{2t_R + t_H} \quad . \qquad (2)$$

Each solid line in Fig. 5 corresponds to the case where t_R is changed with t_H kept constant, and each dashed line corresponds to the case where t_H is changed with t_R kept constant. For t_H=0, the plateau does not appear. The value of $(da/dN)_{scc}=(1/f)(da/dt)_{plat}$ is shown by dash-dot line in Fig. 5 as the crack growth rate in the stress corrosion cracking where $(da/dt)_{plat}$ is the crack growth rate at the plateau in log (da/dt) versus log ΔK diagram.

With respect to the effect of t_R for specific value of t_H, da/dN shows considerably time dependent behaviour, increasing with

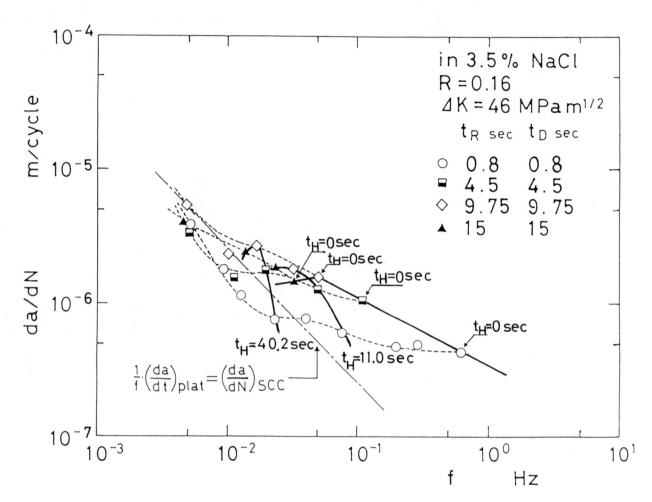

Fig. 5. The relation between da/aN and frequency f as affected by t_H and t_R. $K_{Ii} > K_{Iscc}$.

increase of t_R. With further increase of t_R, da/dN reaches the maximum, say, at t_R=9.75 sec and then tends to decrease. This characteristics may correspond to unstable equilibrium process due to the balance between the accelerating effect by anodic dissolution and deceleration effect by the formation of a non-conductive surface film as t_R increased.[2]

On the other hand, with respect to the effect of t_H for specific value of t_R, in the range of shorter t_H, that is, in higher frequency region, da/dN is not so much affected by t_H or frequency, although it increases slightly with increase of t_H. From this characteristics it is inferred that da/dN is not only time dependent, but also strongly cycle dependent. However, when t_H increases furthermore, then da/dN accelerates remarkably, and t_R does not affect da/dN so much for this region. That is, da/dN approaches to have only time dependent characteristics.

From the results and the analysis, it can be concluded that time effect by stress wave in higher frequency range is not governed by overall effect, that is, time integral of the stress wave per cycle. The stress wave shape per cycle consists of the two different stages* in series, that is, stress rising stage and the subsequent stage under constant stress. However, t_R has the both time effect as cyclic effect, that is, as included in time gradient of the value of stress during stress rising stage and also the time effect due to a different effect at this stage. On the other hand, t_H has only the time effect due to the different effect pertaining to the latter effect of t_R stage.

EFFECTS OF STRESS RISE TIME t_R AND STRESS HOLD TIME t_H ON CRACK PROPAGATION RATE UNDER THE CONDITION OF $K_{Ii} < K_{Iscc}$.——Fig. 6 shows the results of the effect of stress hold time t_H on the plot of the logarithm of da/dN against the logarithm of stress range intensity factor ΔK for stress rise time t_R=0.81 and 4.5 sec,

---

* Stress decreasing stage may also concern time effect, but in the present paper it was not discussed as described in the section: TESTING CONDITIONS.

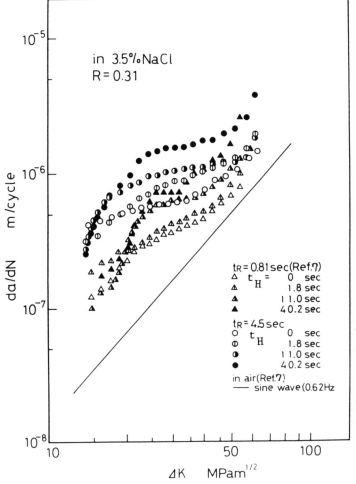

Fig. 6. da/dN versus ΔK with parameters, $t_H = 0 \sim 40.2$ sec and $t_R = 0.81$ and 4.5 sec, respectively. $K_{Ii} < K_{Iscc}$.

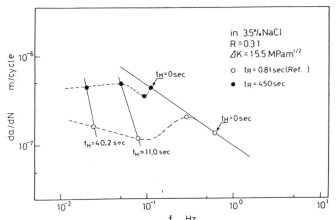

Fig. 8. The relation between da/dN and frequency f as affected by t_H and t_R. $K_{Ii} < K_{Iscc}$.

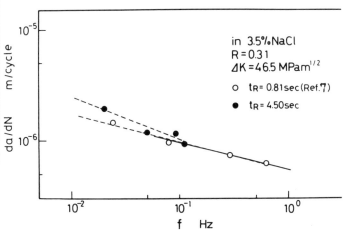

Fig. 9. The relation between da/dN and frequency f as affected by t_H and t_R. $K_{Ii} < K_{Iscc}$.

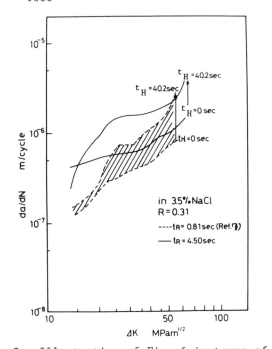

Fig. 7. Illustration of Fig. 6 in terms of band.

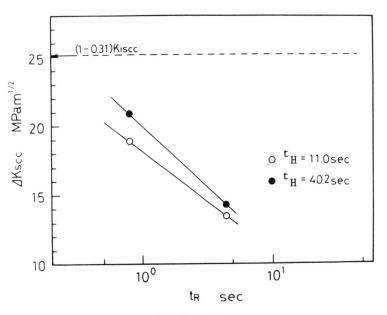

Fig. 10(a). ΔK_{scc} versus t_R.

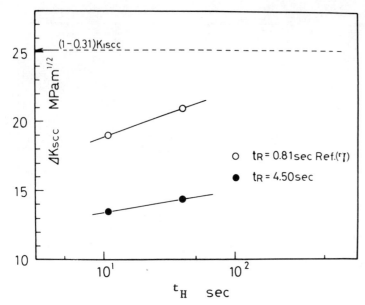

Fig. 10(b) ΔK_{scc} versus t_H.

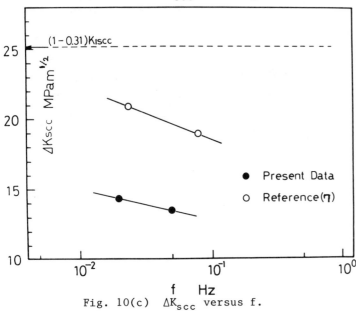

Fig. 10(c) ΔK_{scc} versus f.

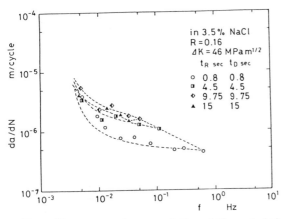

Fig. 11. The comparison of Eqs.(3) and (4) with the experimental data.

respectively. The results are shown schematically in Fig. 7, which show the characteristics more clearly. From Figs. 6 and 7 it can be seen da/dN is considerably accelerated with increase of t_R in smaller range of ΔK, whereas in higher range of ΔK, da/dN is not so much affected by t_R. Figs. 8 and 9 show the effect of t_R and t_H on da/dN for $\Delta K=15.5$ MPa $m^{\frac{1}{2}}$ and 46.5 MPa $m^{\frac{1}{2}}$, respectively. In Figs. 8 and 9 solid line shows for $t_H=0$, and dashed line shows the effect of t_H on da/dN with t_R as parameter, $t_R=0.81$ sec and 4.5 sec, respectively.

For lower value of ΔK, say, $\Delta K=15.5$ MPa $m^{\frac{1}{2}}$ the following can be seen from Fig. 8: For specific value of t_H, with increase of t_R, that is, with decrease of frequency, da/dN is accelerated and shows strongly time dependent characteristics. On the other hand, for specific value of t_R, da/dN is not so much affected by t_H and shows not only time dependent, but also cyclic dependent characteristics. These characteristics is quite similar for the value in plateau region under the condition of $\Delta K_{Ii} > K_{Iscc}$ described in the previous section.

On the other hand, when the crack extends furthermore and ΔK becomes larger, say, $\Delta K=46.5$ MPa $m^{\frac{1}{2}}$, the effect of t_R and t_H on da/dN is expressed in terms of only the linear summation of t_R and t_H, that is, the overall frequency. This means da/dN is controled by only the overall time integral of the stress wave, that is, the rate-controling process in the stress rising stage and the subsequent stage under constant stress may be the same physical process. The characteristics appears to correspond to that in lower frequecny range for the case $K_{Ii} > K_{Iscc}$ shown in Fig. 5.

It can be seen from Fig. 6 that steeply increasing region is observed on log da/dN curve in the low crack growth rate range of da/dN=10^{-7}~10^{-6} m/cycle for $t_H > 11.0$ sec. Let the value of ΔK at this stage be denoted as ΔK_{scc}, then ΔK_{scc} is less than $(1-R) K_{Iscc}$, where R is stress ratio. Fig. 10(a), (b) and (c) shows ΔK_{scc} versus t_R, t_H and f=1/(2t_R+t_H), respectively. From Fig. 10, we can see that ΔK_{scc} decreases steeply with increase of t_R, and gradualy decreases with decrease of t_H and with increase of frequency.

PROPOSAL OF A NEW EXPERIMENTAL FORMULA FOR da/dN IN TERMS OF f, T_R AND T_H

In the previous paper[8] the experimental formula of da/dN at the plateau in log da/dN versus log ΔK diagram in terms of t_R and t_H has been proposed. However, as found in the present paper, da/dN shows the t_R-characteristics having a maximum at $t_R \simeq 9.75$ sec. Thus, one modified attempt is to use the two equations for 9.75 sec $\geq t_R > 0$ and $t_R \geq 9.75$ sec, respectively as follows:

$$\left[\frac{da}{dN}\right]_{t_H} = \frac{4.99 \times 10^{-4} \; t_R^{0.508}}{1 - 5.64 \times 10^{-2} \; t_H^{0.53}} \qquad (3)$$

$$= [2.00 \times 10^3 \, t_R^{-0.508} - 1.13 \times 10^2 (t_H/t_R)^{0.53} t_R^{-0.508}]^{-1}$$

$9.75 \text{ sec} \geqq t_R > 0$, $186 \text{ sec} \geqq t_H \geqq 0$ and

$$\left[\frac{da}{dN}\right]_{t_H} = \frac{2.25 \times 10^{-3} \, t_R - 0.150}{1 - 5.11 \times 10^{-2} \, t_H^{0.493}} \qquad (4)$$

$$= [4.44 \times 10^2 \, t_R^{0.150} - 2.27 \times 10 (t_H/t_R)^{0.493} t_R^{0.63}]^{-1}$$

$t_R \geqq 9.75 \text{ sec}$, $186 \text{ sec} \geqq t_H \geqq 0$. For comparison with experimental data, the calculated curves by Eqs.(3) and (4) are shown by dotted line in Fig. 11. The results are good in accordance with the data. However, Eqs.(3) and (4) have not the term of $2t_R + t_H$, or the frequency, $f(= 1/(2t_R + t_H))$, and, also does not reduces to the equation of the type $\frac{da}{dN} \propto \frac{1}{f}$ corresponding to $t_H \to \infty$.

Therefore, in the present paper we proposed the following formula of da/dN which has also the term of the frequency, f and, also, satisfies both condition at $t_H = 0$ and $t_H \to \infty$, in terms of t_R and t_H. The equation is:

$$\frac{da}{dN} = A\left(\frac{2t_R + t_H}{t_o}\right) \log_{10}\left(C - \alpha \cdot \frac{2t_R}{2t_R + t_H}\right) \qquad (5)$$

,where to, A, C, and α is possitive, and A is a function of stress. C, $\alpha > 1$ and $(2t_R + t_H)/t_o > 1$. Eq.(5) is at least, qualitatively in good agreement with the experimental characteristics shown in Fig. 8. Furthermore, if we take C=10 and $\alpha \simeq 6.5$, then Eq.(5) satisfies both the conditions at $t_H/t_R \gg 1$ including $t_H \to \infty$ and at $t_H/t_R \ll 1$ including $t_H = 0$. That is, Eq.(5) reduces to

$$\frac{da}{dN} = A\left(\frac{1}{ft_o}\right) \qquad (6)$$

for $t_H \to \infty$. Also, Eq.(5) reduces to

$$\frac{da}{dN} = A\left(\frac{1}{ft_o}\right)^{0.544} \qquad (7)$$

for $t_H = 0$, which is exactly in accord with the experimental data (corrosion fatigue) plotted in Fig. 5. Furthermore, it can be shown that Eq.(5) has a minimum for certain value of t_H in t_H-characteristics. An example calculated by Eq.(5) assuming $A = 10^{-4}$ m/sec and to$= 10^2$ sec, is shown in Fig. 12. The curve has the characteristics well in accord with the experimentals as shown, for instance, in Figs. 5 and 8.

THEORETICAL CONSIDERATIONS ON THE CAUSE OF DEPENDENCE OF da/dN ON BOTH T_R AND FREQUENCY AND ON BOTH CYCLE AND TIME.

Next, let us consider theoretically why the time effect consists of the frequency and t_R or t_H, independently for the fracture under

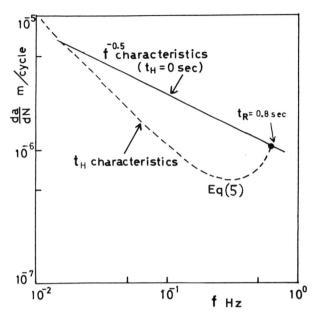

Fig. 12. A characteristics of the curve calculated by Eq.(5).

corrosive environment. The fracture of this kind can be treated as a stochastic process. A transition probability per unit time, μ may be in general written as:

$$\mu = f(\sigma, t). \qquad (8)$$

For the case the stress wave as shown in Fig. 13, σ in Eq.(8) is expressed as:

$$\sigma = \dot{\sigma} t \simeq \frac{\sigma_a}{t_R} t \qquad \text{for } t < t_R,$$

and $\qquad (9)$

$$\sigma = \sigma_a \qquad \text{for } t \geqq t_R.$$

where $\dot{\sigma}$ = applied stress application rate. σ_a is taken as $\sigma_{max} - \sigma_{min}$ for convenience. The transition probability per unit cycle, μ_o is related with μ as [10)]

$$\mu_o = \mu/f \qquad (10)$$

where f = stress frequency.

Here we consider the stages I and II are differnt mechanicaly and physicaly as far as time effect is concerned. That is, in the stage I, the time effect included in σ in Eq.(8) is involved in addition to the term t in Eq.(8). On the other hand, in the stage II, since σ in Eq.(8) is kept constant, only the term t in Eq.(8) has the effect. That is, the time effect in the stage I includes the two different effects including the one in the stage II. This feature is shown in Fig. 13 by double hatched lines. Therefore, μ_o is in general written as:

$$\mu_o = F_o (\sigma, t_R, 2t_R + t_H) . \qquad (11)$$

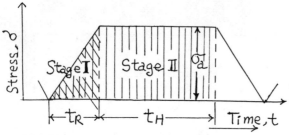

Time effect 1 involved in Stage 1.
Time effect 2 involved in both Stages 1 and 2

Fig. 13. Schematic illustration of how the stages I and II are different with respect to time effect.

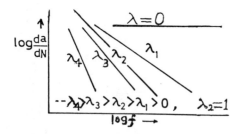

Fig. 14. Schematic illustration of log da/aN verssus log f. For $\lambda=0$, dependent of cycle only. For $\lambda=1$, dependent of time only. For $1 > \lambda > 0$ and $\lambda > 1$, dependent of both cycle and time.

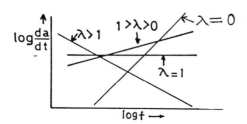

Fig. 15. Schmatic illustration of log da/dt versus log f.

da/dN may simply be regarded to have linear relation[11] to transition probability μ_0 per unit cycle. Therefore, using Eq.(11) da/dN may be written as:

$$da/dN = AF_0(\sigma, t_R, 2t_R + t_H). \qquad (12)$$

Thus we can see Eq.(5) experimentaly proposed is a special form of the type of Eq.(12).

For the case of fatigue ($t_H=0$), as can be seen from Eq.(7), the following equation is obtained in general:

$$\frac{da}{dN} = \frac{A}{f^\lambda} \qquad . \qquad (13)$$

And, since da/dt = f da/dN, then we get:

$$\frac{da}{dt} = \frac{A}{f^{\lambda-1}} \qquad . \qquad (14)$$

log da/dN versus log f and log da/dt versus log f are schematicaly shown in Figs. 14 and 15, respectively. log da/dt versus log f characteristics was shown previously[8] and interpreted in terms of the transition of the mechanism with decrease of frequency in the light of Fig. 15.

From Eqs.(13) and (14) it can be seen that da/dN or da/dt is dependent of cycle only for $\lambda=0$, and dependent of time only for $\lambda=1$, and dependent of both cycle and time for $1 > \lambda > 0$ or $\lambda > 1$.

The line of considerations shown herein is general, and, therefore, the principle may be applied to the crack growth rate at high temperatures and other similar so-called time-dependent problems.

CONCLUSIONS

From the experiments and analysis on the crack propagation in 3.5 per cent NaCl solution of low alloy Cr-Mo steel under the stress wave with hold time, the following conclusions are obtained:

I.   On the effects of stress rise time t_R and stress hold time t_H on crack propagation rate under the condition of $K_{Ii} > K_{Iscc}$:

(1)   With respect to the effect of t_R for specific value of t_H, da/dN shows considerably time dependent behaviour, increasing with increase of t_R. With further increase of t_R, da/dN reaches the maximum and then tends to decrease.

(2)   With respect to the effect of t_H for specific value of t_R, in the range of shorter t_H, da/dN is not so much affected by t_H or frequency, although it increases slightly with increase of t_H. From this characteristics it is inferred that da/dN is not only time dependent, but also strongly cycle dependent, that is, the effect of t_R and t_H on da/dN is non-linear.

(3)   However, when t_H increases furthermore, then da/dN accelerates remarkably, and t_R does not affect da/dN so much. That is, da/dN approaches to have only time dependent characteristics.

(4)   In higher frequency range, time effect by stress wave is not governed by the overall effect, but da/dN is dependent of both cycle and time. In lower frequency range, da/dN approaches to have only time dependent characteristics.

II.   On the effect of stress rise time t_R and stress hold time t_H on crack propagation rate under the condition of $K_{Ii} < K_{Iscc}$:

(1)   In smaller range of ΔK, the characteristics of da/dN with respect to t_R and t_H is quite similar for the case $K_{Ii} > K_{Iscc}$.

(2)   In the higher range of ΔK, the effect of t_R and t_H on da/dN is expressed in terms of

only the linear summation t_R and t_H, that is, the overall frequency. The characteristics appears to correspond to that in lower frequency range for the case $K_{Ii} > K_{Iscc}$.

(3) The effect of t_I and t_H on ΔK_{scc} is also non-linear. ΔK_{scc} decreases steeply with increase of t_R, and graduary decreases with decrease of t_H and with increase of frequency.

III. A new experimental formula of crack propagation rate under corrosive condition is proposed in non-linear terms of both frequency f and stress rising time t_R or stress hold time t_H. This formula satisfies the case for $t_H=0$ and for $t_H \to \propto$ as well as any combination of t_R and t_H.

IV. From a line of consideration as stochastic process, it is explained why the time effect consists of, in general, both frequency f and t_R or t_H, independently, for fracture under corrosive environment. In corrosin fatigue ($t_H=0$), as a special case, da/dN and da/dt are characterized in terms of frequency dependence. The principle is general one, and, therefore, it may be applied to the crack growth rate at high temperatuires and other similar so-called time-dependent problems.

REFERENCES

1. Endo, K. and K. Komai, J. Soc. Mat. Sci., Japan 14, 827 (1965) (In Japanese)
2. Borsom, J. M., Proc. Int. Conf. Corrosion Fatigue, NACE-2, 424 (1972)
3. Seilines, R. J. and R. M. Pellowx, Met. Trans. 3, 2525 (1972)
4. Dawson, D. B. and R. M. Pellowx, Met. Trans. 3, 723 (1974)
5. Kawai, S. and K. J. Koibuchi, Soc. Mat. Sci., Japan, 27, 967 (1978) (In Japanese)
6. Atkinson, J. D. and T. C. Lindley, Metal Sci., 444, (1974)
7. Yokobori, T., H. Kuwano, and H. Takizawa, Trans. Japan Soc. Mech. Engrs. A47, 689-697, (1981) (In Japanese)
8. Yokobori, T., A. T. Yokobori, Jr. and N. Takasu, Proc. 4th Int. Conf. Mech. Behaviour, 2, 967-983, (1983)
9. Gross, B. and J. E. Srawley, NASA Tech. Note, D-3092, (1965)
10. Yokobori, T., J. Phys. Soc. Japan, 10, pp.368-374, (1955)
11. T. Yokobori, "Physics of Strength and Plasticity," p.332, The Orowan Anniversary Volume, MIT Press (1969)

** In Fig.8 the curves for $t_H = 11.0$ sec and 40.2 sec as parameter are shown also by solid lines.

EFFECT OF NIOBIUM ADDITION ON INTERGRANULAR STRESS CORROSION CRACKING RESISTANCE OF Ni-Cr-Fe ALLOY 600

K. Yamauchi, I. Hamada
Y. Sakaguchi, T. Okazaki, T. Yokono
Babcock-Hitachi K.K.
Hiroshima, Japan

Y. Fujiwara, R. Nemoto
Nippon Yakin Kogyo Co., Ltd.
Kawasaki, Japan

ABSTRACT

Intergranular stress corrosion cracking (IGSCC) resistance of Ni-Cr-Fe alloy 600 in 289°C pure water containing 8 ppm O_2 can be improved by suitable heat treatment which promotes healing effect. Such effect, however, is easily degraded by high temperature heat cycles during welding because of dissolution of chromium carbides in heat affected zones (HAZ) of weldments. In this study, the effect of niobium addition on IGSCC resistance of Ni-Cr-Fe alloy 600 and its related alloys, i.e., the stabilization of carbon by niobium, was investigated focusing upon their weldments. To characterize the degree of stabilization by niobium addition, the authors introduced a stabilization parameter, $\bar{N}$, which is related with the number of stabilizing elements per carbon atom of the alloys. We found that the stabilization parameter control could provide the alloys with excellent resistance to IGSCC, and that the critical $\bar{N}$ to be controlled be higher than about 12. The $\bar{N}$ controlled alloys and their weldments, even though followed by post weld heat treatment (PWHT) and/or low temperature sensitization (LTS) heat treatments, were highly resistant to intergranular corrosion (IGC) in a modified ASTM G28 test. They were also fully immune to IGSCC in some accelarated tests under tight crevice in 289°C pure water with 8ppmO_2. The other properties of the alloys such as strength and weldability were compatible with or superior to commercial alloy 600 and its related alloys.

METALLURGICAL CHARACTERISTIC of IGSCC of ordinary aloy 600 in oxygenated high temperature pure water was typically revealed by large heat-to-heat variations in IGSCC susceptibility, which was one of the main results of our study for these ten years.

Although some studies of grain boundary segregation effects of impurities, such as sulpher, phosphorus, boron etc. were reported [1-5], our various test results indicated that the major cause of such variations could be attributed to the degree of sensitization and to the degree of its recovery (healing), which depend on precipitation and growth of chromium carbides along grain boundaries and in grains.

From this viewpoint, we, therefore, could improve IGSCC resistance of ordinary alloy 600 by using some suitable heat treatments, which produced healing states. These heat treatments are, recently, well known as thermal stabilization or thermal treatment for alloy 600 and its related alloys [2, 6-11].

However, one problem to be solved still remains, which is IGSCC susceptibility of weld HAZ of Ni-Cr-Fe alloy 600 family, even though the base metal was in a healing state before welding. Weld HAZ is sensitized by welding with or without further thermal effects (PWHT, LTS etc.). This sensitization results from the dissolution of chromium carbides, which can dissolve during welding at temperatures higher than about 1000°C. The resultant free carbon in the weld HAZ becomes the cause of sensitization. Another ways of improvement, therefore, are to be taken, because the weld process cannot be avoided in any, actual fabrication.

Yamauchi et al. investigated on the improvement of IGSCC resistance of weld metals of Types 82 and 182, which was reported in 1984 [12]. As described in the paper, weld metals themselves have had the same problem of weld sensitization. The conclusion, there, was that the stabilization parameter ($\bar{N}$) introduced by them was useful for improvement of IGSCC resistance of weld metals. In this study, the same parameter $\bar{N}$ was applied to produce SCC resistant materials in weld HAZ of alloy 600 and its related alloys.

The purposes of this study are first to examine the effects of niobium, titanium, carbon and nitrogen on IGSCC of weld HAZ of alloy 600, and then to develop and evaluate SCC resistant, new alloy 600 as structural material with enough strength and weldability. The test materials of the alloy 600 were made according to the specification of ASME Code Sec. II, except addition of stabilizing elements. Ni-Cr-Fe alloys 690 and 625 were also included for comparison.

For confirmation of the new alloy 600, some large pipes of actual size (194 mmOD, 26mm thick) were produced with final alloy chemistry, and were examined with regard to strength and weldability as well as long term IGSCC resistance in 289°C pure water containing 8ppmO$_2$ under creviced conditions.

EXPERIMENTAL

ALLOY DESIGN - The design of experimental alloys were performed as shown in Table 1. The modification of Ni-Cr-Fe alloy 600 can be divided in three categories with some essential variables. On the other hand, although only ordinary alloys 690 and 625 had been evaluated concerning to SCC properties by several workers [13-16], in the present study, these alloys were modified or controlled as shown in the table.

Low Carbon - According to the data of Scarberry et al. [17], fairly low carbon seems to be necessary to prevent the sensitization of Ni-Cr-Fe alloy 600. Coriou et al. showed that low carbon alloys revealed IGSCC susceptibility in a simulated steam generator environment [18]. Recently, Bandy and Van Rooyen, based on the SCC data in a simulated steam generator environment, suggested that low carbon alloys were more susceptible to IGSCC compared with relatively high carbon alloys because the former had lower activation energy [19]. In this study, low carbon alloys less than 0.010wt% were made, which were relatively

pure alloys compared to ordinary, commercial alloy 600.

Stabilizing Elements - Ordinary alloy 600 normally contains titanium of around 0.3wt%. Herein, niobium and/or titanium were added to alloy 600 for stabilization of carbon, although Copson and Economy once tried the addition of these stabilizing elements, and found that they were less effective [20]. Their result is discussed later.

Stabilization Parameter - The parameter $\overline{N}$ was previously introduced in the study on IGSCC of Inconel weld metals by Yamauchi et al. [12]. The $\overline{N}$ implies the number of stabilizing elements per carbon atom in the material itself, which is expressed by the following equation ;

$$\overline{N} = \frac{\dfrac{Ti}{47.9} + \dfrac{V}{50.9} + \dfrac{Zr}{91.2} + \dfrac{Nb}{92.9} + \dfrac{Hf}{178.5} + \dfrac{Ta}{180.9}}{\dfrac{C}{12.0}}$$

$$= 0.13 \cdot \frac{2Ti+2V+Zr+Nb+0.5Hf+0.5Ta}{C} \qquad (1)$$

where Ti, V, Zr, Hf and Ta indicate their individual percent in the materials. For Inconel weld matals, niobium is the main stabilizing element, and titanium is the second. Others are nearly zero. Therefore, Eq.(1) could be expressed as follows ;

$$\overline{N} = 0.13 \cdot \frac{Nb + 2Ti}{C} \qquad (2)$$

Their results showed that the Inconel weld metals of higher $\overline{N}$ values than 12 were resistant to IGSCC in oxygenated high temperature pure water.

In this study, the parameter was adopted to Ni-Cr-Fe alloys 600 and 690. The stabilization parameter control, however, was performed only using niobium addition and carbon adjustment. The $\overline{N}$, therefore, is expressed in the following Eq.(3).

Table 1   Design of Experimental Alloys

Alloy	Modification	Essential Variables
Alloy 600	Low C	< 0.010% C
	Nb and/or Ti Addition	0.5%
		1 ~ 2%
		2 ~ 3%
	$\overline{N}$ controlled	$\overline{N} \gtrsim 12$
Alloy 690&625	$\overline{N}$ controlled	$\overline{N} \gtrsim 12$

$$\bar{N} = 0.13 \frac{Nb}{C} \tag{3}$$

The target of the $\bar{N}$ for their alloys was set in higher than 12. Herein, the $\bar{N}$ controlled Ni-Cr-Fe alloys are temporarily referred to as Ni-Cr-Fe alloys 600M and 690M.

TEST MATERIALS - Each of experimental alloys (Lab. melts) was produced with 6 kg in weight. The ingots were forged by hot-working, whose final size were 70mm width, 10mm thickness and about 1000mm length. The plates were annealed, and were machined for weld. Butt weldments were made using the $\bar{N}$ controlled electrode of Type 182 [12].

For demonstration test, two large melts were made in shop using the same process with commercial products of ordinary Ni-Cr-Fe alloy 600. The forged and annealed materials were machined in pipes of 194mm OD with 26mm thick. The weldments for tests were obtained by butt-welding with horizontal welding position using the $\bar{N}$ controlled electrode of Types 82 or 182 [12].

TEST PROCEDURES - Tests used in this study are as follows;

Mechanical Test - A primary requirement as structural material is sufficient strength. For screening of the materials, we set a target of strength which was compatible with ASME Code. For this, tensile test at RT was done for the base metals of alloy 600M. Tensile test at 288°C and Vickers hardness test were, additionally, applied.

Weldability Test - The weldability of experimental materials was examined using liquid penetration test and bend test. Furthermore, for the pipes of shop melt of the alloy 600M, longitudinal Varestraint test was done using some plates with 10mm thickness cut out from the pipes.

IGC Test - For detection of IGC suscepitility of the test materials, a boiling ferric sulfate-sulfuric acid test standardized in ASTM G28 was modified and used by us as previously described [12], which was named modified ASTM G28 test. This modified test, briefly speaking, have two modifications to the standard test specified [21].

The first modification was that the amount of ferric sulfate to be added to 636ml of a 50wt% sulfuric acid solution was changed from 25g of the standard to 50g.

The second modification was the evaluation procedure of results. The corrosion loss of weight per unit time is used in the standard G28 test. In the modified ASTM G28 test, the maximum IG penetration depth after 24h immersion was employed. For this, the specimen tested was lightly bent, and then cut into several pieces parallel to its longitudinal direction. The depth of attacks, the maximum perpendicular distance from the initial surface of each piece to the IGC front, was measured. Further, the maxi-

mum IGC data among the data of the several pieces was used as the result of IGC evaluation of materials.

Figure 1 shows an example obtained using the modified ASTM G28 test. Two time-temperature-sensitization (TTS) diagrams resulted from a ordinary Ni-Cr-Fe alloy 600 with 0.07 wt% carbon, herein, one is the TTS diagram of the material annealed at 1100 °C for 1 h and the other is the TTS diagram of the material heated at 840°C for 2h. These results showed that the test had good detectability of the sensitization behavior of the material.

For screening of IGSCC resistant materials in 289°C pure water containing 8ppm O_2, we applied a target of IGC resistance less than 0.5mm/day, whose basis is discussed in the later section.

SCC TESTS - Three SCC tests were employed in 289°C pure water containing 8ppmO_2, which were a creviced U-bend (CU-bend) test, creviced bend beam (CBB) test and creviced slow strain rate (C-SSRT) test.

The CU-bend specimens were machined to a dimension of 75 x 15 x 2mm^3. The crevice of

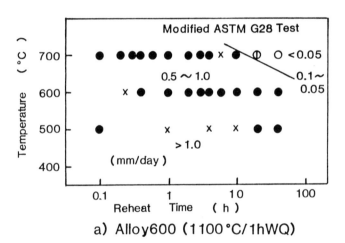

a) Alloy600 (1100°C/1hWQ)

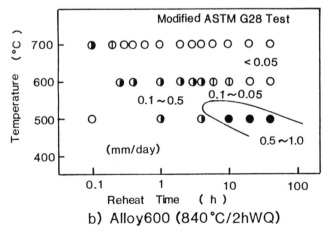

b) Alloy600 (840°C/2hWQ)

Fig. 1   Time-Temperature-Sensitization Diagram of Inconel 600

specimen was first formed by spot-welding a thin foil strip of Type 304 stainless steel of $0.1 \times 15 \times 30mm^3$ size, on the area to be the apex of a U-bend specimen. Then, the specimen was bent to U shape with 8 mm bent radius. This procedure produced a tight crevice on each CU-bend. The calculated strain on the apex of CU-bend specimen was about 12.5%.

The specimen for C-SSRT tests had a gage length of 20 mm. The crevice of each specimen was formed also with Type 304 stainless steel foil of 0.1mm thick spirally overlapped on the gage section by spot-welding, or graphite fiber wool pressed against the gage section with holders. The strain rate used for testing of the developed alloys was 1.7×10^{-7}/s, while faster strain rates, 4.2×10^{-7}/s for ordinary alloy 600 and 8.3×10^{-7}/s for Type 304 stainless steel as a reference of evaluation of IGSCC susceptibility.

The CBB test which was originally developed by Akashi [22], was applied in this study with the environmental condition and the immersion time modified. The holders used were made of ordinary alloy 600, and graphite fiber wool was employed as a crevice former. The specimen size was $50 \times 10 \times 2mm^3$. The applied strain was 1 %.

RESULTS

SCREENING OF TEST ALLOYS – Table 2 shows a summary of the screening test results of forty experimental alloys along with ordinary commercial alloy 600, which includes the tensile test data of base metals and the IGC test results of weld HAZ, partially with the SCC test results C-SSRT and CBB. The heat treatment condition of specimens for tensile test was as-received (AR), and for IGC and IGSCC test was PWHT followed by LTS, whose condition were shown in the table.

Low Carbon Alloy 600 – Both heats B1 and C1 of alloy 600 with low carbon were highly resistant to IGC, but their strength properties did not pass the requirement of ASME Code. When nitrogen was added to the low carbon alloy 600, it was effective to enough strength as seen in heat C3. But, this heat failed in IGC resistance of weld HAZ.

Niobium and Titanium Addition – In low niobium and titanium addition series of around 0.5wt%, all heats of D to F2 of alloy 600 showed high resistance to IGC in weld HAZ, but their strength did not pass the requirement.

Further, in the case of medium addition of niobium of 1.0 to 1.6wt%, most heats of alloy 600 could pass the requirements of strength. However, there were not IGC resistant alloys in weld HAZ except only one heat J2. The good properties of heat J2 seems to result from low carbon and high

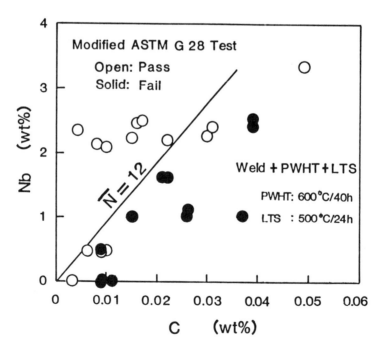

Fig. 2   Pass and Fail Expression of IGC Resistance in Weld HAZ of Niobium Added Alloy 600

Table 2 Screening Test Results of Experimental Alloys

Materials			Heat No.	Chemical Composition (wt%)					Tensile Test 1)	IGC Test 2)	SCC Test 3)	Total Evaluation
				C	Ti	Nb	N	Others				
Alloy 600	Ordinary		A1	0.05	–	–	–	–	o	x	–	x
			A2	0.04	–	–	–	–	o	x	–	x
			A3	0.06	–	–	–	–	o	x	x	x
			A4	0.07	–	–	–	–	o	x	x	x
	Low C		B1	0.003	–	–	–	–	x	o	–	x
			B2	0.009	–	–	–	–	x	x	–	x
			C1	0.009	–	–	–	–	x	o	–	x
			C2	0.011	–	–	–	–	x	x	–	x
			C3	0.004	–	–	0.046	–	o	x	–	x
	Nb and/or Ti Addition	Low	D	0.004	0.52	–	–	–	x	o	–	x
			E1	0.006	–	0.47	–	–	x	–	–	x
			E2	0.009	–	0.48	–	–	x	o	–	x
			E3	0.010	–	0.48	–	–	x	o	–	x
			E4	0.009	–	0.48	–	–	x	o	–	x
			F1	0.005	0.53	0.48	–	–	x	o	–	x
			F2	0.008	0.60	0.46	–	–	x	x	–	x
		Median	G	0.026	–	1.01	–	–	x	x	–	x
			H1	0.015	–	1.00	–	–	o	x	–	x
			H2	0.019	–	1.01	0.038	–	o	x	–	x
			H3	0.026	–	1.01	–	–	o	x	–	x
			H4	0.037	–	1.02	–	–	o	x	–	x
			I1	0.021	–	1.61	–	–	o	x	o	x
			I2	0.022	–	1.62	–	–	o	x	o	x
			J1	0.016	0.24	0.99	–	–	x	o	–	x
			J2	0.021	0.26	1.62	–	–	o	o	o	o
		High	K	0.022	2.03	–	–	–	o	x	x	x
			L1	0.031	–	2.50	–	–	o	o	o	o
			L2	0.030	–	2.34	–	–	–	x	–	x
			L3	0.049	–	3.34	–	–	o	o	o	o
			L4	0.039	–	2.51	–	–	o	x	o	x
			L5	0.039	–	2.50	–	–	o	x	o	x
	$\overline{N}$ controlled ($\overline{N} \geqslant 12$)		P1	0.012	–	1.63	0.039	–	o	o	o	o
			p2	0.010	–	2.10	–	–	o	o	o	o
			P3	0.012	–	2.11	0.045	–	o	–	o	–
			P4	0.016	–	2.58	–	–	o	o	o	o
			P5	0.022	–	2.19	–	–	o	o	o	o
			P6	0.024	–	2.19	0.054	–	o	o	o	o
			P7	0.008	–	2.13	–	–	o	o	o	o
			P8	0.017	–	2.66	–	–	o	o	o	o
Alloy 690	Ordinary		M1	0.01	0.23	–	–	–	o	o	o	o
			M3	0.022	–	–	–	–	x	o	x	x
	$\overline{N}$ controlled		N	0.042	–	3.08	–	–	o	o	o	o
Alloy 625	$\overline{N}$ controlled		Q1	0.03	0.26	3.59	–	8.5Mo	o	o	o	o
			Q2	0.039	–	3.56	–	9.0Mo	o	o	o	o

1) ; Base Metal, AR, RT Tention Test Results   PWHT ; 600°C/40h   o ; Pass
2) ; Weld HAZ, PWHT + LTS, Modified ASTM G28 Test Results   LTS ; 500°C/24h   x ; Fail
3) ; Weld HAZ, PWHT + LTS, CBB, and C-SSRT Test Results

niobium, because heat J2 is close to the high N materials described below.

High titanium addition of 2.03wt%, as seen in heat K, is harmful to IGC resistance not only in the weld HAZ but also in the base metal itself. The cause was not identified so for.

Contrary to this, all of heats of L series, i.e., alloys of high niobium addition from 2.03 to 3.34wt%, were resistant to IGC in the base metal, and especially both heats of Ll and L3 were highly resistant in the weld HAZ. All of the heats of L series showed enough strength. If the lower carbon contents of the heats other than Ll and L3 were lower, they would become enough IGC resistance in weld HAZ.

$\bar{N}$ Controlled Alloy 600 - All heats of P series were controlled by the stabilization parameter $\bar{N}$, which had higher than 12. As seen in the table, they were high resistant to IGC and sufficient strength, and were immune to IGSCC in 289°C pure water containing 8 ppm O_2 with crevice. Therefore, we need only the $\bar{N}$ control for realization of a new Ni-Cr-Fe alloy 600 with IGSCC resistance and sufficient strength.

Figure 2 represents pass and fail expression of IGC resistance in the weld HAZ of ally 600. The data plotted are, mainly, those of niobium addition type, with titanium less than 0.5 wt% and nitrogen less than 0.03 wt%. As seen in the figure, a line can be drawn to separate the data of pass and fail, which corresponds the $\bar{N}$ values of 12. Herein, the alloys of the $\bar{N}$ values higher than 12 are the Ni-Cr-Fe alloy 600M.

Alloy 690, 690M and 625 - One of the two heats of ordinary alloy 690 did not satisfy the strength of ASME Code case requirement, and was susceptible to SCC in weld HAZ of the vicinity to fusion line of the weldment. Although only the heat N was examined, the Ni-Cr-Fe alloy 690M was very promissing as structural material. Alloy 625 itself has a large amount of niobium as alloying element specified in ASME Code case. Therefore, when the $\bar{N}$ of alloy 625 can be controlled by reducing in carbon content, the material would become sufficient. A lot of heats and testing will be needed to characterize the both IGSCC resistance and strength of the $\bar{N}$ controlled alloys 690M and 625.

Ni-Cr-Fe ALLOY 600M - Two heats of Ni-Cr-Fe alloy 600M were made succesfully in shop as ordinary, commercial alloy 600. Their $\bar{N}$ values were 22 and 76 (Heat A and B). The various properties of shop made alloy 600M with some experimental alloy 600M and commercial, ordinary alloy 600 were evaluated with strength, weldability, IGC resistance and IGSCC resistance.

Strength - Figure 3 presents the tensile test results of base metals at room temperature. All of alloy 600M tested showed enough strength as well as ordinary commercial alloy 600. Figure 4 compares the tensile test results at RT and 288°C for the alloy 600M and ordinary alloy 600. The results were also enough strength as well as ordinary alloy 600. We, further, confirmed that the strength of weldments of the alloy 600M were superior to those of ordinary alloy 600, Therefore, Ni-Cr-Fe alloy 600M can be considered as structural material to be identical with alloy 600 of ASME Code Sec. II.

Tested at Room Temperature

Material		Y.S. (MPa) 100 200 300	T.S. (MPa) 400 500 600	El. (%) 30 40 50	Hardness (HB) 1) 140 160 180
Alloy 600M	Heat A				
Alloy 600M Lab. Melts	P3				
	P4				
	P5				
	P6				
	P7				
	P8				
Ordinary Alloy 600	I1				
	I2				
	I3				
	I4				
	I5				

1) : The Data were Converted from Hv unit into HB unit   ⋮ : ASME Code Sec. II

Fig. 3   Comparison of Strength Properties Between Developed Alloy 600M and Ordinary Alloy 600

16

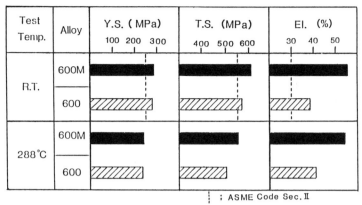

Fig. 4   Strength Properties of Alloy 600M at RT and 288°C

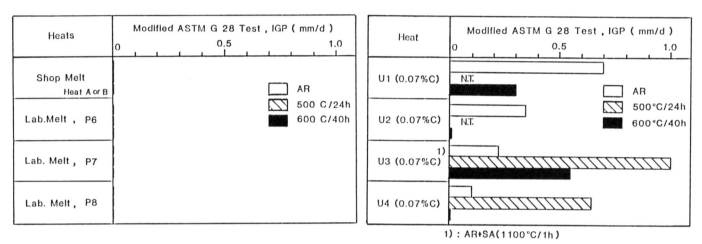

Fig 5   Heat-to-Heat Variations in IGC Resistance of Alloy 600M Base Metals

Fig. 6   Heat-to-Heat Variations in IGC Behaviours of Ordinary Alloy 600 Base Metals

1) : AR+SA(1100°C/1h)

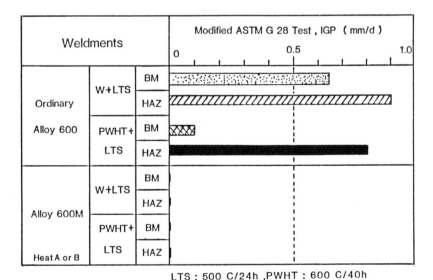

LTS : 500 C/24h , PWHT : 600 C/40h

Fig. 7   Comparison of IGC Resistance between Alloy 600M and Ordinary Alloy 600

Weldability - We performed longitudinal Varestraint tests of the alloy 600M base metal. When the results were evaluated with the relation between total crack lengths of fusion line cracks and longitudinally augmented strain, as a result, the alloy 600M was found equal or superior compared with ordinary, commercial alloy 600. In the actual weldability test as butt-welding of the pipes of the alloy 600M, both the non-destructive and desructive tests showed that the weldability was no problem.

IGC Resistance of Base Metals - The effect of heat-to-heat variations in IGC resistance of base metals of alloy 600M was examined using the modified ASTM G28 test. Figure 5 showed the IGC Test results of alloy 600M. The heat treatment used were AR, PWHT and LTS, whose conditions were shown in the figure. For comparison, Figure 6 presents similar data of ordinary alloy 600. There was little variations among heats of the alloy 600M. Furthermore, Ni-Cr-Fe alloy 600M did possess very stable resistance or immunity to sensitization heat treatments employed.

IGC Resistance of Weld HAZ - Their weldments were similarly tested, which were heated in the conditions of LTS and/or PWHT plus LTS. Figure 7 shows the comparison of

IGC behavior of the alloy 600M and ordinary alloy 600. The weld HAZ of the alloy 600M were excellent resistant to sensitization in all cases of the heat treatment tested.

C-SSRT Test Resutls - Figure 8 depicts the test results of the weldments of the alloy 600M. The heat treatment condition of each specimen was PWHT plus LTS. Although the relatively low strain rate was applied with crevice in 289 °C pure water containing 8 ppm O_2, these weldments were resistant to IGSCC. Contrary to this, the weldments of ordinary alloy 600 and Type 304 stainless steel as reference showed IGSCC susceptibility in spite of the use of faster strain rates.

However, in order to assure the performance in actual use, we will need the data of the incubation time until SCC growth, because the time mainly dominates SCC resistance of the material. The related test data are described below.

CBB Test Results - Figure 9 shows the test results of the weldments of various

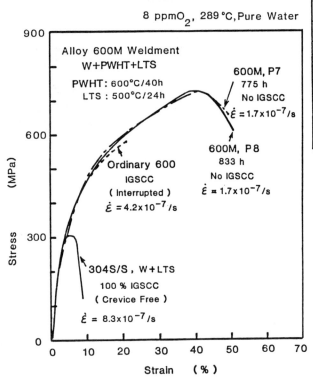

Fig. 8   Creviced SSRT Test Results of Alloy 600M Weldments

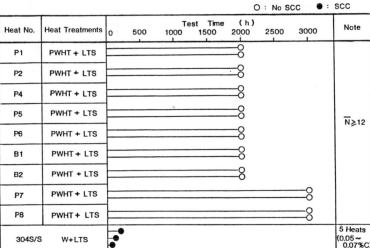

Fig. 9   CBB Test Results of Weldments of Alloy 600M Lab. Melts

Fig.10   CBB Test Results of Weldments of Alloy 600M Shop Melt

18

heats of the alloy 600M. The specimens were heated in PWHT plus LTS condition. Note here that only LTS was applied to Type 304 stainless steel. As seen in the figure, while the Type 304 stainless steel showed a very short incubation time to IGSCC, it was found that the alloy 600M had very long incubation time to IGSCC occurrence so far. This result means that the alloy 600M can have a high factor of improvement to IGSCC occurrence compared with the reference materials of Type 304 stainless steel.

Herein, note that there were no heat-to-heat variations in SCC resistance in the incubation times among the 9 heats of the alloy 600M.

Figure 10 shows the other CBB test result of the alloy 600M concerning to the effect of heat treatments. This result indicates that longer testing time was necessary to observe IGSCC in the alloy 600M for each heat treatment tested, so that their incubation times are not detected within the test time.

CU-bend Test Results - Figure 11 shows the results of the CU-bend test of the alloy 600M. The specimens of the weldments and base metals were heated in various conditions of PWHT and/or LTS. The test duration was prolonged to long immersion more than three thousand hours compared with the short incu-

bation time of Type 304 stainless steel. All of the specimens of the alloy 600M were fully immune to SCC. Note that the results include the samples heated in severer LTS condition, i.e., 500°C/48h, which were also resistant to IGSCC.

Additionally, the $\overline{N}$ controlled Inconel weld metals in the weldments of the alloy 600M shown in Figures 8 to 11 were also resistant to SCC.

DISCUSSION

EFFECT OF NIOBIUM - In 1968, Copson and Economy [20] performed a double U-bend test of some alloy 600 modified by the addition of stabilizing elements in a ammoniated 316°C water as a simulated pressurized water reactor (PWR) environment. They found that the alloy 600 with 1.2 wt% niobium and 0.07 wt% carbon were susceptible to IGSCC under the creviced condition.

Based on the present results, their data may be attributed to the low stabilization parameter $\overline{N}$ of 2.2, which is far less than 12. To be emphasized is that niobium is effective to produce SCC resistance of materials, only when the addition of niobium was controlled by the $\overline{N}$.

DEGREE OF IGSCC RESISTANCE - The degree of SCC resistance, in other words, the factor

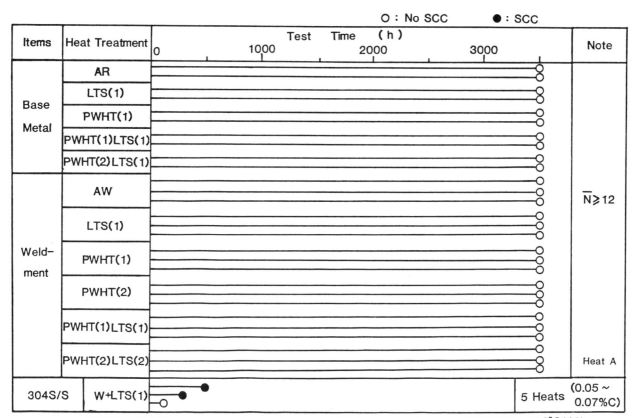

Fig.11 Creviced U-bend SCC Test Results of Alloy 600M Shop Melt

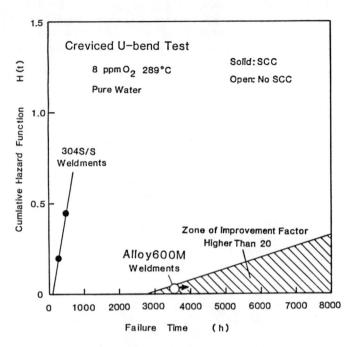

Fig. 12 Relation between Cumulative Hazard Function and Failure Time

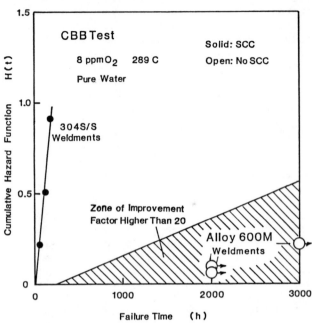

Fig. 13 Relation between Cumulative Hazard Function and Failure Time

of improvement in the IGCC resistance of the Ni-Cr-Fe alloy 600M can be discussed in a statistical way, and the present author's approach to the present SCC life data will be shown below.

Some studies on statistical analyses using a log-normal distribution of SCC life data boiling water reactor (BWR) piping materials were successfully done [23, 24], and it was shown that the alternate materials need to have the factor of 20 improvement to the IGSCC incubation time as a reference of Type 304 stainless steel.

An exponential distribution was also found useful for the analyses of SCC life data [25, 26]. According to Yamauchi et al. [26], of twenty sets of available SCC life data for analyses, the nineteen groups could be fitted by the distribution. Further, they interpreted the fitting of the data to the exponential distribution as implying that SCC life phenomenon is of a first order reaction system, and hence the factor controlling SCC life is mainly activation energy of the process. This approach seems to be essential in discussing the physical meaning of the improvement factor.

Figures 12 and 13 show the relation between the cumulative hazard function and the failure time of the SCC test results of CU-bend and CBB respectively, where the failure rate was assumed to be constant to the test time because of the application of the exponential distribution as dealt in the paper [26]. The shaded zones in both figures

express that the factor of improvement are higher than 20 to the reference of Type 304 stainless steel. Those zones were drawn with the mean failure times and the failure rates, more than twenty times and less than one-twentieth of those of the Type 304 stainless steel, respectively. The open marks with arrows show the data of the Ni-Cr-Fe alloy 600M weldment, no IGSCC. Obviously, all of the present data of the alloy 600M weldments are in the zone of concern; the improvement factor higher than 20 was achieved.

CRITERION OF IGC - The maximum IG penetration revealed by the modified ASTM G28 test can be used as a criterion for screening of IGSCC resistant materials in 289°C, pure water, as with Inconel weld metals [12]. For the present study, the criterion of the IGC resistance is again less than 0.5mm/day. The validity of the criterion was also shown by the resultant SCC data in present study.

The creterion seems to have the following meaning. In the modified ASTM G28 test, the growth of IG penetration in test materials can be expressed as a function of log t, i.e., as A x log t + B, where t is time. Figure 14 shows an example of such data of alloy 600. Note that the behavior is similar to the relation between SCC incubation period and its growth.

Therefore, for comparison of IGC susceptibility between two materials, the time ratio, F_r (= $t2/t1$) at a given IG penetration depth (P_i) gives more reasonable results than the IGP ratio, F_i (= $P1/P2$) at 24 hours immer-

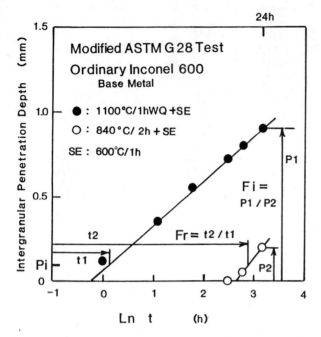

Fig. 14   Relation between IG Penetration Depth
and Test Time of Ordinary Alloy 600 by
Modified ASTM G28 Test

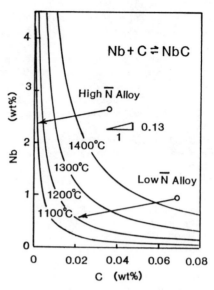

Fig.15 Thermodynamic Relation of
NbC Formation at High Temperature

sion of the standard test.   Therefore, when
the time ratio can be assumed to be nearly
equal to the factor of improvement in SCC
test results, the improvement factor of the
alloys satisfied the criterion can be deduced
as follows.

When the time ratio of the alloy 600 of
IGC of 1.0mm/day to the reference Type 304
stainless steel is Fm based on the data SCC
test, and when the time ratio of the alloy
600 of IGC of 0.5mm/day to the alloy 600 of
IGC of 1.0mm/day is Fr based on the IGC test,
the total factor of the alloy (0.5mm/day) to
the reference is possibly expressed as Fm x
Fr.   As a trial assessment, if Fm is more
than ten and if Fr is five, the total factor
provides more than fifty.   As a interpretation of the criterion of IGC, these discussion is probable.   Furthermore, the factor
of the new alloy 600M, by applying of thus
discussion, obviously become very high value.

LTS RESISTANCE - In this study, the LTS
condition of 500°C/24h was used.   If we use a
relatively low value of activation energy,
i.e. 40 kcal/mol for diffusion or reaction of
the related elements, the LTS condition
becomes nearly equal to a metallurgical state
at 288°C for 58 years.   Furthermore, while
extramely severer, a heat treatment of
500°C/48h was also used, it was concluded
that the LTS problem would never actually
appear in the alloy 600M.

MECHANISM OF STABILIZATION PARAMETER
CONTROL - For modeling the mechanism of the
stabilization parameter control, it seems to
be important to clear why the $\overline{N}$ value higher

than 12 was needed for the prevention of the
sensitization in the weld HAZ of alloy 600.
Although the detail mechanism will be
reported, its main points are described here.

First, a critical carbon content for
immunity to sensitization was about 0.005wt%
for alloy 600 heated by PWHT with or without
LTS as shown in Table 1.

Second, by applying the solubility curve
of niobium carbides in steel [27], we could
consider that niobium alloys was reduced in
lower content compared with that of low
niobium alloys, even though heated at elevated
temperatures higher than 1000°C by formation
of niobium carbides as shown in Figure 15.

Third, the process of reduction in free
carbon and niobium in matrix, as shown with
the arrows in the figure, can occur with the
slope of 0.13, because of the relation of
atomic ratio of niobium and carbon in NbC
formation.

Fourth, the process will be promoted by
elevated temperature, prolonged time and
increased density of NbC nuclei in grain.

Therefore, in the alloy 600M having high
$\overline{N}$ values, free carbon can reduce to the critical content for immunity to sensitization
of PWHT at 600°C.   Especially, the alloys of
the $\overline{N}$ values more than 12 could be sufficiently stabilized only by weld heat effects during
welding.   This stablization model was shown
to be possible by a theoretical analysis [28].

Additionally, a large number of very
fine particles of NbC in grains may play a
role as trapping sites of non-metallic
impurities.

CONCLUSIONS

The present study of the effect of niobium addition on the IGSCC resistance of of Ni-Cr-Fe alloy 600 is concluded as follows.

1. Niobium addition is effective for improvement of resistance to IGSCC and IGC of alloy 600, when the addition is controlled by the stabilization parameter $\bar{N}$. Contrary to this, high titanium addition was less effective.
2. The critical value of stabilization parameter for full resistance to IGSCC and IGC, is higher than about 12 for the weld HAZ followed by PWHT and/or LTS, as well as for base metals.
3. New alloy 600M, which was developed by stabilization parameter control, has the high degree of resistance to IGSCC in 289°C pure water containing 8 ppm O_2.
4. Alloy 600M did not show heat-to-heat variations in IGSCC and IGC resistance, and was immune to sensitization heat treatments tested.
5. The strength of alloy 600M satisfied the requirement specified in ASME Code Sec. II of alloy 600. The weldability was also compatible or superior to commercial alloy 600.
6. The stabilization parameter was also applicable to alloys 690 and 625.

ACKNOWLEDGMENT

The authors wish to gratefully acknowledge Dr. I. Masaoka, S. Kuniya and S. Hattori of Hitachi Research Laboratory of Hitachi, Ltd. for useful discussion.

REFERENCES

1. McIlree, A.R. and H.T. Michels, P.E. Morris. Corrosion, 31, 441-448 (1975)
2. Was, G.S. and H.H. Tischner, R.M. Latanision. Metallurgical Transaction A, 12A, 1397-1408 (1981)
3. Airey, G.P. Metallography, 13, 21 (1980)
4. Mulford, R.A. Corrosion, Metallurgical Transaction A 14A, 865-870 (1983)
5. Guttemann, M. and Ph. Domoulin, N. Tan-Tai, P. Fontaine. Corrosion, 37, 416-425 (1981)
6. Kowaka, M. and H. Nagano, T. Kudo, Y. Okada. Nuclear Technology, 55, 394-404 (1981)
7. Domain H.A. and R.H. Emanuelson, L.W. Sarver, G.J. Theus, L. Kutz. Corrosion, 33, 26-37 (1977)
8. Airey, G.P. and A.R. Vaia, R.G. Aspden. Nuclear Technology, 55, 436-448 (1981)
9. Caramihas, C.A. and D.F. Taylor. Corrosion, 40, 382-385 (1984)
10. Crum, J.R. Corrosion, 38, 40-45 (1982)
11. Bulischek T.S. and D. Van Rooyen. Corrosion, 37, 597-607 (1981)
12. Yamauchi, K. and I. Hamada, T. Okazaki, T. Yokono, A. Nishioka. Fifth International Conference on Pressure Vessel Technology, ASME, 2, 599-616 (1984)
13. Sedriks, A.J. and J.W. Schults, M.A. Corodovi. Corrosion Engineering (Boshoku Gijutsu), 28, 82-95 (1979)
14. Page R.A. Corrosion, 39, 409-421 (1983)
15. Clark W.L. and J.C. Danko, G.C. Gordon. the Eelctrochemical Society, Corrosion Problems in Energly Conversion and Generation, 410-419 (1974)
16. Berry, W.E. and E.L. White, W.K. Boyd. Corrosion, 29, 451-469 (1973)
17. Scaberry, R.C. and S.C. Pearman, J.R. Crum. Corrosion, 32, 401-406 (1976)
18. Coriou, H. and L. Grall, C. Machieu, M. Pelas. Corrosion, 22, 280-290 (1966)
19. Bandy, R. and D. Van Rooyen. Corrosion, 40, 425-430 (1984)
20. Copson, H.R. and G. Economy. Corrosion, 55-65 (1968)
21. Annual Book of ASTM Standard, G28, 10, 902-907 (1981)
22. Akashi, M. and T. Kawamoto. Corrosion Engineering (Boshoku Gijutsu), 32, 9-15 (1983)
23. Kass, J.N. and J.C. Lemaire, R. Davis. J. Alexander, J. Danko. Corrosion, 36, 686-698 (1980)
24. Kuniya J. and S. Hattori, I. Masaoka, R. Sasaki, H. Ito. Corrosion Engineering (Boshoku Gijutsu), 31, 261-267 (1982)
25. Akashi, M. and T. kenjyo, S. Matsukura, T. Kawamoto. Corrosion Engineering (Boshoku Gijutsu), 33, 628-634 (1984)
26. Yamauchi, K. and Y. Sakaguchi, S. Kimura, H. Yamasaki. Corrosion Engineering (Boshoku Gijutsu), No. 32 meating in Sapporo, B-205, 202-205 (1985)
27. Narita. R & D Kobe Steel Engineering Report, 18, 170-190 (1968)
28. Yamauchi, K. Unpublished Results.

CORROSION-FATIGUE CRACKING
IN FOSSIL-FUELED BOILERS

H. J. Westwood, W. K. Lee
Ontario Hydro Research Division
Toronto, Ontario, Canada

ABSTRACT

Corrosion-fatigue in boiler tubing is a major cause of forced outages in conventional (fossil-fuelled) electricity generating plant and similar damage in larger, thick-section boiler components is also of concern. The phenomenological aspects are described and the current theoretical understanding of the crack initiation and growth processes is reviewed. A need is identified for materials data on crack initiation and for examination of the extent to which crack growth data for nuclear systems may be relevant to fossil-fuelled boiler components which operate in significantly different environments.

ROUGHLY 70% OF NORTH AMERICA'S electricity is generated in conventional power stations using large watertube boilers to extract the energy from fossil fuels (coal, oil or gas), producing high pressure superheated steam to drive turbo-alternators. Typical steam conditions are 538°C (1000°F) and 16.2 MPa (2350 psi).

Whilst a generating unit can be forced out of service for many reasons, the perennial main cause of forced outages is boiler tube failures and, since a single tube failure can cause a three day outage costing perhaps $100,000 per day, the problem is a serious one. Among the various types of boiler tube failure, waterside-initiated cracking in carbon steel economizer and waterwall tubes has become very common, to the extent that a major EPRI project to determine the root cause has recently been initiated. The damage has variously been described as stress-induced corrosion, corrosion cracking or corrosion-fatigue but, in view of an apparent association with transient or cyclic loading, the last term is now preferred.

Similar forms of damage have also been encountered in larger, thick section boiler components such as headers and drums, necessitating safety-related defect assessments to be undertaken and, in some cases, expensive replacements to be made.

This paper will briefly review the phenomenological aspects of the problem as it affects boiler tubing and also outline some recent experience of similar damage in larger pressure vessels. The current state of understanding of the damage mechanisms will then be reviewed, firstly as pertaining to the tubing problem and then in the thick section component context. Areas of needed research will be identified and some aspects of experimental techniques will be considered.

CORROSION-FATIGUE IN BOILER TUBING

An earlier paper [1] and a more recently produced Boiler Tube Failure Manual [2] presented detailed descriptions of various cases of corrosion-fatigue failure which had been encountered in Ontario Hydro boilers. For present purposes, it will suffice to reiterate briefly the essential characteristics of the damage and note the circumstances under which the problems are typically encountered.

MATERIALS AND OPERATING TEMPERATURES – Economizer tubes, which perform the final preheating of the boiler feedwater and are located in the rear pass of the boiler, are usually made of SA192 low carbon or SA210 medium carbon steel and run at temperatures in the range ~250–300°C. The steam-generating water wall tubes form the furnace walls of the boiler and are usually made of SA210. Maximum metal temperatures relate to the saturation temperature and are typically in the range 350–400°C.

DAMAGE CHARACTERISTICS – Corrosion-fatigue damage consists of crack initiation and growth from the waterside surface of the tube. Multiple cracking is usually observed with one crack ultimately becoming dominant and causing a through-wall penetration. Figure 1 shows a cross-section through a severely damaged 64 mm o.d., 6.4 mm wall thickness water wall tube. The tube bore surface is shown in Figure 2, in which the cracks are readily discernible.

Fig. 1 – Severe cracking in water-wall tube

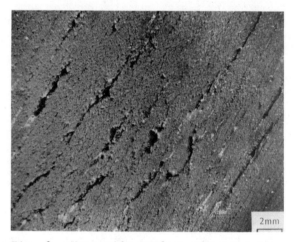

Fig. 2 – Waterside surface of same tube

In many cases, cracks are less obvious on visual inspection though some oxide grooving is usually apparent. Non-destructively, radiography has so far proved to be the most effective method for detecting the damage.

In section, the cracks are usually wide, at least near the waterside surface, are oxide filled and often show an irregular, bulged profile. Sometimes, however, the cracks may be relatively narrow or may change to a narrow profile at some point in growth; the cracks in Figure 1 show this characteristic. Simplistically, the crack width may relate to the relative influence of corrosion and stress on the growth mechanism.

Usually, though not exclusively, corrosion-fatigue failures are of the "leak-before-burst" type in which the dominant crack grows until the remaining ligament fails by tensile overload; the half-wall section in Figure 3 typifies such a failure.

Fig. 3 – Through-wall penetration of corrosion-fatigue crack

CIRCUMSTANCES OF CRACKING – Ontario Hydro experience is that, in general, corrosion-fatigue tube failures are associated with some source of stressing in addition to the normal pressure and thermal stress to which boiler tubes are subjected. The predominant locations for corrosion-fatigue cracking have been where thermal expansion has been restricted or constrained during transient conditions such as start-up, load changing, two-shift operation etc. Such locations are where pressure parts (tubes) are connected by welding or various spacer/support devices to non-pressure attachments.

Crack growth is invariably in a direction perpendicular to the maximum tensile stress and, in most cases longitudinal cracking is observed – Figure 1 is such an example. Sometimes, however, bending stresses may be involved, for example, at the intersection of stub tubes with headers or where complex tube geometry is employed; in such cases, crack orientation is circumferential. Finally, corrosion-fatigue failures, some catastrophic, have occurred where no obvious external stressing was apparent, a prime example being tight hairpin bends in economizer sections. In these cases, it has been concluded that residual stresses from the fabrication process were responsible for the damage. An example of such a failure is shown in Figure 4.

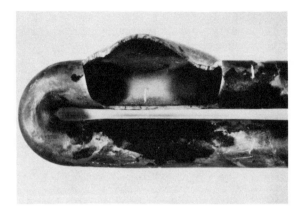

Fig. 4 - Catastrophic failure of economizer hairpin bend

CORROSION-FATIGUE IN LARGE COMPONENTS

In 1981, an investigation of longitudinal corrosion fatigue cracking in economizer header stub tubes led to the disclosure of serious cracking in the header itself. The damage proved to be generic to a number of Ontario Hydro boilers and inquiries revealed that similar problems had also been found in some units in the US.

Figure 5 shows a typical cross-section through the header at the stub tube intersections. Header dimensions were 300 mm o.d., 50 mm minimum wall thickness and the material was SA106B carbon steel. Normal operating temperature was approximately 250°C.

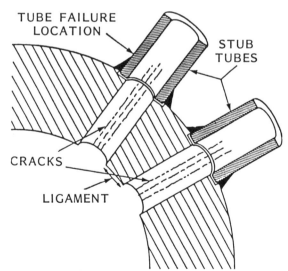

Fig. 5 - Economizer header section at stub-tube intersection

As indicated, cracking continued from the tube down the 32 mm tube hole and, as shown in

Figure 6, fairly profuse cracking of the header inside surface was also present. In general, both circumferential and longitudinal cracking was observed and, as shown, some inter-hole ligament cracks had developed in the circumferential direction. The extent of the damage was revealed by sectioning, one of the worst areas being shown schematically in Figure 7 where the shaded area represents cracking.

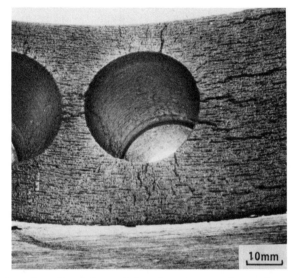

Fig. 6 - Cracking of header internal surface

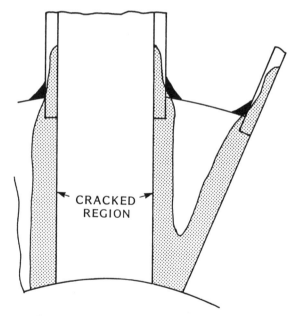

Fig. 7 - Indication of damage extent

In-plant temperature monitoring and stress analysis determined that the damage had been caused by thermal shocks resulting from periodic admission of low temperature feedwater to the hot header during start-up and off-line top-up periods. The resulting thermal stresses were exacerbated by stress concentrations at the edge of the holes and at the ends of the stub tubes. Defect assessment showed that eventual failure would be leak-before-break and, based on a fatigue analysis, modified operational procedures were instituted to reduce the stress levels. Several headers were deemed unfit for continued service and were replaced, the replacement headers being modified to reduce the stress concentrations.

For the present paper, the above brief summary of a serious and costly problem will suffice. The main purpose here is to characterize the actual cracking in the header in terms of the apparent mechanisms involved.

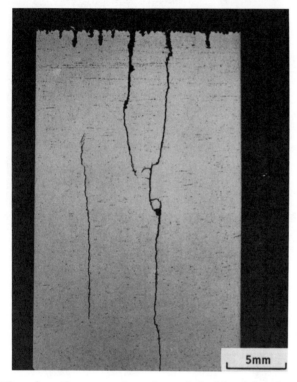

Fig. 8 - Macrosection through half of header wall

Figure 8 shows a macrosection through the header close to one of the tube holes, the waterside surface being at the top. The multiple thick cracks from the surface are strongly reminiscent of those typified for boiler tubing by Figure 1. Also notably similar to the tube crack morphology is the change from wide to narrow cracking, at a depth of ~2.5 mm. In the tube shown in Figure 1, the transition took place at a depth of ~3.7 mm. A microsection through the wide portion of a crack is shown in Figure 9 and, by comparing with Figure 3, the similarity with the tube damage is again apparent.

The major crack shown in Figure 8 extends roughly half way through the 50 mm wall of the header and the morphology at the tip is shown at relatively high magnification in Figure 10. Despite its distance from the

waterside surface, the crack sides are significantly oxidized and the profile has a similar bulged appearance to the initial wide portion of the cracks, albeit on a smaller scale. Also significant is that the crack path is clearly transgranular, again similar to boiler tube corrosion-fatigue cracks.

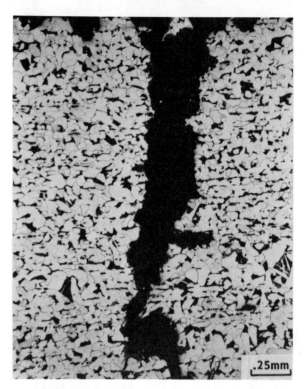

Fig. 9 - Crack details near inside surface

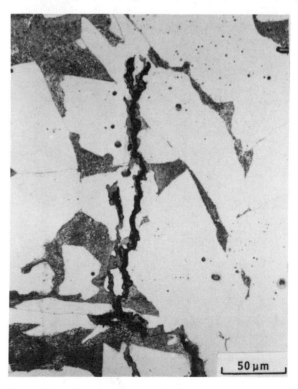

Fig. 10 - Crack tip morphology

26

In summary, the damage in the header showed essentially similar characteristics to the corrosion-fatigue damage now well-characterized in boiler tubing. The main difference, however, lay in the relative proportions of wide and narrow cracking, since it appears that the transition occurs at roughly the same depth in both sizes of component.

PRESENT UNDERSTANDING AND RESEARCH NEEDS

The previous section has described the corrosion-fatigue problem in boilers from a phenomenological viewpoint. It is now appropriate to examine the present theoretical understanding of fatigue damage in high temperature water, to explore the availability of relevant materials data, and to identify areas where new research is needed.

BOILER TUBING - CRACK INITIATION - In contrast with the large pressure vessels in both nuclear and conventional stream-raising plant, boiler tube dimensions are usually in the range of 38-75 mm o.d. with wall thicknesses seldom exceeding 6.3 mm and waterside surfaces normally free of macroscopic defects. Accordingly, crack initiation, rather than growth from in-built defects, is important and may well occupy a large fraction of the corrosion-fatigue life of the tube, although there is no present quantitative information on this point.

Few studies specifically relevant to fossil-fuelled boiler tube corrosion-fatigue have been reported except for some work by Schoch and Spahn [3] who characterized the damage in some detail and also reported some laboratory simulations; some aspects of this work will be considered later.

At this point, it may be appropriate to recall that the successful use of carbon steel for service in high temperature water is crucially dependent on the formation and maintenance of a protective film of magnetite, Fe_3O_4 on the waterside surface. The reaction between iron and oxygen-free neutral or slightly alkaline water at temperatures above about 230°C is complex, probably involving several stages, but can conveniently be summarized as

$$3 \ Fe + 4 \ H_2O \rightarrow Fe_3O_4 + 4 \ H_2$$

During normal boiler operation, the magnetite builds up to a thickness of 10-15 μm; this is about 10^4 times the thickness of the typical "passive" film on iron [3].

The fact that the vast majority of boiler tubes perform satisfactorily attests to the protection normally afforded by the magnetite. Conversely, when damage does occur, disruption of the magnetite is an apparent prerequisite.

In principle, the magnetite can be disrupted by stress and/or water chemistry deficiencies, particularly the presence of excess oxygen. As will be discussed later, oxygen levels can be quite high during start-up periods. However, the fact that all cases of corrosion-fatigue have involved adverse stressing suggests that the water chemistry aspects may be of secondary importance.

In general, an oxide will crack perpendicularly to the metal-oxide interface if the oxide fracture strain is exceeded and it has been shown that this strain is reduced under cyclic stress or temperature conditions [4]. The damage pattern, as typified in Figure 1, consists of parallel cracks in the oxide aligned at right angles to the maximum tensile stress direction. The cracks are equally spaced, with new cracks typically initiating midway between existing cracks - where the oxide strain maximizes after having relaxed around each crack [5].

As-formed under boiler operating conditions, the magnetite is stress-free whereas the base metal will be under pressure-derived tensile stress, possibly enhanced by thermal stress due to the through-wall temperature gradient. During shut-down, these stresses decline or disappear with the result that the magnetite becomes stressed in compression. It has been suggested that the protective layer can be disrupted by the addition of further compressive "peaks" due to transient thermal stressing under start-up conditions [3]. Service experience, however, suggests that normal operating stresses are insufficient to cause damage since all cases of corrosion-fatigue failure have involved some form of additional stressing, whether residual or associated with attachments and/or constraints [1,2].

In some cases, an apparent correlation has been noted between increased incidence of corrosion-fatigue failures and number of unit starts, typically around several hundred. This is typical of low-cycle fatigue, ie, involving significant plastic strain per cycle. The implication is that, not only are external stresses necessary for corrosion-fatigue cracks to initiate, but these stresses must be quite high.

Schoch & Spahn proposed a mechanism of crack initiation involving fatigue failure of the magnetite film, assisted by the presence of structural defects in the thick layer [3]. The corrosive medium then had access to the tube metal where the magnetite reaction occurred, consuming some additional metal and creating an oxide notch. Subsequent load cycles caused the progressive growth of this "chemico-mechanical" notch. This proposed mechanism contrasts with the situation where the much thinner passive films are involved; in this case, damage was thought to initiate, not by cracking of the film, but by slip processes in the metal beneath.

The same workers [3] considered that the cracking involved two damage processes, namely corrosion-fatigue and stress-induced corrosion. The former was defined as the simultaneous action of start-up stresses and the corrosive medium provided by oxygen-free water. The latter was suggested to result from the electrochemical corrosive action taking place after shut-down in aerated water. The successive "pits" often observed along the cracks, as typified in Figure 1, were believed to result from oxygen ingress during shut-down periods. Whilst this shut-down pitting damage explanation may well be valid, as noted earlier, oxygen-free water is not typical of normal start-up conditions, so that pitting during this part of the load cycle may also be possible.

Having effectively nucleated a crack, subsequent propagation is favoured, mechanically due to the stress concentration, and chemically, because the pit will act as a crevice and be anodic relative to the tube surface.

As the crack (or cracks) grow into the wall, environmental access to the crack tip becomes increasingly restricted and the observed amount of scale decreases. This effect is shown both for tubing and heavy section components in Figures 1 and 8. The stress intensity will normally increase as the crack grows so that the later stages of growth may be mechanically-dominated and the crack may be relatively narrow and oxide free. In some cases, where large thermal stresses are involved, growth may be into a declining stress field so that, beyond a certain depth, growth rates might reduce to insignificant levels.

RESEARCH NEEDS IN CRACK INITIATION - As outlined, the mechanistic understanding of the magnetite disruption during fatigue-type loading is by no means complete and, in terms of root cause identification, more effort is needed in this area. Perhaps a more immediate practical need, however, is for data relating crack initiation to the usual mechanical loading parameters in realistic water chemistry environments.

The approach advocated would involve conventional reverse stress low-cycle fatigue tests on smooth bar specimens conducted in autoclaves with flow loop capability for water

chemistry control and monitoring. The specimen surfaces should be pre-conditioned to develop a realistic magnetite layer. Tests should be conducted under strain range control and the endurance (cycles to crack detection) related to strain range. Other mechanical parameters to be varied should be cyclic frequency and wave-form (triangular or trapezoidal).

These tests would establish the material corrosion-fatigue properties. There is also a need, however, for tests on actual tubes under realistically simulated operational conditions.

Concerning water chemistry, as noted, oxygen is the most obvious potentially deleterious water impurity. Accordingly, it is suggested that tests should be run at three oxygen levels representing worst, best and intermediate water conditions. These would be saturation (~8 ppm), normal full load level (~5 pb) and some level between. Other species such as phosphate, chloride or certain organic materials could also be investigated once the oxygen dependency had been established.

Another question to be addressed concerns the relative importance (duration) of the initiation and growth stages. This should be clarified when specific data pertaining both to initiation and crack growth become available. The crack growth aspects will be considered in the next sections.

On a practical note, if the root cause of corrosion-fatigue should be shown to involve water chemistry deficiencies, the scope for remedial measures is limited; oxygen especially is essentially impossible to exclude during shut-down periods. Thus, it may be, as concluded by Schoch & Spahn [3], that the only practical solution may be mechanical, ie, to limit the cyclic stresses to acceptable levels. Data from the work proposed above, in conjunction with plant monitoring and stress analysis, would facilitate the identification of susceptible areas of the boiler where modifications to reduce stress levels might overcome or prevent corrosion-fatigue damage.

LARGE (THICK-SECTION) COMPONENTS - CRACK GROWTH - As noted earlier, in the larger boiler components such as headers, drums, pipework etc, the main corrosion-fatigue concern is with crack growth rather than initiation. This section will therefore examine the availability of crack growth rate data for carbon and low alloy steels in high temperature water environment and then discuss its relevance to the specific fossil-fuel boiler situation.

Although a considerable body of general information exists concerning environmental crack growth (particularly stress corrosion cracking) in carbon steels, little or no data is available on corrosion-fatigue crack growth in the fossil-fuel boiler water environment. Plentiful data is available, however, relevant to the Pressurized Water and Boiling Water (PWR and BWR) nuclear systems in which steam conditions are up to 288°C (550°F) and 8.6-13.8 MPa (1250-2000 psi). These will subsequently be referred to as Light Water Reactor (LWR) systems. This information will now be reviewed briefly in order to identify general trends which should give some indication of material behaviour in an environment which is not radically different.

REVIEW OF AVAILABLE CRACK GROWTH DATA - A review of the important factors controlling corrosion fatigue in LWR water environments is available [6]. Some of the pertinent results will be outlined here and then some subsequent developments will be described.

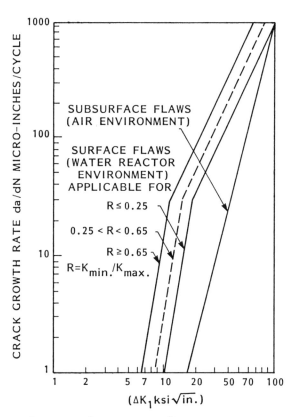

Fig. 11 - ASME "wet and dry" fatigue crack growth data for carbon and low-alloy ferritic steels (Ref 7)

For design purposes, the ASME codes provide "air" and "wet" lines for different R ratio range for crack growth in pressure vessel steels [7]. Bamford's [8] data formed the basis for these recommended crack growth guidelines which are shown in Figure 11. The "bend-over" region suggested that the environ-

mental effect gradually diminished at high stress intensity range and the data band direction pointed towards the air line. The hold time in trapezoidal waveform tests did not result in any significant increase in the crack growth rates, and limited information was available concerning threshold conditions.

More recent data from Scott and Truswell, shown in Figure 12, [9] revealed that, under conditions felt to be more plant-relevant, slower crack growth rates were observed than those reported by other workers. In Scott's tests, a flow rate of 40 L/min was used and turbulent conditions were present. Most other work has been conducted under static or low flow conditions eg, 0.8 L/min. The bend-over or plateau region observed in low-flow testing was attributed by Scott to stress corrosion cracking. The corresponding crack tip strain rate was determined and compared with the dissolution rate using the Faraday's material dissolution equivalent approach. The agreement was within an order of magnitude within the plateau region.

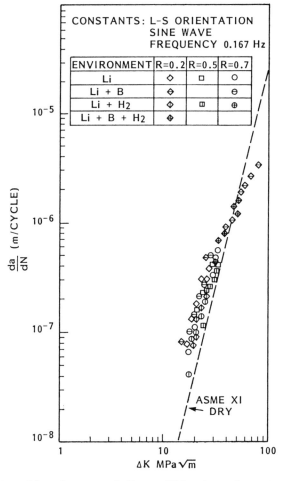

Fig. 12 - Scott and Truswell's data from high flow rate tests (Ref 9)

In an inter-laboratory exchange program [10], Scott's specimens were tested under low-flow conditions and, as shown in Figure 13, crack growth rates were significantly higher than those shown in Figure 12.

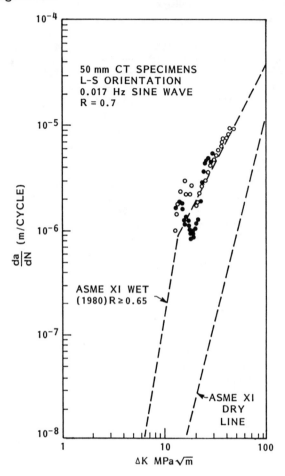

Fig. 13 - Data from low flow rate tests on Scott's specimens (Ref 10)

Parallel work by Van Der Sluys and DeMiglio [11] showed that the crack growth rate was controlled by ΔK, ν, R and was independent of prior loading history. The experiments were carried out in a refreshed autoclave using the constant stress intensity method. The validity of the conventional constant load increasing K test was questioned, in that the crack growth rate may represent either a steady state or a transient behaviour at a particular stress intensity level. Again, the data could be described by the crack tip strain rate equation [9] for the maximum crack growth rate.

In a systematic study by Scott et al [12] on the effect of sulphur, test specimens of "pure", commercial and sulphur-doped steels were tested. The high sulphur content material after fracture showed some sites of pre-existing inclusions, and a small portion

of the fracture surface contained quasi-cleavage facets. There was no evidence that the crack branched and followed a preferential route along specific micro-structures or chemically enhanced surfaces. The change in sulphur content modified the fractographic behaviour, but, as shown in Figure 14, only a small change in macroscopic crack growth rate was measurable. Thus, crack growth retardation through control of the sulphur content may not be a realistic proposition. Fractography also revealed different types of oxide as a result of different testing conditions. Pure magnetite (Fe_3O_4) was found in the high flow loop, whereas Bamford's specimen contained 30% haematite (Fe_2O_3) [8]. Clearly, long-term electrochemical measurement is required to clarify the above observations.

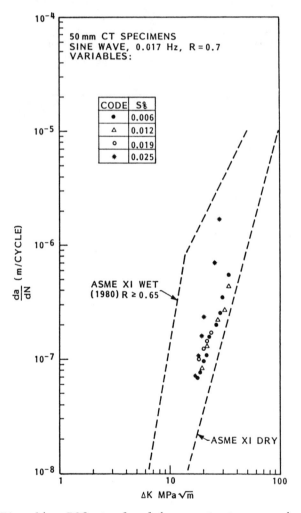

Fig. 14 - Effect of sulphur content on crack growth rates (Ref 12)

Atkinson et al [13] measured the corrosion potential during fatigue tests on A533B steel in a simulated PWR environment over the temperature range 150-290°C using a Ag/AgCl reference electrode. As the oxygen content of the water decreased, there was a transition from high to low potential. A specially designed electrode with a cooling jacket has also been designed and proven for use up to 288°C at 1330 psi [14]. The electrode chemical potential data may be used in conjunction with polarization curves and/or Pourbaix diagrams to infer the susceptibility of a material to corrosion. The balance between the rate of bare metal production and the repassivation kinetics as deduced from electro-chemical potential can help to reveal the crack advance mechanism.

RELEVANCE TO FOSSIL-FUELLED BOILER COMPONENTS - This section will characterize the essential differences between fossil-fuelled and nuclear plant mechanical-environmental parameters in order to identify crack growth research needs relevant to thick section components in conventional boilers.

(1) Material - The main fossil plant material of interest is SA106-B carbon steel (>.25%C) which is widely used for economizer headers and other thick section components in the lower temperature parts of the boiler (excluding the superheater and reheater sections which operate at temperatures in the creep range). This material contains similar amounts of carbon, phosphorous, sulphur and silicon to SA516 grade 70 carbon steel and the SA508 and SA533 low alloy steels used for LWR pressure vessels. The low alloy steels contain additional nickel, chromium and molybdenum (each <1%) and have slightly higher manganese content.

The carbon and low alloy steels were shown to have similar crack growth behaviour in LWR environments [8]. Thus, although data

in SA106B material are needed, evidence to date suggests that small compositional differences are unlikely to have a strong influence on crack growth behaviour.

Information concerning the effect of microstructural variation is limited but there is evidence of faster crack growth rates in weld heat affected zone material [15] and slower rates in weld material [16].

(2) Water Chemistry - Table I gives representative water chemistry data for the two LWR nuclear systems and for a typical fossil fuelled boiler. For the latter, chemistries relevant to economizer and steam drum are given; the differences reflect additions, notably sodium phosphate, made to the drum before the water reaches the steam-generating water walls.

Clearly, there are significant differences in water chemistry as well as in temperature and pressure between the three systems. A common objective in all cases, however, is to control oxygen and pH within specified limits.

For a typical 500 MW fossil-fuelled boiler, on-line monitoring showed that the dissolved oxygen depended on start-up and shut-down conditions and procedures [17]. In general, a relatively high oxygen level was present during fire-up and the level dropped down after turbine synchronization, typically 1 to 2 hours later. For approximately worst case conditions, involving a cold start with drained economizer header, initial oxygen levels of 5000 ppb have been observed; such levels are close to the saturation value of

TABLE I

LIGHT WATER REACTOR AND FOSSIL-FUELED BOILER WATER CHEMISTRIES

	BWR	PWR	Economizer F.F. Feedwater	F.F. Boiler Water
Pressure, MPa	6.9	13.8	18.0	18.0
Temperature, °C	288	288-315	250-300	300-375
pH (room temperature)	6.5-7.0	4.0-10.5	8.8	9-9.4
Oxygen, ppb	50-200	<100	3	3
Hydrogen, ml/kg at STP		0-50		
Lithium Hydroxide, ppm		0.6-6.2		
Boric Acid, ppm		0-13,000		
Chloride, ppm	<0.1	<0.15		<0.75
Fluoride, ppm	<0.1	<0.15		
Ammonia, ppm			0.25-0.40	
Hydrazine ppb			10	
Sodium Phosphate ppm				0.5-3.0
Sodium Hydroxide ppm				<2
Silica ppm				<0.2
Conductivity μmho/cm	<0.1	1-40	<3.0	5-10

~8 ppm. The level typically drops to ~10 ppb by two hours after turbine synchronization and eventually reaches ~5 ppb under steady-state conditions. The important point to note is that the transient fatigue cycles during start-up (see later) are accompanied by the highest levels of dissolved oxygen in the boiler water.

Whilst, in principle, many water chemistry parameters may affect the crack growth behaviour, oxygen would be expected to have the greatest influence and should thus be the initial focus of a laboratory test program. From the above plant information, initial tests should concentrate on the two extremes of oxygen, namely near saturation (~8 ppm) and near oxygen-free (~5 ppb).

For meaningful laboratory testing, it is clear that the water environment should closely simulate plant conditions both chemically and in terms of flow conditions. Concerning the latter, it is not usually possible to incorporate typical plant water flow conditions in laboratory tests involving autoclaves. It has, however, become apparent that use of "static" autoclave systems is not satisfactory so that at least the "refreshed" approach is needed in which the autoclave water volume is changed, typically once per hour. Clearly, the closer the flow conditions can approach those in-plant, the more meaningful the test data will be.

Slight differences in crack tip electrochemical conditions due to flow rate variations between different test rigs have drawn attention to the importance of monitoring the electrochemical potential which, as noted previously, relates to the oxygen level in the water [13]. Further, since crack tip water chemistry may differ significantly from that of the bulk fluid, it is important that the measurements are made as closely to the crack tip as is practical.

(3) Temperature - Depending on the particular component and its location in the boiler, metal temperatures can range from ~250°C up to ~375°C. Some economizer headers run at the lower temperatures whereas steam drums and some piping can run at essentially saturation temperature. LWR-related data have been generated at 288°C, so on this basis alone, may be irrelevant to fossil plant components. Currently, there is little, if any, information concerning temperature effects on crack growth above 288°C. However, since it seems unlikely that higher temperatures would retard crack growth, the use of 288°C data may be non-conservative for many fossil plant components. Clearly, this is an area where research effort is needed.

(4) Pressure - There appears to have been no systematic study on the effect of water pressure on fatigue crack growth. However, comparison between PWR and BWR environments with different system pressure has indicated good agreement, although the water chemistry was different. If, as seems likely, the crack advance mechanism depends on crack tip strain rate and local electrochemical potential, external pressure may be of relatively minor concern.

(5) Loading Frequency - As mentioned before, the highest "bend-over" region in the da/dN versus ΔK plot [8] occurred at a frequency of 1 cpm and, in general, the crack growth rates reached a maximum between 0.1 and 1 cpm. In actual plant operation, to reduce the magnitude of thermal stresses due to through-wall temperature differentials, the start-up/shut-down temperature ramp rates have to be limited to minimize damage.

Since a typical start takes several hours, the overall ramp rate is much slower than the above values. It has been shown, however, that a large number of significant thermal fatigue cycles can occur during the overall loading cycle due to rapid and numerous temperature fluctuations associated with additions of relatively cold feedwater to the system [17]. Thus a start-up does not constitute one but many fatigue cycles and, importantly, the ramp rate of these transient cycles may well fall within the range cited above at which maximum crack growth rates occur. In a laboratory corrosion-fatigue program, the carbon steel should therefore be tested at frequencies in the above range to establish the "worst-case" effect of the water environment.

(6) R (Minimum to Maximum Load) Ratio - R ratio is important as noted in the ASME guidelines [7]. Generally, as R increases, the crack growth rate increases as the crack tip is opened above a critical crack tip opening displacement. However, Van der Sluys and DeMiglio [11] observed that, under constant stress intensity test conditions, the greatest crack growth enhancement in SA508 steel occurred at an intermediate R (R = 0.46) value. Obviously, in a laboratory test program, several R ratios should be examined.

(7) Waveform - Various workers [10,16,18,19,20] have found that, in aqueous environment, fatigue crack growth occurred only during plastic deformation, ie, during load ramping. Atkinson reported that square wave or negative sawtooth loading at 0.1 Hz produced almost the same crack growth rates as in air [18].

Similarly, as noted earlier, a hold time in trapezoidal loading resulted in negligible change in the crack growth rates in Bamford's work [8]. As the typical operational start-up/shut-down essentially involves sawtooth loading waveform, hold time effects may be of secondary importance. A limited test effort should adequately clarify this point.

RESEARCH NEEDS IN CRACK GROWTH - SUMMARY

In contrast to crack initiation, where virtually no data are available, the need in crack growth is to establish to what extent the existing LWR-related data base is relevant to fossil-fuelled boiler components. Since the main differences lie in the environmental aspects -- temperature, pressure and water chemistry -- a limited program of conventional crack growth testing in realistically simulated fossil plant environment should readily indicate whether a more extensive research effort is needed.

SUMMARY AND CONCLUSIONS

(1) The phenomenological aspects of corrosion-fatigue in tubing and thick section fossil-fuelled boiler components have been described.

(2) In tubing, crack initiation appears to be of more significance than in large components in terms of the failure process; crack growth appears to dominate the thick section component endurance.

(3) Theoretical understanding of the crack initiation process is presently limited and there is essentially no relevant materials data; an initial program of low-cycle fatigue tests in representative water environments has accordingly been advocated.

(4) Some exploratory testing is needed to determine the extent to which the large LWR-related crack growth data base may be relevant to fossil-fuelled boiler components which are exposed to a significantly different water environment.

ACKNOWLEDGEMENTS

This work was carried out in connection with a Canadian Electrical Association-funded project sub-contracted to the University of Toronto and directed by Prof. D.W. Hoeppner, now at the University of Utah.

The authors are grateful to Dr. D. Sidey of Ontario Hydro Technical and Training Services Division for permission to refer to unpublished work on plant monitoring and defect assessment and to Drs. G. Bellamy and M.A. Clark of Ontario Hydro Research Division for permission to use Figure 7.

The comments and conclusions are those of the authors and are not implied to represent the official views of Ontario Hydro.

REFERENCES

1. M.D.C. Moles and H.J. Westwood, Proceedings of ASM International Conference on Materials to Supply the Energy Demand, Harrison, British Columbia, Canada, 1980, p 515-537.

2. R.B. Dooley and H.J. Westwood, "Analysis and Prevention of Boiler Tube Failures", Final Report on Canadian Electrical Association Contract 990-650A, 1983.

3. W. Schoch and H. Spahn, "Corrosion Fatigue," NACE-2, Editors O.F. Devereux, A.J. McEvily, R.W. Staehle, 1972, p 52-64.

4. A.G. Crouch and R.B. Dooley, Corrosion Science, Volume 16, 1976, p 341.

5. J.C. Grosskreutz and M.B. McNeil, J. App. Phys., 40 (1969), p 355.

6. K. Torronen and W.H. Cullen, Jr., ASTM STP 770, Editors C. Amzallag, B.N. Leis and P. Rabbe, 1982, p 460-481.

7. ASME Boiler and Pressure Vessel Code, Section XI, Appendix A, 1983.

8. W.H. Bamford, J. of Engineering Materials & Technology, Vol 101, July 1979, p 182-190.

9. P.M. Scott and A.E. Truswell, Proceedings of an IAEA Conference on "Sub-Critical Crack Growth", Freiburg, West Germany, 13-15 May 1981, p 376.

10. P.M. Scott, A.E. Truswell, Proceedings of ASME Pressure Vessel and Piping Conference, Orlando, Florida, Paper PVP82-029, June 1982.

11. W.A. Van Der Sluys and D.S. DeMiglio, Proceedings of Second International Conference on the Environmental Degradation of Engineering Materials, Blacksburg, Virginia, September 21-23 1981.

12. P.M. Scott, A.E. Truswell and S.G. Druce, Proceedings of Corrosion 83, Paper No 133, NACE 1983.

13. J.D. Atkinson, S.T. Cole and J.E. Forrest, Proceedings of an IAEA Meeting

on "Sub-Critical Crack Growth", Freiburg, West Germany, 13-15 May 1981.

14. J. Leibovitz, W.R. Kassen, S.G. Sawochka and W.L. Pearl, Proceedings of Corrosion 83, Paper No 129, NACE, 1983.

15. M. Suzuki, H. Takahashi, T. Shoji, T. Kondo and H. Nakajima, "The Environment Enhanced Crack Growth Effects in Structural Steels for Water Cooled Nuclear Reactors", Institute of Mechanical Engineers, London, 1977.

16. T. Shoji, S. Aiyama, H. Takahashi, M. Suzuki, Corrosion-NACE, Vol 34, No 8, Aug. 1978, p 276-282.

17. D. Sidey, Private Communication, Ontario Hydro, Toronto.

18. J.D. Atkinson, T.C. Lindley, The Influence of Environment on Fatigue, Institute of Mechanical Engineers, London, 1977, p 65-74.

19. J.M. Barsom, Corrosion Fatigue: Chemistry, Mechanics, and Microstructure, NACE, 1972, p 424-436.

20. E.J. Imhof, J.M. Barsom, Progress in Flaw Growth and Fracture Toughness Testing ASTM STP 536, American Society for Testing and Materials, 1973, p 182-205.

THE ROLE OF CREVICES IN COMPARISON TO PITS IN INITIATING STRESS CORROSION CRACKS OF TYPE 310S STEEL IN DIFFERENT CONCENTRATIONS OF MgCl$_2$ SOLUTIONS AT 80°C

Shigeo Tsujikawa, Tadashi Shinohara
University of Tokyo
Tokyo, Japan

Yoshihiro Hisamatsu
Nisshin Steel Co. Ltd.
Tokyo, Japan

STRESS CORROSION CRACKING (SCC) in neutral chloride solutions , which may be more often encountered service environments , requires local anodes[1] as initiation sites which are pits[2,3] or crevices[4] .

Pitting occurs at relatively noble potentials , while crevice corrosion can occur at less noble potentials. This difference is significant in the corrosion protection of stainless steels especially in neutral solutions with lower concentrations of chloride ions. It is corrosion crevices that provide initiation sites for SCC cracks in such less aggressive media.

This paper discusses the roles of crevices compared with pits as initiation sites for cracks based on the SCC tests conducted simultaneously with smooth and creviced specimens in neutral 15-35%MgCl$_2$ solutions at 80 °C.

EXPERIMENTAL PROCEDURE

The material of Type 310S steel obtained in round bar form , 8mm in diameter , was solution treated at 1050 °C for 30min followed by water quenching. The chemical composition in wt% of the steel was as follows : 24.13Cr, 20.15Ni, 0.18Mo, 1.14Mn, 0.027C, 0.66Si, 0.029P, 0.19Cu, 0. 018N, 0.09S, balance Fe.

SMOOTH SPECIMEN - Crack initiation from pits was examined with a smooth specimen of a 8mm dia round bar of wich surface had been polished to #1200 in SiC grit size number , followed by buffing. With respect to sizes of pits introduced to initiate cracks for the smooth specimen , two SCC tests were conducted.

In the SCC test A shown in Fig.1(a) both a stress of 20kgf/mm^2 and a potential , E , were simultaneously applied immediately after passivating at -400mV for 1h. According to the results in the SCC test A at E=-350mV in 35% MgCl$_2$ solution , cracks initiated from relatively small pits of which depths was less than 10μm under applied stress at and above 19.5kgf/mm^2.

In the SCC test B shown in Fig.1(b) , a stress of 20kgf/mm^2 was applied after a polarization procedure to grow pits at -330mV for 20h , which had introduced onto the specimen relatively large pits of which depth attained from 30 to 100μm in 35% MgCl$_2$ solution , for an example. The test B at E= -350mV in 35%MgCl$_2$ solution showed that cracks initiated from pits of 60 - 300μm in depth , h , introduced by prolonged holding time at -330mV under applied stresses , σ , at and above 14kgf/mm^2 , only when the stress intensity $K_I = 1.1\sigma\sqrt{\pi h/Q}$ was larger than $K_{ISCC} \simeq 8$kgf/mm$^{3/2}$, where σ was the applied stress , h was the pit depth and Q was the flaw shape parameter. The stress of 20kgf/mm^2 to be applied to the smooth specimen is larger than threshold stresses of 19kgf/mm^2 in the test A and 13Kgf/mm^2 in the test B.

NOTCHED SPECIMEN - Crack initiation from crevices was examined with a notched specimen to which a notch of the size of a=0.3mm and h=2.2mm as shown in Fig.2 had been spark machined. Copper concentrated inner surface layer of the notch was dissolved away at 1.35V.SCE for 30min in 1N H$_2$SO$_4$ solution and finished by mechanical polishing[5] . The notch introduced to the notched round bar specimen works as a mechanical notch as well as a corrosion crevice. The notched specimen was polarized in similar procedure to that for $E_{R,CREV}$ measurement to introduce crevice corrosion within the notches and then brought to a predetermined potential , when a initial stress intensity ,$K_{I,i}$, from 14 to 23kgf/mm$^{3/2}$ was applied. SCC test C conducted as above for the notched specimen both in 25% and 35% $MgCl_2$ solutions showed that threshold stress intensity for SCC , K_{ISCC} , was nearly equal to 8kgf/mm$^{3/2}$ and that applied stress intensities above 10kgf/mm$^{3/2}$ gave constant crack growth rate , the plateau velocity in l vs K_I characteristic.

REPASSIVATION POTENTIALS FOR PIT AND CREVICE - Repassivation potential for crevice corrosion has been developed in Japan and it is regarded to be a well reproducible electrochemical parameter which covers the repassivation as well as the initiation of crevice corrosion. It is not a unique material property but a crevice characteristic which reflects geometrical , environmental and material factors of a given crevice[6)7)].

A notched specimen was polarized in the following sequence : -400mV(SCE) for 1h, potential sweep in noble direction in steps of 10mV every 10min to the potential where the anodic current , I_t , increased more than 50μA , potential sweep in less noble

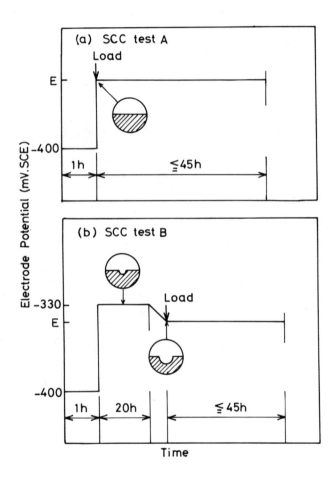

Fig. 1 - Polarization procedures for SCC tests A and B.
Holding specimen potential at -330mV for 20h in the test B introduces comparatively large pits onto the specimen before loading.

After SCC tests specimens were broken in the air to observe initiated cracks under a scanning electron microscope (SEM).

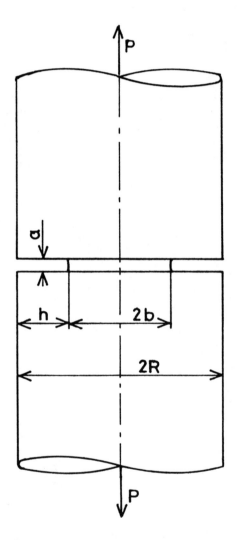

Fig. 2 - Notched round bar specimen. The notch of h=2.2mm in depth and a=0.3mm in width works as a mechanical notch as well as a corrosion crevice to initiate cracks.

direction at the same rate as above at relatively noble potentials and in steps of 10mV every 1h in the range where I_t decreased less than 5μA as shown in Fig.3. Thus , the repassivation potential for the crevice , $E_{R,CREV}$, was determined as the most noble potential at which repassivation occurred during 60min holding. The value in the case of Fig.3 was determined to be −385mV.

Stationary pittig potential , V_C , under no applied load was measured with a smooth specimen , which was kept at −400mV for 1h and polarized to a predetermined potential where increase or decrease of the anodic current was observed for 20h. The value of V_C was determined as the lowest potential above which growing pits initiated.

The repassivation potential for pit , $E_{R,PIT}$, was determined after the similar procedure to that for $E_{R,CREV}$ with additional in-situ measurement for enlargement of pit radius to confirm repassivation of the pit observed , as shown in Fig.4. Determined value of $E_{R,PIT}$ for the pit of 187μm in depth in the case shown in Fig.4 was −365mV and coincided with those for the pits of 75μm and 187μm in depth. Those facts show that $E_{R,PIT}$ does not depend on pit size in such highly concentrated chloride solutions as used in this work.

DISSOLUTION RATE WITHIN LOCAL ANODE − Because pit radius enlarged linearly with time , the growth rate of pit radius , da/dt, was determined as the slope of the regression line. The dissolution rate at pit-bottom ,I_h, could be given by:

$$I_h=(h_0/a_0) \cdot (da/dt) \qquad (1)$$

where h_0 and a_0 were depth and radius of the pit , respectively , measured after the test. The dissolution rate at the crevice was assumed to be equal to that of I_h at corresponding specimen potential and $MgCl_2$ concentration.

EXPERIMENTAL RESULTS

Variations of $E_{R,PIT}$, $E_{R,CREV}$ and V_C with $MgCl_2$ concentration are shown in Fig.5. $E_{R,CREV}$ was less noble than $E_{R,PIT}$ in 15−35% $MgCl_2$ solutions. Relations of dissolution rate determined at pit-bottom , I_h , with electrode potential , E , in different concentrations of $MgCl_2$ solutions are shown in Fig.6. The relationship between I_h and E can be expressed by:

$$E(mV.SCE)=α + β \log I_h(μm/h) \qquad (2)$$

where α and β are constants shown in the table in Fig.6.

CRACKS FROM PITS − Fig.7(a) summarizes the results of SCC test A conducted for 45h , the symbol O indicating no cracks or no localized corrosion , X indicating crack initiation from pits and Δ no cracks from pits which continued to grow.

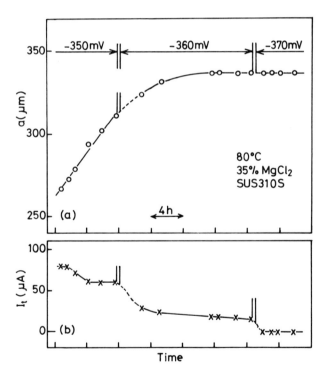

Fig. 4 − Variations of mouth radius of a pit (a) and corresponding total current, I_t , from the specimen (b) with time under successive descents of electrode potentials. Enlargement of pit radius stopped at −360mV, followed by decrease in current down below zero at −370mV. Repassivation potential for the pit, $E_{R,PIT}$, was determined to be −365mV.SCE in this case.

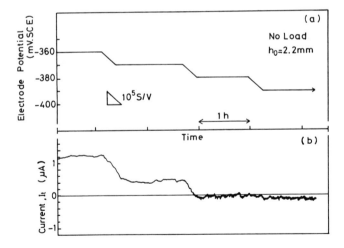

Fig. 3 − Procedure for potential control (a) and corresponding variation of current (b) under no applied stress for the specimen with a notch as corrosion crevice of 2.2mm in depth in 25% $MgCl_2$ solution. The repassivation potential for the crevice, $E_{R,CREV}$, was determined to be −385mV in this case.

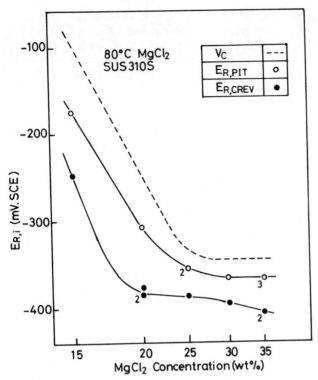

Fig. 5 - Variations of repassivation potentials, $E_{R,i}$, for pits (o, i=PIT) and crevices (•, i=CREV) with $MgCl_2$ concentration. The dashed line shows the stationary pitting potentials, V_C.

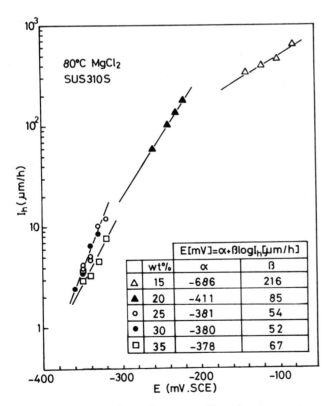

Fig. 6 - Relations between dissolution rate at pit-bottom, I_h, and electrode potential, E, in different concentrations of $MgCl_2$ solutions.

The depths of pits which initiated cracks were larger than 40μm at -350mV in 30% $MgCl_2$ solution and at -340 and -350mV in 35%$MgCl_2$, and smaller than 10μm at -350 and -360mV in 35% $MgCl_2$. Electron micrographs of the former and the latter cases were shown in Fig.8(a) and 8(b), respectively. The results of SCC test B shown in Fig.7(b) are the same as those of SCC test A except an additional occurrence of SCC at -360mV in 30% $MgCl_2$. The lower limit of the potential range for cracking (X) is the potential just more noble than $E_{R,PIT}$ in each solution as shown in Fig.5. This fact means that repassivated surface of the steel could not initiate SCC cracks. The cracking range for Type 310S steel was found to locate in electrode potentials more noble than that for 304 steel and less noble than that for 316 steel[3].

In order to examine meanings of the upper limit of the potential range for SCC in Fig.7 crack growth rates were compared with dissolution rates at pit bottom in 35% $MgCl_2$ solution.

Variations of maximum depth of pits (o) and cracks(x) with testing time in SCC test A are shown in Fig.9. Crack growth rates, l, at the three potentials obtained as the slopes of the regression lines in the figure are 4.2 - 5.7μm/h, which is regarded to be almost equal to that from the crevices. The rates of pit depths as shown in Fig.9 at -350 and -340mV under a applied stress of 20kgf/mm² were 2.0 and 4.0μm/h which were found to be nearly equal to I_h values calculated from Eq.(2) at corresponding potentials under no applied stress. Stresses seem to give no

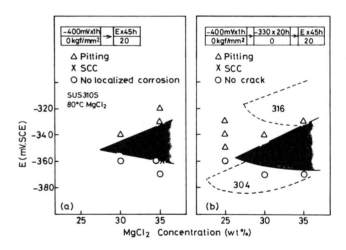

Fig. 7 - The SCC region for smooth specimens in terms of electrode potential and $MgCl_2$ concentration obtained from the SCC tests A (a) and B (b). A stress of σ=20kgf/mm² and a electrode potential of E were loaded for 45h from the beginning of the test A. The test B was conducted after introducing large pits on the specimen at -330mV for 20h.

significant effect on the growth rates of pits[8]. The values of $\dot{1}$ and I_h obtained in 35% MgCl$_2$ solution are shown in Fig.10 against electrode potential, E. The lower limit of potential range for SCC (+) is the one just more noble than $E_{R,PIT}$ and the upper corresponds to the potential, E^V, where I_h is equal to $\dot{1}$. SCC region in terms of electrode potential, E, is written as:

$$E_{R,PIT} < E < E^V \qquad (3)$$

CRACKS FROM CORROSION CREVICES - The results of SCC test C for 45h in 20-35% MgCl$_2$ are shown in Fig.11. The lower limit of potential range for cracking shown by (x) is the potential just more noble than $E_{R,CREV}$ in each solution as shown in Fig.5.

Variations with time of depth of maximum cracks at various potentials in SCC test C in 25% and 35% MgCl$_2$ solutions are shown in Fig.12(a) and Fig.12(b), respectively. Crack

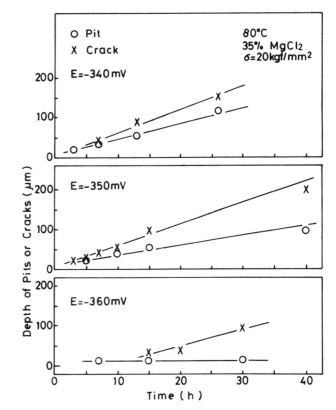

Fig. 9 - Variations of maximum depths of pits (o) and cracks (x) with testing time on the smooth specimens potentiostated at the three electrode potentials.

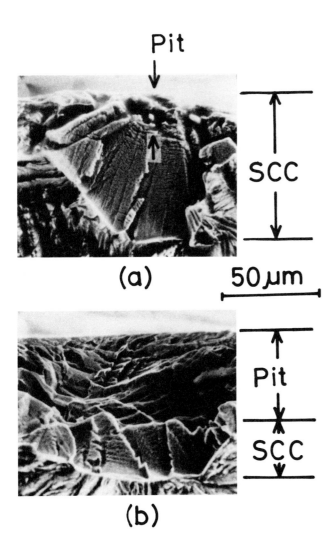

Fig. 8 - SCC cracks observed after a test duration of 15h at the bases of small pit (a) and large one (b) at a electrode potential of -350mV.SCE in 35%MgCl$_2$ solution at 80°C.

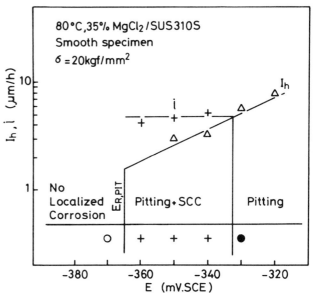

Fig. 10 - Comparison of crack growth rates, $\dot{1}$, with the dissolution rates at pit-bottom, I_h, against electrode potential for the smooth specimens. The lower limit of the potential range for SCC is the one just more noble than $E_{R,PIT}$ and the upper corresponds to the one where I_h is equal to $\dot{1}$.

growth rates , $\dot{l}$, obtained as the slopes of the regression lines are 4.8μm/h in 25%MgCl$_2$ solution and 4.6μm/h in 35% MgCl$_2$ solution. Changes in electrode potential[4][9] and MgCl$_2$ concentration in solution made no significant difference in the crack growth rate.

Comparison of $\dot{l}$ with I$_h$ in 25% and 35% MgCl$_2$ solutions are shown in Fig.13(a) and Fig.13(b) , respectively. The upper limit of the potential range for SCC corresponds to the one where I$_h$ is equal to $\dot{l}$ and the lower is the one just more noble than E$_{R,CREV}$. The SCC region in terms of electrode potential , E , was written as:

$$E_{R,CREV} < E < E^V \qquad (4)$$

where E^V is the potential at which I$_h$ is equal to $\dot{l}$. The potential range for SCC with the notched specimen is enlarged to less noble potentials than that with the smooth specimen shown in Fig.7. This comes from the fact that E$_{R,CREV}$ is less noble than E$_{R,PIT}$, as shown in Fig.5. Corrosion crevices which could continue to grow at less noble potentials than pits could initiate cracks at the less noble potentials.

COMPARISON OF CREVICES AND PITS AS INITIATION SITES FOR SCC CRACKS - Fig.14 shows comparison of SCC region for the smooth specimen with that for the notched specimen in terms of I$_h$ and MgCl$_2$ concentration. The ordinate of each data point is calculated by Eq.(2). I$^*_{h,PIT}$ and I$^*_{h,CREV}$ are the values of I$_h$ at the potential just above

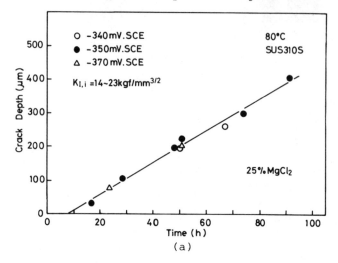

(a)

E$_{R,PIT}$ and E$_{R,CREV}$, respectively. It is shown that crack growth rate , $\dot{l}$, regareded to be equal to the plateau velocity does not depend on MgCl$_2$ concentration , electrode potential and types of crack initiation. The SCC region in terms of I$_h$ was written as:

$$I^*_{h,i} < I_h < \dot{l} \qquad (i=PIT,CREV) \qquad (5)$$

There exist critical MgCl$_2$

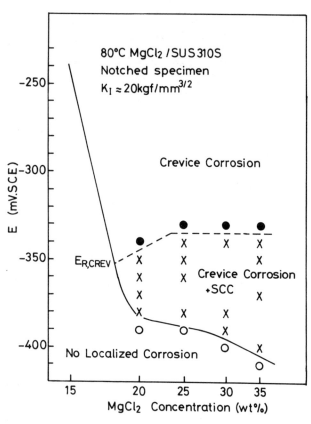

Fig. 11 - The SCC region in terms of electrode potential and MgCl$_2$ concentration for the notched specimen under a stress intensity of 20kgf/mm$^{3/2}$ at 80°C, which is much enlarged to less noble potentials and to lower MgCl$_2$ concentrations than that for the smooth specimen shown in Fig. 7(b).

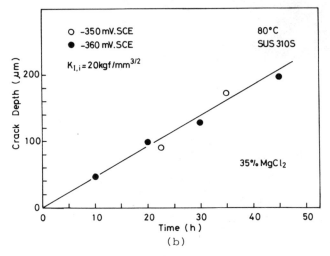

(b)

Fig. 12 - Growth behavior of maximum cracks on the notched specimen in 25% (a) and 35% (b) MgCl$_2$ solutions at different electrode potentials.

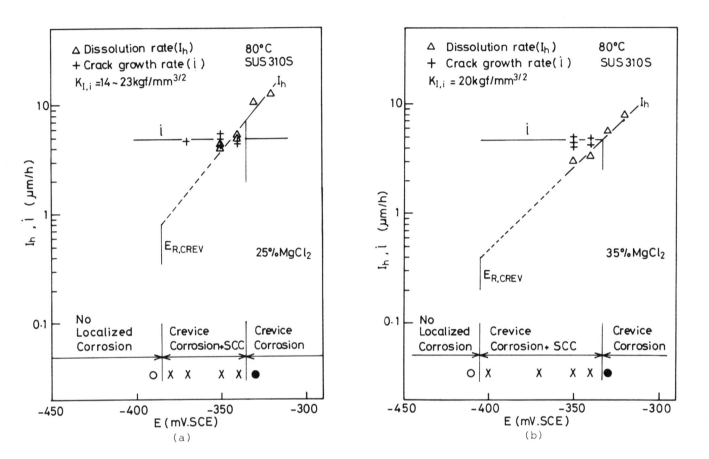

Fig. 13 - Comparison of crack growth rates, $\dot{\mathrm{l}}$, with dissolution rates at crevice-bottom, I_h, against electrode potential for the notched specimens in 25% (a) and 35% (b) MgCl$_2$ solutions. Obtained conclusion is similar to that of Fig. 10 for the smooth specimens, except that the lower limit of SCC range is determined by $E_{R,CREV}$ for crevice corrosion at the base of notch instead of $E_{R,PIT}$.

concentrations, C^*_i (i=PIT,CREV),where $I^*_{h,i}$ is equal to $\dot{l}$ and below which SCC will not occur. C^*_{CREV} is much lower than C^*_{PIT} ,then SCC region for the notched specimen was enlarged to lower $MgCl_2$ concentrations. Small difference between C^*_{CREV} and C^*_{PIT} comes from the geometrical configuration of the crevice used in this work. A through the thickness crevice of 0.25mm in width and 4mm in depth introduced to a tapered DCB specimen of Type 316 steel could initiate SCC cracks in even 0.03% NaCl solution at 80°C [4].

CONCLUSION

SCC tests of Type 310S steel were conducted in neutral 15 - 35% $MgCl_2$ solutions at 80 °C to examine two types of crack initiation , namely from pits in the smooth specimen and from crevices in the notched specimen.

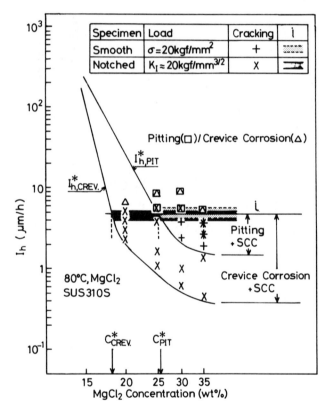

Fig. 14 - Comparison of SCC region for the smooth specimen with that for the notched specimen in terms of dissolution rate at pit/crevice-bottom, I_h , and $MgCl_2$ concentration. The crackings from pits for the smooth specimen occurred in the region of $I^*_{h,PIT}<I_h<\dot{l}$, and those from crevices for the notched specimen occurred in the region of $I^*_{h,CREV}<I_h<\dot{l}$. Intersections of $I^*_{h,i}$(i=PIT, CREV) with crack growth rate, $\dot{l}$, give the critical concentrations of $MgCl_2$, C^*_i (i=PIT, CREV), below which there exists no SCC region.

1) The repassivation potential for pit , $E_{R,PIT}$, and that for the notch as crevice , $E_{R,CREV}$, were determined as a function of $MgCl_2$ concentration.
2) The dissolution rate at pit-bottom , I_h , was found to be expressed as:
$$E = \alpha + \beta \cdot \log I_h$$
where α and β are constants which depend on $MgCl_2$ concentration. Dissolution rate at crevice was taken to be equal to that of I_h at corresponding electrode potential and $MgCl_2$ concentration.
3) The crack growth rate , $\dot{l}$, did not depend on $MgCl_2$ concentration , electrode potential and types of crack initiation under the stressing conditions used.
4) The SCC region in terms of dissolution rate at local anode , I_h , was written as:
$$I^*_{h,i} < I_h < \dot{l} \quad (i=PIT,CREV)$$
or in terms of electrode potential , E , as:
$$E_{R,i} < E < E^V \quad (i=PIT,CREV)$$
where $I^*_{h,i}$ is the value of I_h at just above $E_{R,i}$ and E^V is the potential where I_h was equal to $\dot{l}$. The potential range for SCC of the notched specimen was enlarged to less noble potentials than that of the smooth specimen , because $E_{R,CREV}$ is less noble than $E_{R,PIT}$ or $I^*_{h,CREV}$ is smaller than $I^*_{h,PIT}$. In higher concentrations of $MgCl_2$ solutions, $I^*_{h,i}$ became smaller , then SCC occurred more easily.
5) There existed the critical $MgCl_2$ concentrations , C^*_i (i=PIT,CREV) , below which SCC did not occur. C^*_{CREV} was much lower than C^*_{PIT} , which is another marked feature of crevices compared with pits.

REFERENCES

1. S. Tsujikawa and Y. Hisamatsu, J. Inst. Metals Japan, 41, 829 (1977)
2. R. W. Staehle , F. H. Beck and M. G. Fontana , Corrosion , 15, 373 (1959)
3. M. Kowaka and T. Kudo , Tetsu-to-Hagane, 62 , 390 (1976)
4. S. Tsujikawa , K. Tamaki and Y. Hisamatsu, Tetsu-to-Hagane , 66 , 2067 (1980)
5. T. Nakamura , T. Tsujikawa and Y. Hisamatsu, J. Inst. Metals Japan , 46, 503 (1982)
6. S. Tsujikawa and Y. Hisamatsu , Proc. of the 3rd Soviet-Japanese Seminar on Corrosion and Protection of Metals , 119, Nauka (1984)
7. S. Tsujikawa , Y. Soné and Y. Hisamatsu, Conference on Corrosion Chemistry within Pits , Crevices and Cracks , held at NPL ,London (Oct.,1984)
8. O. Ando , S. Tsujikawa and Y. Hisamatsu, Boshoku Gijutsu, 27 , 580 (1978)
9. A. J. Russel and D. Tromans , Met. Trans. , 10A , 1229(1979)

THE RELATIONSHIP BETWEEN HYDROGEN-INDUCED THRESHOLDS, FRACTURE TOUGHNESS AND FRACTURE MODES IN IN903

Neville R. Moody, Ron E. Stoltz, Mark W. Perra
Sandia National Laboratories
Livermore, California, USA

ABSTRACT

The effect of hydrogen concentration and grain size on sustained load cracking thresholds and rising load fracture toughness has been studied in the high strength Fe-Ni-Co superalloy, IN903. Sustained load crack growth tests in 207 MPa hydrogen gas show that thresholds increase from 30 to 40 MPa-m$^{1/2}$ as the grain size increases from 23 to 200 μm. In contrast, fractures toughness values determined from precracked, gas phase charged samples are independent of grain size. Sustained load cracking in hydrogen gas is typified by an intergranular fracture mode at small grain sizes and a mixed intergranular and slip band fracture mode at large grain sizes. Fracture toughness tests of precharged samples exhibited slip band fracture at all grain sizes. Analysis and modelling showed that matrix carbides govern the fracture process in fracture toughness tests whereas grain boundaries govern the fracture process in slow crack growth tests. The differences between rising load fracture toughness and slow crack growth susceptibility and corresponding fracture modes can qualitatively be related to the differences between hydrogen distributions.

IT IS WELL ESTABLISHED that hydrogen can significantly reduce the resistance to crack initiation and crack growth in many alloy systems. As a result, determination of hydrogen-induced fracture susceptibility is of critical importance to many materials applications in hydrogen and hydrogen-producing environments.

Two measures of crack growth resistance are rising load fracture toughness tests and in-situ slow crack growth tests under sustained load or decreasing load conditions. An extensive data base exists on fracture toughness of bcc steels [1-5] and hydrogen-affected crack growth thresholds in bcc [6-12] and fcc alloys [13-15]. However, corresponding data on fracture toughness as a function of hydrogen concentration is sparse due to experimental difficulties. As a result, the relationship between fracture toughness and slow crack growth thresholds in hydrogen environments is not well established.

It is well documented that hydrogen concentration, temperature, stress state, and yield strength exert strong effects on crack growth susceptibility [7, 8, 16, 17]. Alloy microstructure (second phases, precipitates, inclusions, carbides, etc.) also influences susceptibility [18]. However, the interactions between each of these variables is not well defined. Consequently, recent studies [2-5, 17, 19, 20, 21] seek a comprehensive understanding of fracture for a material through the use of quantitative models based on the micromechanisms of fracture. A similar approach was used in this study to examine the relationship between hydrogen-induced fracture toughness and slow crack growth thresholds in the fcc, Fe-Ni-Co superalloy IN903. The fracture toughness values were higher and exhibited a different grain size dependence than the slow crack growth thresholds at similar hydrogen concentrations. It will be shown through both modelling and observation that fracture initiates at different locations in each type of test. Combined with the difference between hydrogen distributions in each type of test, the difference in properties and fracture modes will be qualitatively related to the possible operation of a rapid hydrogen transport path along grain boundaries.

MATERIALS AND PROCEDURE

IN903, a γ' strengthened, fcc superalloy with the composition given in Table I, was used for this study. Seven 76 mm diameter sections were cut from bar stock of this material, solution-treated at temperatures ranging from 940°C to 1180°C, water quenched and double-aged

Table I. Composition of IN903 (wt.%)

Ni	Co	Ti	Al	Nb	C	Mn	Si	Mo	Fe
37.78	15.25	1.33	0.07	3.07	0.04	0.15	0.7	0.1	bal

Table II. Solution Treatment Temperature and Corresponding Grain Sizes in IN903.

940°C/1h	28 μm
1000°C/1h	49 μm
1060°C/1h	114 μm
1060°C/4h	172 μm
1100°C/1h	200 μm
1120°C/1h	226 μm
1180°C/1h	350 μm

Table III. Correlation Between Charging Pressure and Hydrogen Concentration from Charging at 300°C.

P, atm	f, atm	C_H, appm
1360.5	3143	5000
680.3	1059	2900
340.1	443.3	1876
68.0	74.1	767

at 720°C/8h plus 620°C/8h. The solution treatments resulted in grain sizes ranging from 28 to 350 μm as shown in Table II. The microstructures consist of equiaxed grains with a significant number of annealing twins, matrix and grain boundary carbides, and 20 nm diameter Y' precipitates. The size and volume fraction of Y' did not change with grain size. The size, spacing and volume fraction of matrix carbides was also independent of grain size. However, the size and number density of grain boundary carbides increased with increased grain size.

Four smooth bar tensile samples with 5 mm diameters and 25.4 mm gage lengths were machined from each solution treated section. Charpy bars for fracture toughness tests were also machined from sections with grain sizes up to 200 μm. The Charpy bars facilitated hydrogen charging because of the relatively short diffusion distances required for hydrogen saturation. As shown by Hirth and Froes [22], fracture toughness measured from Charpy bars is in reasonable agreement with K_{Ic} values measured from compact tension samples for materials having yield strengths from 1000 to 1500 MPa and toughness values up to 120 MPa-m$^{1/2}$. WOL samples were used to measure slow crack growth thresholds in hydrogen gas. These 22.2 mm thick samples were machined from material with grain sizes ranging from 49 to 200 μm.

Two tensile samples for each solution treated condition were hydrogen-charged in 1360.5 atm pressure hydrogen gas at 300°C for 60 days. This gave a calculated hydrogen concentration of 5000 appm in each sample when pressure was corrected to give effective hydrogen fugacity [23, 24]. Charpy bars with 28 μm grain sizes were also charged at hydrogen pressures which gave hydrogen concentrations up to 5000 appm as shown in Table III.

The slow crack growth tests were conducted in 207 MPa pressure hydrogen gas at room temperature. These conditions give an estimated surface hydrogen concentration of 5500 appm. Whereas estimation of hydrogen concentrations in precharged samples was calculated directly from permeation data obtained at 300°C [23, 24], estimation of hydrogen concentration at room temperature must include the effects of hydrogen trapping at Y' precipitates. This requires trap site density, lattice concentration and binding energy information [25]. IN903 has a trap site density of 4.55×10^{20} sites/cm^3 and a binding energy of 0.15 eV as determined from permeation data [26]. Because the binding energy for hydrogen trapping at Y' precipitates is 0.15 eV, trapping is not a factor at 300°C but is at room temperature. With a room temperature lattice hydrogen concentration of 2800 appm and equations relating trapped and lattice concentrations to total concentration [25], a maximum hydrogen concentration of 5500 appm was calculated.

The tensile tests were conducted on an Instron Tensile testing machine using an initial strain rate of 3.3×10^{-4} s^{-1}. Fracture toughness tests using the precracked Charpy bars were conducted under three point bend loading as described in ASTM E399 at a loading rate of 42 MPa-m$^{1/2}$/min. Tests at loading rates from 4.2 to 472 MPa-m$^{1/2}$/min showed that fracture toughness values are independent of loading rate below a rate of 190 MPa-m$^{1/2}$/min. The charged samples exhibited a distinct load drop at maximum load corresponding to a pop-in failure. The maximum load with measured fatigue precrack lengths were used to determine fracture toughness. After charging, all charged samples were maintained at 0°C in an ice bath to prevent near surface hydrogen offgassing [27]. Just prior to testing, all samples were brought to room temperature.

Slow crack growth thresholds were determined using bolt loaded WOL samples tested at 298 K in 207 MPa hydrogen gas. The samples were initially loaded to a stress intensity of 70 MPa-m$^{1/2}$ then exposed to hydrogen gas. During the tests, loads were continuously monitored to determine when crack arrest

Table IV. Mechanical Properties of Hydrogen Charged and
Uncharged Tensile Samples.

Grain Size	Yield Strength	Ultimate Tensile Strength	Reduction in Area	Fracture Strain	Fracture Strength
μm	MPa	MPa	%		MPa
Uncharged					
28	1081	1315	37	0.47	2093
49	1088	1327	36	0.44	2078
114	1080	1295	34	0.42	1981
172	1065	1289	32	0.37	1981
200	1054	1276	15	0.17	1490
226	1007	1248	13	0.14	1407
350	1000	1119	4	0.04	1164
Charged					
28	1102	1314	13	0.14	1542
49	1091	1324	13	0.14	1549
114	1082	1257	11	0.11	1427
172	1073	1239	11	0.11	1394
200	1072	1201	6	0.06	1249
226	1052	1166	4	0.04	1200
350	982	1001	3	0.03	1030

* Data for grain sizes up to 172 μm were previously reported [28].

occurred. After four months with no detectable crack growth, the samples were removed from the gas environment, allowed to outgas several weeks and then pulled to failure. The threshold stress intensities were determined from the crack lengths and loads at crack arrest. More complete experimental details are given elsewhere [13].

RESULTS

The tensile test results are given in Table IV. (The results for grain sizes up to 172 μm have been previously reportd [28].) As these results show, yield strength is essentially independent of grain size up to 200 μm. At larger grain sizes, yield strength exhibits a small decrease with increasing grain size. In samples with grain sizes larger than 49 μm, the ultimate tensile strengths exhibit a grain size dependence. The fracture stress (calculated at the sample center-line with the Bridgeman correction) and fracture strain exhibit a two stage grain size dependence. Whereas a weak grain size dependence is observed up to grain sizes of 172 μm, a stronger grain size dependence is observed at larger grain sizes.

Tensile sample fracture occurred by several different failure modes which exhibit a complex dependence on hydrogen concentration and grain size. Fracture in small-grained uncharged tensile samples occurred by microvoid coalescence with large dimples at matrix carbides as shown in Fig. 1. In uncharged samples with grain sizes larger than 172 μm, fracture occurred by a combination of intergranular and ductile intergranular failure which became totally intergranular at a grain

size of 350 μm. In contrast, fracture in hydrogen charged tensile samples with grain sizes up to 172 μm occurred primarily by slip band failure with some twin band failure. Some intergranular failure was also observed in 114 and 172 μm grain size samples. At grain sizes larger than 172 μm, hydrogen-induced failure was totally intergranular.

The fracture toughness of IN903 without hydrogen was 95 MPa-m$^{1/2}$ and independent of grain size as shown in Fig. 2. Addition of 5000 appm reduced the fracture toughness to 48 MPa-m$^{1/2}$. The fracture toughness of the hydrogen charged material was also independent of grain size. The fracture toughness of 28 μm grain size samples for hydrogen concentrations up to 5000 appm is shown in Fig. 3. This figure shows that the greatest decrease in fracture toughness occurred with the addition of 770 appm hydrogen. Further hydrogen additions reduced fracture toughness to a lesser degree.

Fracture in all uncharged three point bend fracture toughness samples occurred by microvoid coalescence with primary dimple formation at matrix carbides (Fig. 4a). In contrast, fracture in all samples charged with 5000 appm hydrogen occurred primarily by slip band failure with some twin band failure (Fig. 4b). Fracture in 28 μm grain size samples with hydrogen concentrations ranging from 770 to 5000 appm also occurred by slip band failure as shown in Fig. 5. However, the slip band facets exhibited more local plasticity at lower hydrogen concentrations. The local plasticity arises from tear ridge and void formation along slip band intersections.

The slow crack growth thresholds were lower than the fracture toughness values and

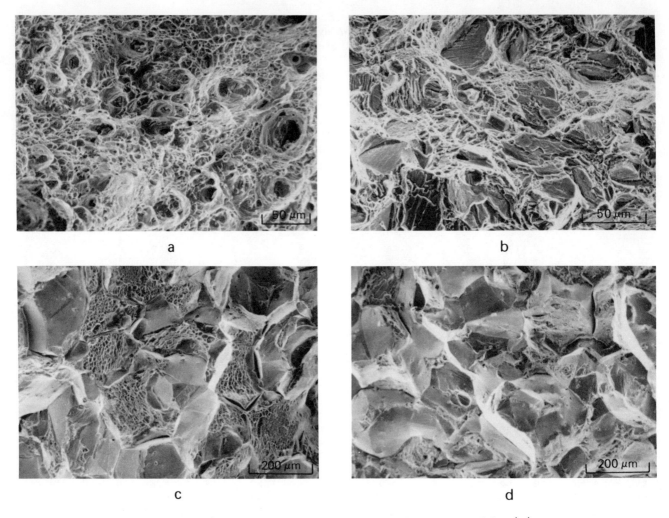

Fig. 1.   Fracture in 43 μm grain size tensile samples occurred by (a) microvoid coalescence in uncharged samples with primary void formation at matrix carbides and by (b) slip band fracture in charged samples.   In contrast, fracture in 226 μm grain size samples occurred by (c) a mixed mode of intergranular and ductile intergranular fracture in uncharged samples and by (d) intergranular fracture in charged samples.

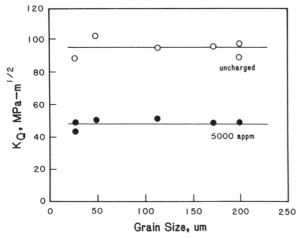

Fig. 2.   Addition of 5000 appm hydrogen reduced the fracture toughness of IN903 from 95 to 48 MPa-m$^{1/2}$.   Fracture toughness of both the charged and uncharged samples was independent of grain size.

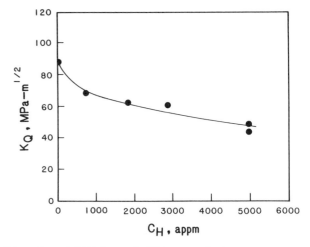

Fig. 3.   Addition of 770 appm hydrogen caused a significant drop in fracture toughness. Addition of more hydrogen led to further decreases in fracture toughness.

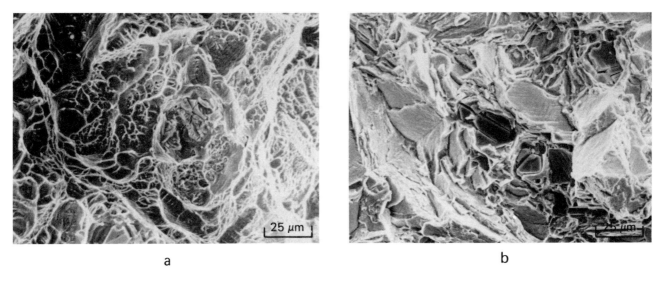

a

b

Fig. 4. Fracture in all (a) uncharged fracture toughness samples occurred by microvoid coalescence with primary void formation around matrix carbides and in all (b) charged samples by slip band failure.

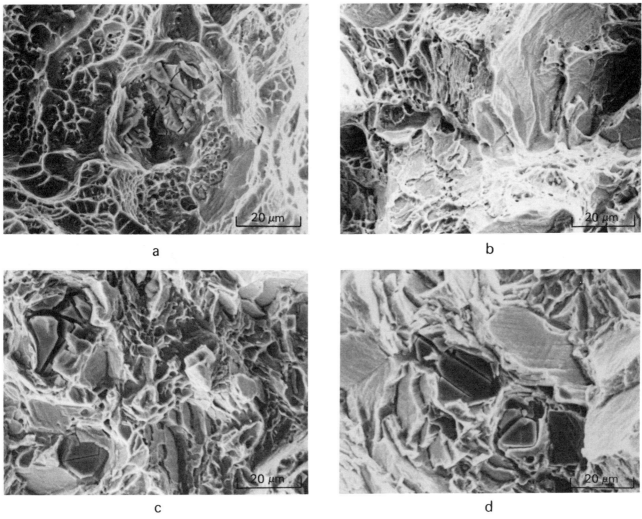

a

b

c

d

Fig. 5. Fracture in 28 μm grain size samples changed from (a) microvoid coalescence to slip band fracture with addition of (b) 770, (c) 2900 and (d) 5000 appm hydrogen. The amount of local plasticity on the slip band facets decreased with increased hydrogen concentration.

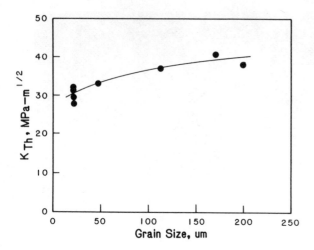

Fig. 6. IN903 slow crack growth thresholds increased from 30 MPa-m$^{1/2}$ in 23 µm grain size samples to 40 MPa-m$^{1/2}$ in 200 µm grain size samples.

exhibited a moderate grain size dependence. As shown in Fig. 6 (which includes small grain size IN903 data from previous studies [13, 29]), the thresholds increased from 30 MPa-m$^{1/2}$ in 23 µm grain size samples to 40 MPa-m$^{1/2}$ in 200 µm grain size samples.

Fracture in all slow crack growth samples occurred intergranularly at stress intensities near the initial stress intensity. However, fracture became more complex during stage I crack growth near the crack arrest front (near threshold). Near the crack arrest front, slow crack growth fracture (Fig. 7) occurred intergranularly in the small grain size samples. As grain size increased, fracture occurred by mixed mode of intergranular and slip band failure. The fracture surfaces in Fig. 7 show that intergranular fracture was the predominant failure mode in samples with grain sizes up to 49 µm. However, slip band failure was the predominant failure mode in larger grain size samples.

DISCUSSION

The IN903 results show that fracture toughness values are higher than slow crack growth thresholds and exhibit different responses to variation in grain size. This is similar to previous work on 4340 [2, 3, 17, 30, 31] in which different responses of fracture toughness and slow crack growth thresholds to grain size were observed (Fig. 8). In IN903, the fracture toughness tests are rising load tests on material with a uniform hydrogen concentration. In contrast, the IN903 slow crack growth tests are decreasing load tests on material with a steep hydrogen concentration gradient. The resulting fracture toughness and slow crack growth threshold values differ in magnitude and grain size dependence as shown in Fig. 9. The corresponding fracture modes also differ significantly. The difference in fracture toughness and slow crack growth threshold behaviors will be analyzed below by comparison of fracture modes and application of fracture models. These results will be compared to the different hydrogen distributions that exist in each type of test to explain why the two measures of crack growth resistance differ.

FRACTURE TOUGHNESS - Fracture in the uncharged samples occurred by microvoid coalescence. The large voids in uncharged samples at matrix carbides indicate that the matrix carbides are fracture initiation sites (Fig. 4a). The interconnecting regions of many small dimples shows coalescence of primary voids was truncated by second generation void

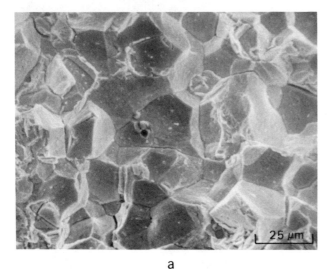

a

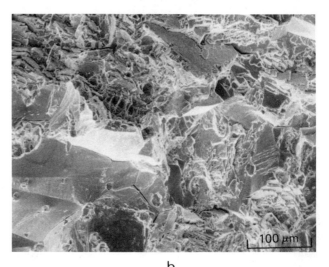

b

Fig. 7. Fracture near threshold in the (a) 23 µm grain size samples was predominantly intergranular. Slip band fracture was the predominant fracture mode in the large grain size samples as shown in (b) for the 172 µm grain size sample.

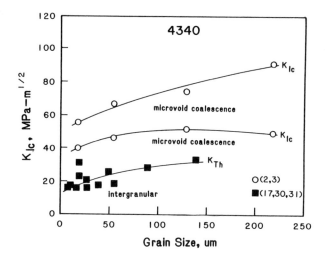

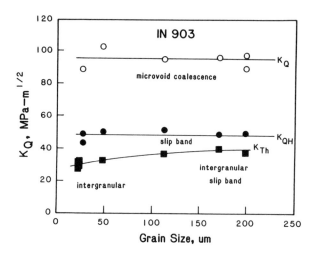

Fig. 8. Fracture toughness and slow crack growth thresholds of 4340 exhibit different behaviors with respect to grain size. No distinct correlation exists between behavior and fracture mode.

Fig. 9. Comparison of fracture toughness of charged and uncharged samples and slow crack growth thresholds shows that the threshold values are lower than the fracture toughness values and exhibit a different grain size dependence.

formation. This process is characteristic of strain localization. Although slip band fracture appears brittle (Fig. 4b), close examination of the fracture surfaces and orientation of the facets show it is also a ductile fracture process which initiates at fractured matrix carbides. In hydrogen charged fracture toughness samples, slip band fracture occurs by microvoid formation at slip band intersections. Subsequent failure occurs along interconnecting coplanar slip band segments. This is identical to hydrogen-induced slip band fracture in IN903 tensile samples which initiates at fractured matrix carbides [28].

In order to support the assertion that the fracture process in uncharged and charged samples initiates at matrix carbides, a model was sought which continuously described the fracture process over all hydrogen concentrations. Several ductile fracture models assume that fracture occurs when the equivalent plastic strain, $\bar{\varepsilon}_p$, ahead of a crack tip equals or exceeds the critical fracture strain, ε_f^*, over a characteristic fracture distance, ℓ^* [20]. A theoretical ductile fracture model which describes fracture toughness with this failure criteria is given as follows [4, 20]:

$$K_{Ic} \approx 6\sqrt{\sigma_0 E \varepsilon_f^* \ell^*} \qquad (1)$$

This equation was derived from the underlying relationship

$$\delta_{Ic} \sim \varepsilon_f^* \ell^* \qquad (2)$$

where δ_{Ic} is the critical crack tip opening displacement.

Solution of eq.(1) requires yield

strength, σ_0, elastic modulus, E, crack tip plane strain fracture strain, ε_f^*, and characteristic fracture distance, ℓ^*, data. All IN903 fracture toughness samples had the same elastic modulus and similar yield strengths. The crack tip plane strain ε_f^* are unknown. However, uniaxial ε_f^* as a function of hydrogen concentration corresponding to the hydrogen concentrations used in this study are known [33].

Use of uniaxial fracture strains instead of crack tip plane strain ε_f^* requires that two assumptions apply: (1) that the relationship between uniaxial fracture strains ahead of a plane strain crack tip is described by a unique function of stress state and (2) that hydrogen lowers the fracture strains by a constant value for any given stress state. Work on the superalloy JBK-75 suggests that these two criteria are met [34].

Fig. 10 shows that a linear relationship exists between fracture toughness of charged and uncharged samples and the corresponding square root of the uniaxial fracture strains. Furthermore, extrapolation of the results through zero indicates that fracture is hole growth dominated. Most importantly, the linear relationship shows that the characteristic fracture distance is constant. This characteristic fracture distance can be determined from eq.(1). Solution of eq.(1) with the yield strength, elastic modulus and uniaxial fracture strain data gives ℓ^* equal to 2.6 μm. To compare the calculated characteristic fracture distance with microstructural features requires solution of eq.(1) for ℓ^* with crack tip plane strain ε_f^* values. A relationship between uniaxial and plane strain ε_f^* can be obtained from the work

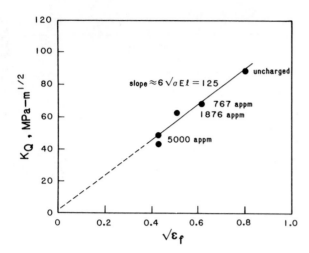

Fig. 10. The linear relationship between fracture toughness and uniaxial fracture strain indicates that the characteristic fracture distance is the same in charged and uncharged samples.

of Gerberich and coworkers [35] on IN718 and 4340. Fracture in IN718 and 4340 is similar to fracture in uncharged IN903 samples; it occurs by primary void growth at inclusions or carbides which is terminated by second generation void formation. From the ratio of primary void size to the initiating inclusion size, Gerberich et.al. [35] found that ε_f^* under plane strain conditions ahead of a crack tip is a factor of 12 lower than corresponding uniaxial values. Solution of eq.(1) with the IN903 uniaxial ε_f^* values reduced by a factor of 12 gives ℓ^* equal to 31 µm. This result, combined with the observation that fracture toughness is independent of grain size and

exhibits a linear relationship with $\sqrt{\varepsilon_f^*}$, shows

that the matrix carbide spacing of ~ 20 µm (3D nearest neighbor spacing) governs fracture toughness in both charged and uncharged samples.

SLOW CRACK GROWTH THRESHOLDS - In slow crack growth tests, fracture occurs by both intergranular and slip band failure. Fractographic observations indicate that failure occurs first by intergranular fracture of the most favorably oriented grain boundaries with subsequent slip band failure in the regions between intergranular facets. This is supported by Fig. 11 which shows the fracture modes along the crack arrest fronts for the 23 and 172 µm grain size samples. Examination of fracture along the paths of crack arrest shows that the points of furthest hydrogen-induced crack growth occurred intergranularly in all samples. The areas which exhibit ductile fracture were unfailed regions at crack arrest.

Slow crack growth thresholds have been accurately modelled in several alloy systems with local critical stress models when fracture occurs intergranularly [3, 10, 17, 36]. A local critical stress model states that fracture occurs when the applied normal stresses equal or exceed the fracture stress over the characteristic fracture distance. Unfortunately, this characteristic distance cannot be derived from first principles [20]. However, previous observations on fcc superalloys with yield strengths greater than 1000 MPa show that the fracture process zone size is defined by the grain size [29]. In a recent study of these fcc alloys [37], hydrogen-induced crack growth thresholds, yield strengths, and the critical distances for sustained crack growth were related to the

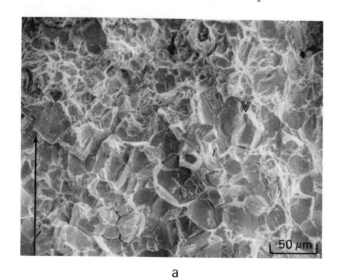

a

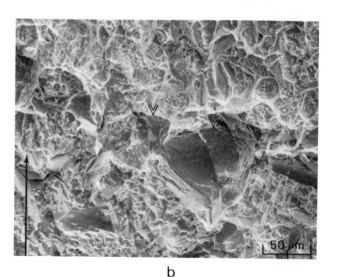

b

Fig. 11. The furthest extent of hydrogen-induced crack growth in small grain size samples (a) 23 µm and large grain size samples (b) 172 µm is always defined by intergranular fracture.

crack tip stress distribution equations given by Tracey [38] and Schwalbe [39] as follows

$$\frac{\sigma yy}{\sigma_0} = \frac{0.3}{X + 0.1} \left(\frac{0.04}{X}\right)^{n/(n+1)} \qquad 0 < X \leq 0.04$$

$$\frac{\sigma yy}{\sigma_0} = \frac{0.3}{(X + 0.1)} \qquad 0.04 \leq X \leq 0.0728 \qquad (3)$$

$$\frac{\sigma yy}{\sigma_0} = \frac{1}{(2\pi(X-0.02))^{1/2}} \qquad 0.0728 \leq X$$

$$X = x/(K/\sigma_0)^2$$

σyy is the stress normal to the crack plane, σ_0 is the yield strength, K is the applied stress intensity, n is the work hardening coefficient, and x is the distance from the crack tip of interest taken as the critical distance for fracture in this study. At threshold, σyy equals the fracture stress (σ_f) at the characteristic fracture distance from the crack tip. To predict thresholds with a local critical stress model, accurate critical fracture stress values are required. In previous studies of bcc steels, this value was obtained from notched bars tested under slow deflection rates in four point bend tests [21]. Use of precharged IN903 samples tested under the same conditions as described in previous a study[21] showed that fracture occurred exclusively by slip band failure. However, hydrogen charged IN903 tensile samples with grain sizes of 200 μm and larger, failed intergranularly. From these tensile tests, intergranular fracture stresses are obtained with the following relationship to the grain size (μm).

$$\sigma_f = 349.5 + 12744d^{-1/2} \qquad (4)$$

Solution of eq.(4) for grain sizes corresponding to those of slow crack growth threshold samples gives the values shown in Table V. These are reasonable values based on a recent study of slow crack growth thresholds in fcc alloys [37]. In that study of stainless steel and superalloys with grain sizes of 20 to 60 μm, the relationship between thresholds and fracture process zone size were accurately described using a fracture stress of 2800 MPa and working hardening coefficient of 0.15. Most interestingly, fracture stresses

Table V. Extrapolated fracture stresses from large grain size sample data

grain size	σ_f*, MPa
23	3007
49	2170
114	1543
172	1321
200	1251

extrapolated from large grain tensile sample results are in good agreement with fracture stresses calculated from eq.(3) as shown in Fig. 12. Only the extrapolated fracture stress for the 23 μm grain size samples varies from that calculated from corresponding threshold values. No correction of the calculated fracture stresses to account for hydrogen accumulation in the crack tip stress field is necessary because the small partial molar volume of hydrogen in fcc materials results in no significant hydrogen accumulations [40].

The agreement between extrapolated fracture stresses and fracture stresses calculated from threshold values show that slow crack growth thresholds can be quantitatively described with a local critical stress model.

TEST METHOD EFFECTS - The application of simple models with microstructural fracture criteria based on fractographic observations show that two distinct failure processes govern hydrogen-induced crack growth in rising load (fracture toughness) and decreasing load (slow crack growth) tests. The slow crack growth thresholds are significantly less than the fracture toughness values at similar surface hydrogen concentrations. In 23 μm grain size samples, slow crack growth thresholds are only 60% of the corresponding fracture toughness values. Although slow crack growth tests involve a moving crack front, the slow crack growth thresholds and intergranular failure cannot be attributed to the moving crack tip stress and strain field distributions because fracture in the slow crack growth tests always initiates intergranularly from the static precrack.

However, a significant difference does exist between hydrogen distributions in the two tests. Fracture toughness was determined on

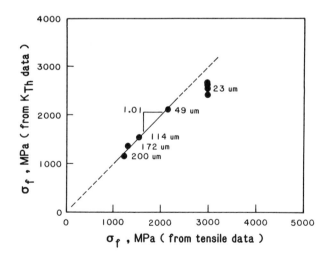

Fig. 12. Good agreement exists between fracture stresses calculated from eq. (5) with the characteristic distance equal to the grain size and fracture stresses extrapolated from tensile data of hydrogen charged large grain size samples which failed intergranularly.

STRESS CORROSION CRACKING AND CORROSION FATIGUE OF STEELS IN HOT WATER

Markus O. Speidel
Institute of Metallurgy
Swiss Federal Institute of Technology
Zurich, Switzerland

ABSTRACT

Stress corrosion cracking of quenched and tempered steels in hot water is a phenomenon which has only recently been observed in steels with yield strengths from 230 to 1000 MN/m^2. In nuclear reactor pressure vessel steels with about 430 MN/m^2 yield strength, SCC is transgranular and has a threshold K_{ISCC} of about 20 MN·m$^{-3/2}$. The maximum crack growth rate is about 7×10^{-8} m/s at 288°C. This threshold and this crack growth rate are identical to the threshold and maximum crack growth rate derived for the time-based component of corrosion fatigue crack growth observed in reactor pressure vessel steels exposed to nuclear reactor coolant at 288°C. The conclusion is drawn that corrosion fatigue crack growth under these conditions is just the superposition of stress corrosion cracking and fatigue crack growth.

STRESS CORROSION CRACKING, (SCC), of quenched and tempered steels can occur in pure, hot water [1]. It has long been known that high strength steels are susceptible to stress corrosion cracking in water at ambient temperature [2]. It is, however, a discovery of the 1980's that steels with yield strengths as low as 230 MN/m^2 can exhibit stress corrosion cracking in water. [1] [3] [4] [5] [6] The fact that stress corrosion cracking of such a wide variety of steels in such a simple environment has been observed so late may be due to the exceedingly slow growth rates of stress corrosion cracks in steels of low strength. [1] [3] [4] [5] Water temperatures higher than ambient and up to 350°C, however, may result in much faster stress corrosion crack growth rates [4] [5]. It is for this reason that both, real SCC service failures and anticipated failures in power generating equipment [3] [7] have drawn attention to SCC of steels in hot water.

In the present paper we attempt to use a specific part of the SCC-information so gained and apply it to the well known phenomenon of environment assisted fatigue crack growth in nuclear pressure vessel steels [8] [9] [10]. Stress corrosion cracking may be defined as environment-assisted subcritical crack growth under <u>constant</u> load or slowly raising load. Corrosion <u>fatigue</u>, in contrast, is defined as environment-assisted subcritical crack growth under <u>cyclic</u> loads.

STRESS CORROSION CRACKING

There is no doubt that nuclear reactor pressure vessel steel (American designation A533B, German designation 20MnMoNi55) can exhibit transgranular stress corrosion cracking in water at 288°C. This is evident from figure 1. The phenomenon has been described by various authors between 1982 and 1985, and this literature is summarized in [3].

The effect of stress intensity on the growth rates of such stress corrosion cracks is shown in figure 2. Note that there is a threshold stress intensity, K_{ISCC}, around 20 MN·m$^{-3/2}$, below which cracks are too slow to be observed so far. At higher stress intensities there is a stress-independent "plateau" stress corrosion crack growth rate. The fastest cracks, according to figure 2, propagate with a velocity of about 7×10^{-8}m/s. This crack growth rate of RPV steel in 288°C water is significantly faster than the crack growth rate in austenitic stainless steels in similar water. SCC of welded austenitic stainless steel in 288°C water, however, has resulted in many cracks in recirculation pipes in boiling water reactors. The fact then, that apparently so far no nuclear reactor pressure vessels have exhibited SCC is not due to the impossibility of SCC, nor due to the slowness of such cracks; it is probably rather due to the careful limitation of stresses (both, residual and service stresses) in nuclear reactor pressure vessels.

A further examination of figure 3 indicates that it is not entirely impossible that future, more extensive studies might reveal a K_{ISCC} even lower than 20 MN·m$^{-3/2}$ for RPV steel in 288°C water.

CORROSION FATIGUE

Fatigue crack growth rates are primarily influenced by the cyclic stress intensity range, ΔK and the modulus of elasticity, E. They can be reasonably well predicted by the two equations given in figure 5. The two equations for fatigue crack growth differ only by a factor of 3. They have been developed for fatigue crack growth rates in vacuum and in air, respectively. [1] [12]. The two equations bracket nicely experimental fatigue data developed ten years later [13]. Both, prediction and experimental data, however, are intended and valid only for the case where no environmental assisted time-base crack growth occurs in addition to fatigue itself. This means, the frequency must be high enough and the stress ratio R must be low enough. These conditions are fulfilled for the data shown in figure 5.

Lower frequencies and higher stress ratios, however, can result in much faster corrosion fatigue crack growth, as indicated in figure 6. Figure 7 is intended to help analyzing the data in figure 6.

Note that there is a threshold ΔK_{TH} for environment assisted fast (corrosion-) fatigue crack growth. This threshold, according to figure 6, depends on the stress ratio R. If we plot all the threshold values ΔK_{TH} so far reported [8] [9] [10] versus the stress ratio R, figure 8, we find that the lower boundary is given by a line that passes through $\Delta K_{TH} = 20$ MN·m$^{-3/2}$ at R = 0. This corresponds directly to the threshold stress intensity $K_{ISCC} = 20$ MN·m$^{-3/2}$ which we have found for stress corrosion cracking (figure 2). As both types of environment-assisted cracking start at the same stress intensity, the thought may be permitted that they are essentially the same.

If then for the moment we assume that corrosion fatigue of RPV steel in nuclear reactor coolant at 288°C is identical with stress corrosion cracking (figure 2), but now under cyclic loads, we can make two predictions. The first is shown schematically in figure 9, indicating that each time the frequency is lowered by a factor of 10, the "plateau" corrosion fatigue crack growth rate (figures 7 and 8) should go up by a factor of 10. This is indeed born out experimentally as shown

in figure 10 [15] and 11.

The second prediction is also indicated in figure 11: if corrosion fatigue in this particular case is just stress corrosion cracking under cyclic loads, then the maximum (plateau) corrosion fatigue crack growth rate at each frequency should (within a factor of two) be identical to the stress corrosion crack growth plateau velocity (figures 2, 3 and 4) times the cyclic frequency. This prediction is again substantiated by the experimental data obtained by a number of authors [9].

CONCLUSION

Corrosion fatigue crack growth (environment-assisted cyclic crack growth) of reactor pressure vessel steel in nuclear reactor coolant (PWR water) at 288°C can be considered as the superposition of fatigue and stress corrosion cracking.

This provides one more reason for a much more detailed investigation of stress corrosion cracking of steels in hot water.

REFERENCES

1) Markus O. Speidel, "Stress corrosion cracking and corrosion fatigue-fracture mechanics", in "Corrosion in Power Generating Equipment", M.O. Speidel and A.Atrens, editors, Plenum Press, New York and London,1984, pp 85-132

2) Hugh L. Logan, "The stress corrosion of metals", John Wiley, New York, 1966

3) Markus O. Speidel, "Stress Corrosion Cracking of Nuclear Pressure Vessel Steel", Proc. IAEA Specialists' meeting on "Subcritical Crack Growth", Sendai, Japan, May 15-17, 1985, to be published as a NUREG report, USA, 1986.

4) Markus O. Speidel and Ruth M. Magdowski, "Stress Corrosion Cracking of Steam Turbine Steels - an Overview", Proc. Second International Symposium on Environmental Degradation of Materials in Nuclear Power Systems-Water Reactors, Monterey, California, USA, Sept. 9-12, 1985

5) Markus O. Speidel and Ruth M. Magdowski, "Environmental Cracking in Steam Turbines", Proc. CORROSION/86, March 17-21, 1986, Houston/Texas, USA

6) Markus O. Speidel, unpublished work, ETH Zurich Switzerland, 1985

7) Corrosion in Power Generating Equipment,
   Markus O. Speidel and Andrejs Atrens, editors,
   Plenum Press, New York and London, 1984

8) Proceedings of the International Atomic Energy
   Agency Specialists' Meeting on Subcritical
   Crack Growth, Freiburg, 1981, NUREG/CP-0044
   MEA-2014, Vol I and II, 1983

9) IAEA Specialists' Meeting on Subcritical
   Crack Growth, May 1985, Reprots of the Research Institute for Strength and Fracture
   of Materials, Tohoku University, Senday,
   Japan. Vol. 19, 1985

10) Proceedings of IAEA Specialists' Meeting on
    Subcritical Crack Growth, Sendai, 1985.
    to be published as NUREG report, USA, 1986.

11) H.C. Cowen, P. Hurst and G.J. Lloyd, "The
    JCCGR inter-laboratory round robin on slow
    strain rate stress corrosion testing of
    PWR forging", in 9), pp 120-136.

12) Markus O. Speidel, "Fatigue crack growth at
    high temperatures", in "High-temperature
    materials in gas turbines", P.R.Sahm and
    M.O . Speidel, editors, Elsevier, Amsterdam
    London, New York, 1974, pp 250-255.

13) R.L. Jones, "Results of ICCGR cyclic crack
    growth round robin test programs", in 10)

14) B. Tomkins and P.M. Scott, "Some thoughts
    on establishing design and inspection codes
    for corrosion fatigue", in 8), pp 329-342

15) J.D. Atkinson, P.M. Scott and D.R.Tice,
    "A review of the UK research programme
    on corrosion fatigue crack propagation
    in pressure vessel steels exposed to PWR
    environments", in 9), pp 166-173, and in 10).

Fig. 1 Transgranular stress corrosion crack in nuclear reactor pressure vessel steel

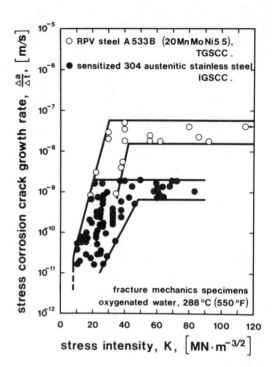

Fig. 3 Comparison of the growth rates of stress corrosion cracks in steels commonly used for nuclear reactor pressure vessels and for recirculation piping

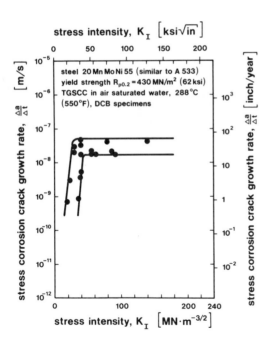

Fig. 2 Effect of stress intensity on the growth rates of transgranular stress corrosion cracks in nuclear pressure vessel steel

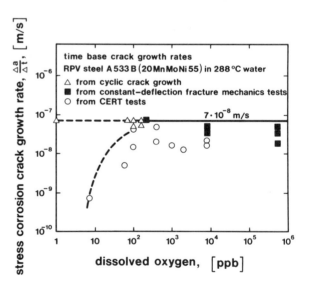

Fig. 4 Effect of dissolved oxygen on time based crack growth (stress corrosion cracking and corrosion fatigue) in reactor pressure vessel steel exposed to 288°C high temperature water

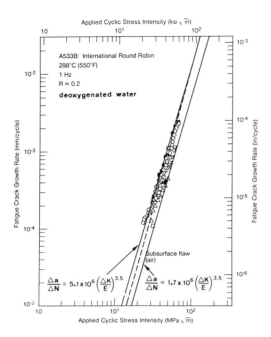

Fig. 5 Fatigue crack growth rates in reactor pressure vessel steel measured in 288°C deoxygenated water at low R ratio and relatively high frequency of 1 Hz. Dashed line corresponds to crack growth in air

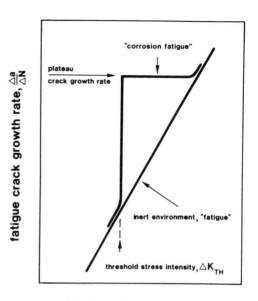

cyclic stress intensity range, △K

Fig. 7 Schematic analyzing the corrosion fatigue data of figure 6

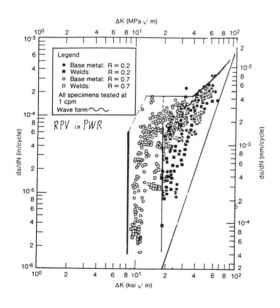

Fig. 6 Corrosion fatigue crack growth rates in reactor pressure vessel steels in PWR primary water. Effect of stress intensity and R ratio. Tested at the relatively low frequency of 1 cycle per minute

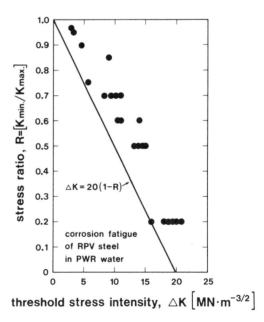

threshold stress intensity, △K [MN·m$^{-3/2}$]

Fig. 8 Threshold stress intensity ΔK_{TH} for the onset of high corrosion fatigue crack growth rates according to figures 6 and 7

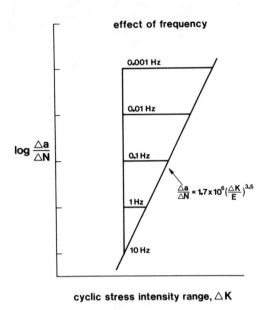

Fig. 9   Schematic predicting the effect of
frequency on corrosion fatigue crack
growth rates when corrosion fatigue is due
to the superposition of fatigue and stress
corrosion cracking

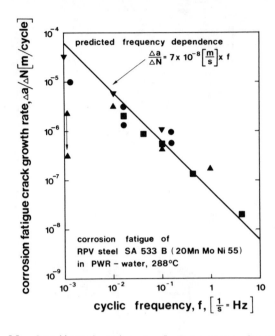

Fig. 11   Predicted and actual frequency dependence
of the corrosion fatigue crack growth
rate in reactor pressure vessel steel
exposed to PWR water at 288°C. Prediction
based on the assumption that corrosion
fatigue results from the superposition of
fatigue and stress corrosion cracking.

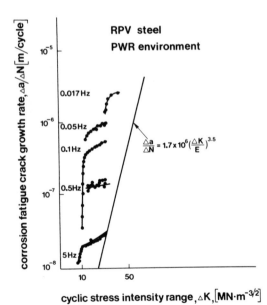

Fig. 10   Effect of frequency on the corrosion
fatigue crack growth rates in reactor
pressure vessel steel exposed to PWR
water at 288°C. Prediction based on the
assumption that corrosion fatigue results
from the superposition of fatigue and
stress corrosion cracking

ELECTROCHEMICAL AND HYDROGEN PERMEATION STUDIES OF IRON BINARY ALLOYS CONTAINING Cr, Mo, Si, Cu and Mn

James A. Smith
Naval Research Laboratory
Washington, D.C., USA

A study was made of the effect of Cu, Cr, Mn, Mo, and Si additions on the electrochemical and hydrogen absorption behavior of iron. Utilizing the permeation technique and a computer driven electrochemical potential stepping and data monitoring system, it is shown that in 1N H_2SO_4 solutions permeation rates, ip values, decrease with respect to the base iron material and decrease with increasing alloy content within alloy groups. The binary alloys show the following order of decrease in ip values: Mo<Mn<Si<Cu<Cr. The experimentally observed permeation rates differ by at least an order of magnitude in 0.6N NaCl solutions (pH 6.3) when compared to those in 1N H_2SO_4 solution (pH 0.4). However the order of ranking remains essentially the same. Electrochemical values of Ecorr and Icorr are also shown to increase with increasing alloy content. The exceptions are Cr and Si alloys in which Icorr values decrease.

These trends can be explained by noting the inhibition of cathodic hydrogen evolution processes and anodic metal ionization processes by the surface absorption of solution anions and/or by the formation of protective films.

THE ENTRANCE OF HYDROGEN into iron-base alloys is objectionable because of the deleterious effect it has on material mechanical and physical properties (1). Although hydrogen may be already present in a particular alloy, it is rarely present in sufficient amounts to cause the embrittlement associated with the delayed failure phenomenom. However, additional hydrogn may come from external sources, eg. cathodic protection schemes or welding operations (2,3). It is this additional hydrogen that is of particular importance in the embrittlement of high strength alloys (4).

There is general accord among investigators that adsorption is the precursor of all heterogeneous reactions at the metal/solution interface (5-7). The nature of surface adsorbed "complex" primarily determines whether metal dissolution, passivation, absorption or desorption occurs.

In iron-base alloys the basic process that causes hydrogen adsorption is shown in Figure 1.

Cathodically absorbed hydrogen atoms are produced on the iron surface by the Volmer reaction.

$$H_3O+ + e- = Had + H_2O$$

or

$$H_2O + e- = Had + OH-$$

The absorbed ions may combine to form hydrogen molecules by the Tafel reaction:

$$2Had = H_2$$

or form hydrogen molecules by the Heyrovsky reaction:

$$H_3O+ + Had + e- = H_2 + H_2O$$

or

$$H_2O + Had + e- = H_2 + OH-$$

The present study is part of a program to examine the role of hydrogen in stress corrosion cracking. The purpose of the study is to characterize experimental material as to composition and as to electrochemical and permeation behavior.

MATERIALS AND EXPERIMENTAL WORK

The apparatus used for the permeation studies, an adaptation of the Devanathan and Stachurski cell (8), is shown in Figure 2. The cell utilizes two matching electrochemical chambers with the working electrode mounted on Teflon O-rings between the two chambers.

Alloy membranes are machined from vacuum melted stock material, rolled to the desired thickness, annealed at 650 C in vacuum for 2 hrs and furnace cooled. Membranes are then cut to provide specimens with input/output diameters of 19 mm , abraded through 600 grit silicon carbide paper, degreased and washed with distilled water. The exit side of the membranes is plated with a thin layer of palladium (420 A) deposited from a mildly alkaline electroplating formulation called "Sel-Rex Palladex VI". The palladium provides a uniform, low porosity layer which increases permeation efficiency and prevents surface attack under the imposed anodic potential. Alloy compositions are given in Table 1.

The charging solutions are respectfully 0.6N NaCl (pH 6.3) and 1N H_2SO_4 (pH 0.4). Chemicals are reagent grade quality and the expreriments are conducted at 25 C under conditions of deaeration.

Measurements are conducted in the following manner. In the exit cell the permeation current is measured in 0.2N NaOH at +100 mV. When the cell background permeation current decreases to less than 1 uA, the charging solution is introduced into the charging cell and the change in permeation current with time is recorded. After establishing a constant value for the permeation (approximately 12 hours), polarization studies are initiated.

The electrochemical polarization experiments are ideally suited to computer control since the potential can be altered as a function of time and the resulting current responses accurately measured. In these tests the potential on the alloy membranes is controlled by a Model 173 PAR potentiostat/galvaniostat which in turn is controlled by a Apple IIe computer system with 128K memory, dual floppy disk storage, and 12 bit input-output facilities including digital to analog (DAC) and analog to digital (ADC) converters. Communication between the operator and computer is via Basic computer language and between computer and cell by an in-house Basic program (9).

The system automatically sets the potential to the desired value, applies potential pulses in staircase form to the electrode and samples the current which flows during the pulses. Experimental polarization and current data together with permeation data are then recorded as a function of time by a interactive data acquisition system. These data are subsequently stored on a disc for later processing.

RESULTS

Table 2 gives permeation rate (ip) data characteristic of iron and iron-base alloys in 0.6N NaCl. The Fe-Cu and Fe-Cr alloy membranes had 80% and 96% reductions in their respective permeation rates compared to that of pure iron. Manganese, molydbenum, and silicon binary alloys, however, showed increases in the permeation rate. At the Ecorr potentials all alloys showed decreases in permeation current densities relative to pure iron. These results may be explained by suggesting that the oxide surface films of Fe-Mo, Fe-Mn and, Fe-Si undergo cathodic reduction at the high cathodic charging potentials. Cathodic reduction has been observed at this laboratory in furnace produced surface oxides on pure iron membranes.

Electrochemical polarization data for the 0.6N NaCl solutions are not included because of their dubious results. Faulty data results from these near neutral solutions when externally applied currents or voltages are supplied in ramp or staircase fashion. The resulting anodic and cathodic polarization curves, E vs. log i, will not produce true Tafel slopes because of mass transfer effects caused by the transport of reactants (eg. H+ ions) to or products (eg. M+ ions) from the membrane surface. These reactions tends to proceed more sluggishly than the charge-transfer reactions and thus contribute to the overall corrosion rate.

Fe-Cu - Figure 3 shows the influences of copper additions to iron on cathodic current density (i_c), anodic current density (i_a) and permeation current density (ip) studied in 1N H_2SO_4 at different cathodic potentials. It is seen that both Icorr and Ecorr increase with increasing copper additions. Of particular interest is the very large reduction in ip. The modestly thick surface layer is identified as Cu_2O by X-ray diffraction analysis.

Fe-Cr - The Tafel behavior of both anode and cathode reactions are strongly affected by chromium additions. Figure 3 shows Ecorr trends similar to those of Fe-Cu alloys. However, Icorr decreases for the Fe-25%Cr binary alloy . It is also noticed that the critical current density for passivation, as expected, decreases with Cr content.

Fe-Mn - The effects of manganese in one and three percent additions are seen in Figure 4. The cathodic Tafel slope of the one percent alloy and the pure iron are practically superimposed while that for the three percent alloy is transposed upward approximately 1/2 decade but remains roughly parallel to the pure iron Tafel slope. The ip values at the initial charging potential for the one percent alloy are comparable to those of pure iron. At Ecorr the ip values are decreased by 100 to 150 mA/M2 compared to pure iron.

Fe-Mo - As seen in Figure 5 the anodic and cathodic Tafel slopes show little change from the values for pure iron but Icorr and Ecorr both increase. Permeation current density values similarly shows little influence from molybdenum additions.

Fe-Si - The presence of silicon in iron results in significant changes in the cathodic Tafel slope as seen in Figure 6. Significant reductions in permeation current density also occur.

COMPARISON OF ALLOY ADDITIONS

The electrochemical and permeation data for the alloys tested in 1N H_2SO_4 are summarized in Table 3. From this table together with plots of ip vs. alloy composition given in Figures 8 and 9 several important factors emerge.

1. Most alloy additions increase Ecorr.
2. Significant increases in Icorr were noted in most alloy systems. The exceptions are Si and Cr additions which show decreases.
3. Additions of Mo produce minimal influence on the ip value for iron while Mn, Si, Cu and Cr produce progressivly greater decreases in the permeation values. Similar trends are observed for ip at the Ecorr potentials.

DISCUSSION

The influence of the different alloying elements on the electrochemical polarization and permeation behavior of pure iron is explainable on the premise that the absorption of hydrogen is primarily determined by conditions or reactions at the environment/metal surface. Alloy additions may affect both the cathodic and anodic processes either singulary or together.

Consider a typical permeation experiment in which the production of hydrogen increases with increasing cathodic potential. This produces an increase in the fugacity of the hydrogen at the metal/solution interface and also an increase in the concentration of subsurface hydrogen. It is this subsurface hydrogen that controls the rate with which hydrogen passes through a binary alloy. This suggests that the permeation rate is a simple function of fugacity and the available number of surface sites. As either or both of these increase / decrease, the ip value correspondingly responds. In the real world, as shown in these experiments, various surface phenomenoa tend to control the passage of hydrogen from one state (adsorbed) to another state (absorbed). Thus both the cathodic hydrogen evolutaion process and the anodic metal ionization process are essentially determined by the inhibition of the respective processes by surface adsorption of solution anions or by the formation of protective films.

Smialowska suggests that adsorption of specific ions occurs near the potential of zero charge (10). In the absence of adsorbable substances the potential of zero charge (PZC) for iron is approximately -611 mV in 1N H_2SO_4. Thus the PZC for pure iron occurs slightly cathodic to its corrosion potential of -492 mV,S.C.E.. Iron surfaces consequently are positively charged at potentials ranging from -611 mV to -492 mV, and at potentials more positive than -492 mV. In 0.6N NaCl, near neutral solutions, the Ecorr value is -550 mV

and iron surfaces are positively charged from -611 mv and at all potentials positive to it. If it is assumed that there is no essential difference in the mechanism for permeation of hydrogen in acidic or basic solutions (11), then at Ecorr a competitive adsorption process exists between hydrogen adatoms and hydroxyl and chloride ions in 0.6N NaCl solutions. In 1N H_2SO_4 solutions the competitive adsorption would occur between hydrogen adatoms and sulphate ions. The data suggest that the inhibitive properties of the hydroxyl ions are greater than that of the sulphate ions as shown by their lower ip values at Ecorr. At initial charging potentials of -1400 mV used in the 0.6N NaCl solutions, previous studies on a series of furnace produced oxide surface layers showed that cathodic reduction of the oxide surface layers occured. The oxide films of the iron binary alloys at the high initial charging potential utilized in the 0.6N NaCl solutions appear to all undergo cathodic reduction with the exception of Fe-Cr and Fe-Cu alloys. As a result of this cathodic reduction the ip values for these alloys approach those seen in acid solutions.

CONCLUSIONS

1. In 1N H_2SO_4 solutions ip decreases with alloy content. The order of maximum decrease is: Mo < Mn < Si < Cu < Cr
2. Permeation rates in 0.6N NaCl differ by an order of magnitude from those in 1N H_2SO_4 solutions. However,the order of ranking at Ecorr remains essentially the same as in the H_2SO_4 solutions.
3. Ecorr and Icorr show positive increases with alloy additions in both 0.6N NaCl and 1N H_2SO_4 solutions. Exceptions are Cr and Si alloys where Icorr decreases.
4. The diminished permeation rates in near neutral 0.6N NaCl solutions are suggestive of competitive adsorption by hydroxyl ions and hydrogen adatoms.
5. Fe-Cr and Fe-Si binary alloys show reductions in ip values suggestive of the formation of protective films.
6. The Fe-Cu alloys show reductions in ip considered caused by the formation of a surface layer composed mainly of Cu_2O.

REFERENCES

1. M. Smialowski, "Hydrogen in Steel", Pergamon, London (1962)

2. B.F. Brown, Metallurgical Review, 13, No. 129, 171-183 (1968)

3. W. Beck, E.J. Jankowsky and P. Fisher, Naval Air Development Center Report No. NADC-MA-7140, (1971)

4. A.R. Troiano, "Hydrogen in Metals", I.M. Bernstein and A.W. Thompson (Eds.), American Society for Metals (1974)

5. T.P. Hoar, J. Electrochem. Society. 117, 17C-22C (1970)

6. B.E. Conway, "Theory and Principles of Electrode Processes", Ronald Press, New York (1965)

7. J. O'M. Bockris and A.K.N. Reddy, "Modern Electrochemistry", Vol. 2, Plenum Press, New York (1970)

8. M.A.V. Devanathan and Z. Stachwiski, Proc. Roy. Soc. A270, 90 (1962)

9. D.A. Meyn, P.G. Moore, R.A. Bayles, and P.E. Denny., in "Automated Test Methods for Fracture and Fatigue Crack Growth, ASTM STP 877", 27-43, W.H. Cullen, R.W. Landgrof, L. Kaisand and J.H. Underwood, Eds., American Society for Testing and Materials, Philadelphia, PA, (1985)

10. M. Jesionek and Z. Szklarska-Smialowska, Corrosion Science, 23, No. 2, 183-187 (183)

11. T. Zakroczymski, Z. Szklarska-Smialowska and M. Smialowska, Werkstoffe und Korrosion, 27, 625-630 (1976)

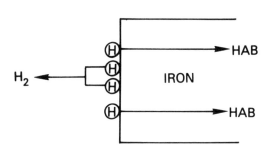

Figure 1.   Cathodically Polarized Iron

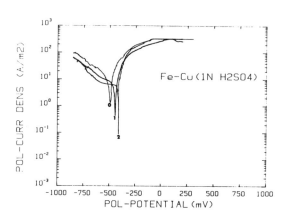

Figure 3.   Polarization behavior of the Fe-Cu binary alloys in 1N H2S04
0-pure iron, 1-low alloy, 2-high alloy

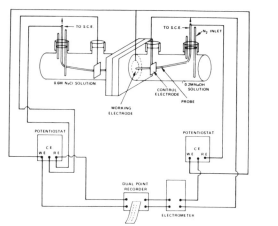

Figure 2.   Schematic of Hydrogen Permeation Apparatus

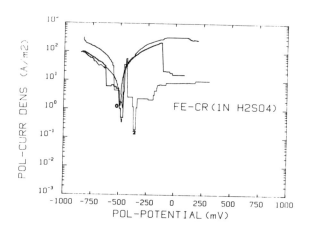

Figure 4.   Polarization behavior of the Fe-Cr binary alloys in 1N H2S04
0-pure iron, 1-low alloy, 2-high alloy

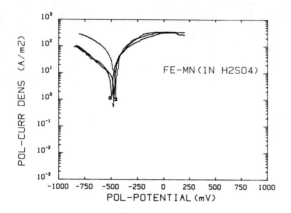

Figure 5.   Polarization behavior of the Fe-Mn
            binary alloys in 1N H2SO4
            0-pure iron, 1-low alloy, 2-high
            alloy

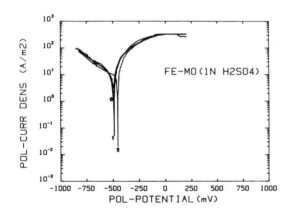

Figure 6.   Polarization behavior of the Fe-Mo
            binary alloys in 1N H2SO4
            0-pure iron, 1-low alloy, 2-high
            alloy

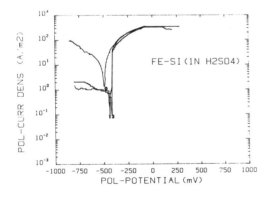

Figure 7.   Polarization behavior of the Fe- Si
            binary alloys in 1N H2SO4
            0-pure iron, 1-low alloy, 2-high
            alloy

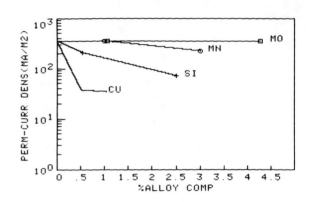

Figure 8.   Permeation behavior at initial
            charging potential (-800 mV,S.C.E)
            versus alloy content for Fe-Cu,
            Fe-Cr, Fe-Mn, Fe-Mo and
            Fe-Si binary alloys

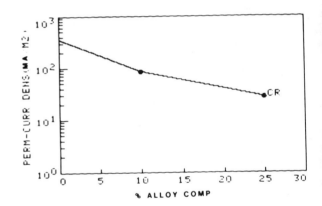

Figure 9.   Permeation behavior at initial
            charging potential (-800 mV,S.C.E)
            versus alloy content for Fe-Cr
            binary alloys.

Alloys
| Alloy | Wt % of Element | |
Element	1	2
Cu	0.52	1.04
Cr	9.90	24.30
Mn	1.00	3.00
Mo	1.06	4.27
Si	0.54	2.50

Table 1.   Concentration of alloying element in
            binary iron alloys

Alloy	E_{corr} (mV,S.C.E.)	i_p MA/M²	$i_2@E_{corr}$ MA/M²	D_H M²/sec
Fe	-549	109	18	1.393×10^{-9}
Fe-0.52%Cu	-495	35	3.7	1.706×10^{-10}
Fe-1.04%Cu	-480	34.3	3.47	1.874×10^{-10}
Fe-10%Cr	-482	11.4	1.1	2.084×10^{-11}
Fe-25%Cr	-237	8.1	.4	1.433×10^{-12}
Fe-1.0%Mn	-457	290	14.8	2.454×10^{-10}
Fe-3.0%Mn	-457	251	10	1.810×10^{-10}
Fe-1.06%Mo	-491	334.4	12.7	1.389×10^{-10}
Fe-4.27%Mo	-488	342	9.0	1.212×10^{-10}
Fe-0.54%Si	-400	119	7.0	1.774×10^{-10}
Fe-2.5%Si	-457	180	7.1	2.022×10^{-10}

Table 2. Corrosion potentials and poermeation behavior values for iron binary alloys in 0.6N NaCl , pH 6.3

Alloy	E_{corr} (mV,S.C.E.)	I_{corr} A/M²	Tafel Slopes (mV/dec) BA	Tafel Slopes (mV/dec) BC	i_p MA/M²	$i_p@E_{corr}$ MA/M²	D_H M²/sec
Fe	-492	4.0	42	113	355	341	3.603×10^{-9}
Fe-0.52%Cu	-444	4.2	58	175	36.0	34.6	6.424×10^{-10}
Fe-1.04%Cu	-410	5.8	45	–	33.6	29.1	5.244×10^{-10}
Fe-10%Cr	-466	0.5	–	–	86.6	47.7	3.332×10^{-11}
Fe-25%Cr	-352	0.2	–	–	27	19	3.561×10^{-12}
Fe-1.0%Mn	-476	6.0	40	65	355	250.8	5.840×10^{-10}
Fe-3.0%Mn	-457	7.1	67	–	305	170	4.440×10^{-10}
Fe-1.06%Mo	-486	6.5	74	218	347.9	323	3.208×10^{-10}
Fe-4.27%Mo	-452	7.2	76	–	345.6	319	2.301×10^{-10}
Fe-0.54%Si	-437	2.6	64	–	209	147	2.385×10^{-10}
Fe-2.5%Si	-415	0.3	120	–	70	61	2.658×10^{-10}

Table 3. Electrochemical and permeation behavior values for iron binary alloys in 1N H2SO4 , pH 0.4

67

STRESS CORROSION CRACKING BEHAVIOR
OF 40CrMnSiMoVA
ULTRA-HIGH STRENGTH STEEL

Jin, Shi; Liu, Xiaokun
Xi'an Shaanxi
People's Republic of China

ABSTRACT

The stress corrosion cracking (SCC) behaviors of
three heats of new type low alloy steel 40CrMnSi-
MoVA (σ_b=1800 - 2100 MPa) in both 3.5% NaCl solution and distilled water are investigated using
BL - WOL type of specimens. The influences of
environment temperature and cathodic polarization
on fracture are also observed. The data show that
the crack growth rates independent of stress intensity $(da/dt)_I$ of three heats in 3.5% NaCl solution are significantly different. At E_{corr}
(-640mv, SCE) and 35°C, $(da/dt)_{II}$ of Heat 1
($\sigma_{0.2}$=1575MPa), Heat 2 ($\sigma_{0.2}$= 1430MPa) and Heat 3
($\sigma_{0.2}$=1370MPa) are 14, 4.2 and 1mm/Hr. , respectively. However, K_{ISCC} of three heats under same
conditions are almost the same — 14 - 14.3MPa$\sqrt{m}$,
about one fourth or one fifth of their K_{IC} . The
cathodic polarization promotes both crack nucleation and propagation. For specimens of Heat 3
coupled to zinc, K_{ISCC}= 12.1MPa$\sqrt{m}$,and $(da/dt)_I$ =
1.9mm/Hr. . The environment temperature has
little influence on crack growth. The SCC characteristic in distilled water is similar to that in
3.5% NaCl solution. No crack branching is observed. SEM fractographic examination shows that the
intergranular failure is typical for specimens of
all heats. Quasi-cleavage can be seen only in the
front region of crack tip. The mechanism of SCC
of 40CrMnSiMoVA is considered to be mainly
hydrogen assisted cracking.

40CrMnSiMoVA is a new type of low alloy ultra-
high strength steel with tensile strength of
1800 - 2100MPa and yield strength of 1400 - 1600
MPa depending on heat treatments, and has been
utilized for manufacturing main components of
structures. However, cracking failures which
cause numerous serious problems are often observed with this steel during manufacturing and
operating processes. In addition to the lack of
toughness, high susceptibility to SCC has been
considered and experimentally proved to be one
of the main causes of such cracking [1]. Obviously, it is significant to investigate the effect of environments on crack growth of 40CrMnSi-
MoVA steel for ensuring the structural reliability and evaluating the remains of strength and
service life of components in practical condition.

EXPERIMENTAL MATERIALS AND METHODS

The vacuum arc remelted 40CrMnSiMoVA steel
was used for tests after composition analysis and
metallurgical structure examination at both high
and low magnifications. The composition of
40CrMoSiMoVA is listed in Table 1.
Stress corrosion cracking data were obtained
using BL - WOL type of plane - strain precracked
specimens with side grooves. The specimens in
thickness of 20mm were machined from a thicker
wrought plate in TL orientation as usual. Pre-
cracks of specimens in length of 1.5 - 2mm were
made using high frequency fatigue machine. Three
heats of specimens were investigated. The heat
treatments and mechanical properties of these
heats are shown in Table 2.
3.5% NaCl solution and distilled water were
employed as the typical simulated environments
for SCC testing. The crack length were measured
before and after SCC, and was monitored during
testing by direct optical observation of the
specimen with a microscope of low magnification.
The fractographic examination was made using both
optical and scanning electronic microscope (SEM)
of JEOL JSM - 35C Type.

RESULTS AND DISCUSSION

Table 3 and Fig. 1 - 5 show the SCC data obtained from three heats of specimens in different
conditions.
EFFECT OF HEAT TREATMENT - It can be seen
from Table 3 and Fig. 1 that the effect of heat
treatment is significant on crack growth rates

but negligible on threshold stress intensity factors K_{ISCC} in 3.5% NaCl solution. The crack growth rate independent of stress intensity $(da/dt)_I$ at 35°C and $E_{corr} = -640mv$ (SCE) is 14mm/Hr. for Heat 1, 4.2mm/Hr. for Heat 2 and 1mm/Hr. for Heat 3. The value of the former is one order of magnitude higher than that of the latter. The K_{ISCC} of three heats under the same conditions, however, are almost the same: 14.3MPa√m̄ for Heat 1 and 14MPa√m̄ for Heat 2 and 3. The main difference among three heats is the temperature of isothermal heat treatments. The effect of isothermal temperatures on $(da/dt)_I$, just like that of tempering temperature (2), are obviously related to the resulting yield strength. The relationship between $(da/dt)_I$ and yield strength is plotted in Fig.2. It shows that the crack growth rate increases with increasing yield strength, which is in agreement with the general trend for SCC resistance of other high strength steels (3).

The influence of isothermal treatment on SCC resistance can also be interpreted with microstructural changes. Since the different isothermal treatments produce different microstructures, the amount of austenite is considerably increased from Heat 1 to Heat 3, and in general, martensite may provide preferential and easy paths for crack growth, the remains of austenite retard the crack propagation (2,3,4), $(da/dt)_I$ of Heat 1 should be higher than that of the others.

However, both metallurgical structures and yield strenth do not appear to influence K_{ISCC} of 40CrMnSiMoVA significantly.

EFFECT OF ENVIRONMENT - It can be found from Table 2 and Table 3 that the values of threshold stress intensity factors of three heats in 3.5% NaCl solution are about one fourth or one fifth of their fracture toughness K_{IC} in air, showing that 40CrMnSiMoVA is relatively susceptible to SCC. However, no significant changes in both crack growth rates and K_{ISCC} values are observed as the environment is changed from 3.5% NaCl to distilled water (see Fig.3 and Table 3, $(da/dt)_I$ changes from 0.75mm/Hr. to 0.71mm/Hr.; K_{ISCC} are the same —14MPa√m̄). Similar effects have been observed with other high strength steels (2), and considered to be probably related with the occlusive cell reaction occuring at the crack tip region.

EFFECT OF TEMPERATURE - The influence of temperature on SCC kinetics of Heat 3 in 3.5% NaCl solution is shown in Fig.4. The $(da/dt)_I$ increases from 0.75mm/Hr. to 1mm/Hr. as the temperature of solution is enhanced from 15° to 35°C. K_{ISCC} for all heats at both temperature, however, are the same, as shown in Table 3.

EFFECT OF ELECTROCHEMICAL POTENTIAL - The open circuit potential of 40CrMnSiMoVA in 3.5% NaCl solution with deaeration is -850mv (SCE). The corrosion potential of this steel in same solution without deaeration is -640mv (SCE), according to the polarization curve, it just falls into the passive - transpassive region (1). After

coupling to zinc, the potential is cathodically polarized to -940mv (SCE). The SCC behaviors of 40CrMnSiMoVA at different potential are different. As shown in Fig.5, the cathodic polarization promotes both crack nucleation and crack propagation. $(da/dt)_I$ increases from 1mm/Hr. to 1.9mm/Hr. and the K_{ISCC} decreases from 14 to 12MPa√m̄ , as the electrochemical potential moved from -640 to -940mv. The important effect given by cathodic polarization on the susceptibility to SCC indicates that hydrogen induced cracking can be considered as the mechanism of SCC with 40CrMnSiMoVA steel.

FRACTOGRAPHIC EXAMINATION - Fig.6 is the appearance of a typical crack - arrest BL - WOL specimen of 40CrMnSiMoVA after SCC testing. Crack propagated along the side groove. No crack branching is observed. Fig. 7 shows the macro - fractographic picture of a crack - arrest specimen broken apart after testing by mechanically fracturing. The fractured plane is perpendicular to the applied stress. The tongue shape of the crack tip is typical, which indicates that SCC is more severe in the central part than on the surface of specimens, and is generally considered to be resulted from the plain strain load condition and occlusive cell effect at the crack tip.

SEM examination indicates that the micro - fractography of specimens of all heats has the same characteristics, as shown in Fig. 8 - 11. There always exists intergranular failure,typical for HI - SCC of high strength steel, on the fracture surfaces, independently of what kind of environment and electrochemical potential applied. Quasi - cleavage can be seen only on the fracture surface in front region of crack tip. This region apparently suffered from environmental influence.

CONCLUSION

1. 40CrMnSiMoVA steel is relatively sensitive to SCC. The K_{ISCC} is only one fourth or one fifth of its K_{IC}.
2. Heat treatment gives significant influence on SCC behavior of 40CrMnSiMoVA steel. $(da/dt)_I$ decreases from 14 to 1mm/Hr. as isothermal treatment is changed from 190°C, 60' to 310°C, 9'.
3. Environments and temperatures have little influences on SCC behavior of 40CrMnSiMoVA. However, cathodic polarization promotes both crack nucleation and propagation in corrosive environments.
4. The mechanism of SCC of 40CrMnSiMoVA is considered to be mainly hydrogen assisted cracking .

REFERENCE

1. S.Jin and X. Liu, "Susceptibility to Stress Corrosion Cracking of 40CrMnSiMoVA Ultra - high Strength Steel", Technical Notes of Northwestern Polytechnical University, Xian, China, (1984)

2. B.F.Brown, "Stress - Corrosion Cracking in High Strength Steel and in Titanium and Aluminium Alloys", Naval Research Laboratory, Washington D.C., (1972)
3. P. Mcintyre, Mechanics and Mechanisms of Crack Growth, Proceedings of a Conference, Cambrige APr. 4 - 6, P. 130 - 155, (1973)
4. E.H. Phelps, "A Review of the Stress Corrosion Behavior of Steels with High Yield Strength", (1969)

Table 1. The Composition of 40CrMnSiMoVA

Element	C	Si	Mn	Cr	Mo	V	P	S
Content (%wt)	0.4	1.45	0.90	1.44	0.50	0.10	0.017	0.002

Table 2. Mechanical Properties of 40CrMnSiMoVA

Heat No.	Heat Treatment* (Isothermal)	E (MPa)	σ_b (MPa)	$\sigma_{0.2}$ (MPa)	δ (%)	ψ (%)	K_{Ic} (MPa√m̄)
1	190°C,60'	195089	1982	1576	10.6	40.9	55.5
2	280°C,60'	195677	1878	1430	11.5	45.0	54.3
3	310°C,9'	195677	1879	1372	11.8	46.6	66.0

* Specification : 920°C,20' —— Isothermal heated —— Cooled in 60°C water —— Tempered at 260°C, 4 Hrs.

Table 3. Stress Corrosion Cracking Behavior of 40CrMnSiMoVA

Heat No.	Environment	K_{Iscc} (MPa√m̄)	$(da/dt)_{II}$ (mm/Hr.)	Number of specimens
1	3.5%NaCl, 15±1°C	14.3	——	3
	3.5%NaCl, 35±1°C	14.3	14.00	3
2	3.5%NaCl, 15±1°C	14.0	——	3
	3.5%NaCl, 35±1°C	14.0	4.20	3
3	3.5%NaCl, 15±1°C	14.0	0.75	4
	3.5%NaCl, 35±1°C	14.0	1.00	3
	Distilled Water 15±1°C	14.0	0.71	3
	3.5%NaCl, 35±1°C, Coupled to Zn	12.1	1.90	2

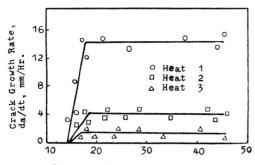

Fig. 1 SCC kinetic behavior of different heats of 40CrMnSiMoVA in 3.5% NaCl solution.

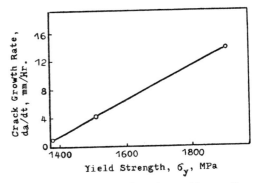

Fig. 2 Influence of yield strength on the crack growth rate of 40CrMnSiMoVA.

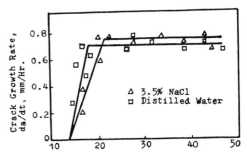

Fig. 3 SCC kinetics of 40CrMnSiMoVA (Heat 3) in 3.5% NaCl solution and distilled water at 15°C.

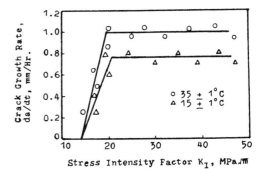

Fig. 4 SCC kinetics of 40CrMnSiMoVA (Heat 3) in 3.5% NaCl solution at 15°C and 35°C.

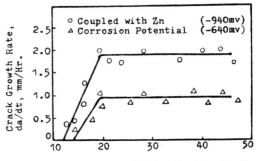

Fig. 5 Influence of electrochemical potential on SCC behavior of 40CrMnSiMoVA (Heat 3) in 3.5% NaCl solution at 35°C.

Fig. 6 Appearance of crack-arrest BL-WOL specimen of 40CrMnSiMoVA.

3   2   1

Fig. 7 Macro-morphology of BL-WOL specimen of 40CrMnSiMoVA broken after SCC testing in 3.5% NaCl solution.
1, 2, 3 -- Heat number.

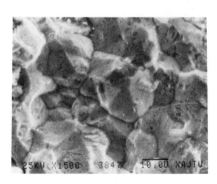

Fig. 8 Fractography of 40CrMnSiMoVA (Heat 2) in 3.5% NaCl at 35°C.

Fig. 9 Fractography of 40CrMnSiMoVA (Heat 3) in 3.5% NaCl at 35°C.

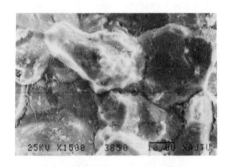

Fig. 10 Fractography of 40CrMnSiMoVA (Heat 3) coupled to zinc in 3.5% NaCl at 35°C.

(a)

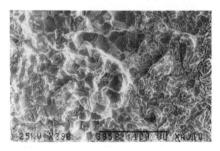

(b)

Fig. 11 Fractography of 40CrMnSiMoVA (Heat 3) in distilled water at 15°C.
(a)-- SCC region;
(b)-- front region of crack tip broken apart by mechanical fracturing.

INVESTIGATION OF SUSCEPTIBILITY OF POST-TENSIONED SYSTEMS TO ENVIRONMENTAL ASSISTED CRACKING

Ellen G. Segan, Dawn R. White
U.S. Army Construction Engineering Laboratory
Champaign, Illinois, USA

Darrell F. Socie, Dan L. Morrow
Peter Kurath
University of Illinois at Urbana
Champaign, Illinois, USA

ABSTRACT

A 1983 study of a post-tensioned system found that the critical stress intensity factor for mode 1 environmental assisted cracking (K_{1EAC}) of material conforming to ASTM-A-416-74 was 16 to 33 MPa-m$^{1/2}$ (15 to 30 ksi-in$^{1/2}$). Failure of a single wire by steady-state crack growth was estimated to occur in 16 to 166 hours.

A 1985 study of post-tensioned bars which failed during installation found that K_{1EAC} of material conforming to ASTM-A-722 was on the order of 16 MPa-m$^{1/2}$ (15 ksi-in$^{1/2}$); the bars failed in less than 24 hours after being grouted.

The failure of the bars occurred very much in the way predicted in the prior study. The normal accepted practice of assuming the steel to be immune to failure in the environment was followed. The results of these studies show that environmental assisted attack should be considered in the design of systems incorporating high strength steels in concrete environments.

HIGH-STRENGTH STEELS EMBEDDED IN CONCRETE have been assumed to be safe from failure by environmental assisted cracking (EAC) for many years. However, recent experience with failures of steel in concrete have led to renewed research to investigate the mechanisms of failure of steel in concrete environments.

RESEARCH ON EAC OF POST-TENSIONED SYSTEMS - Post-tensioned tendons have been used to reinforce concrete structures since the 1930s. The durability performance of post-tensioned systems was surveyed by Schupack[1] in 1978. It was noted that the reliability of post-tensioned structures had been excellent over the approximately 50 years of experience with the systems. Twenty-eight failures of stress-relieved steel were reported and it was noted that the failures could be attributed to poor design, introduction of corrosive environments, or poor construction of the systems. No catastrophic failures of post-tensioned permanent structures in the western world were observed. it was noted that there had been several instances of catastrophic failure of wrapped structures which could be attributed to improper construction detail and material selection. The failure rate of post-tensioned tendons was noted to be about 7×10^{-4} percent. It was concluded that post tensioned tendons which are properly detailed and constructed offer excellent durability against corrosion.

Many post-tensioned designs have been utilized in service. These include stress-relieved wires and strands, high-tensile steel strips,[2] partially prestressed post-tensioned beams with unbonded tendons,[3,4] and other techniques[5]. Research on the behavior of steel in concrete lead to various conclusions with regard to whether there are corrosion related problems with the use of post-tensioned steel in concrete. Although the potential problems which can be encountered when using high strength steel in concrete have been understood for many years, it has generally been assumed that adequate design and construction procedures should alleviate any potential problems.

The following results led to the conclusion that protecting high-strength steel from corrosion in concrete environments can be accomplished with appropriate design and construction considerations.

1. Concrete environments inhibit corrosion of steel in the presence of moisture, abundant free oxygen and most common ions except chloride and bisulfide.[6,7]

2. Corrosion of high strength steel (ultimate tensile strength = 1800 MPa (260 ksi)) in concrete leads to damage with flat-bottom or layer type geometry. Failures of long term test specimens result from overload failure resulting in loss of net cross-section due to corrosion.

Hydrogen only plays a significant role in the fracture process at the time when the tensile strength is approached.[8,9]

3. Corrosion of steel in concrete can be minimized by externally applied coatings. Additions to concrete can control corrosion of steel. Accelerated corrosion of steel in concrete is often associated with the presence of oxygen which can result from cracks in the concrete.[10]

4. Corrosion of steel in concrete can be controlled by control of the size and geometry of cracks in the concrete.[11]

Post-tensioned systems often use high strength materials stressed to a large fraction of their ultimate tensile strength. This can lead to problems which are not normally encountered at the lower stresses at which most metal systems operate. The following behavior in high strength steels has led to concern about the long range behavior of post-tensioned systems. High strength steels have been shown to be susceptible to EAC in Cl^-, SO_4^{2-}, S^{2-}, NO_3^-, NH_4^+, O^{2-}, H_2, H_2S, and low pH solutions.[12-18] Attempts have been made evaluate the behavior of high strength steel in concrete and other environments. In 1962, Unz[19] suggested that electrolytic stress corrosion tests be used to evaluate the suitability of high strength steel bars for use in underground or submersed structures. In 1974, Gilmour and Walker[20] suggested use of a fracture mechanics approach to assess the susceptibility of prestressing tendons to stress corrosion cracking (SCC). It was found that the susceptibility of the tendons to SCC increased with increasing tensile strength. In severe environments, the apparent critical stress intensity factor of high strength steels was found to be as low as 20% of the apparent critical stress intensity factor in air.

Stress relaxation of high strength bars was investigated by Krishtal and Postnikov[21] in 1976. It was found that internal friction increases prior to the onset of microscopic cracking and that crack opening occurs prior to local deformation. This suggests that the onset of cracking can be measured by measuring load relaxation.

In more recent investigations, efforts to evaluate the usefulness of high strength steels based on laboratory tests of SCC resistance were made. McGuinn and Elices[22] investigated the behavior of cold drawn prestressed tendons in $Ca(OH)_2$, cement paste, and NaCl environments at various pHs. Significant batch to batch differences in SCC susceptibility of the materials were observed, these effects were particularly pronounced in Cl^- environments. Ellyin and Matta[23] investigated the behavior of high-strength steels used for prestressing. A criterion to assess susceptibility to SCC based on mechanical properties was suggested. It was found that loss of ductility was not proportional to the loss of tensile strength of corroded wires. Furthermore, under constant deformation and load test and for stresses greater than 25% of the ultimate tensile strength (UTS), the degree of stress appeared to be more important than the aggressiveness of the solution.

SIGNIFICANT FAILURES OF POST-TENSIONED STRUCTURES - Prior to 1980, there were no catastrophic failures of post-tensioned structures due to EAC. Although tendon failures occurred and several wrapped structures failed, no catastrophic failures were noted.[1,19]

However, in 1980 the Berlin Congress Hall post-tensioned roof collapsed.[24] The structure was built in 1957 and failed after 23 years of service. The failure resulted in loss of life and property. The failure occurred due to the failure of eight tension elements and the partial failure of two which were caused by previous failures of the tension rods. This was aggravated by the fact that seven of the elements which had been fracturing for some time lay next to each other in one section of the eastern roof. Successive failures of the tension elements eventually resulted in increased load on the remaining elements until the south arch became structurally unstable and collapsed. The failure was attributed to hydrogen induced SCC of tendons. The cracking occurred in region where corrosion inducing conditions arose because of structural conditions and the finishing of the ring beam joint. It was noted that the tension elements were insufficiently surrounded by concrete and that even if high quality joint material had been used, protection against corrosion would have been insufficient.

Another instance of failure of a structure long after post-tensioning occurred in 1981 when a 26 year old bar in the Grundfors hydropower station failed.[25] The bar was ejected from the structure at high speed. In this case there were no injuries. It was found that the anticorrosion protection of the bars was deficient and unsatisfactory, and the material was sensitive to cracking. The primary cause of failure was intergranular stress corrosion cracking with the design safety factor was judged to be too low. The critical load on the tendon at failure was estimated to be 230 kN (52 kips) and the critical crack length was found to be 5 mm. The tendon was originally loaded to 300 kN (67 kips).

EVALUATION OF POST-TENSIONING SYSTEM ON THE JOHN DAY LOCK AND DAM

BACKGROUND - During the summer of 1981, the U.S. Army Corps of Engineers instituted remedial repair of the south lock wall of the John Day Lock located in Oregon and Washington. The repair was made to reduce lock wall movement due to the presence of cracks near its base. 45 post-tensioned tendons stressed to 70 percent of their ultimate tensile strength were installed in the south wall of the lock as shown in Figure 1. Each tendon conforming to ASTM-A-416-74 is made of 37 stranded cables;

each cable contains six 5 mm (0.2 in) diameter wires. The tendons were grouted in place; grease was used as a corrosion inhibitor on the individual cables.

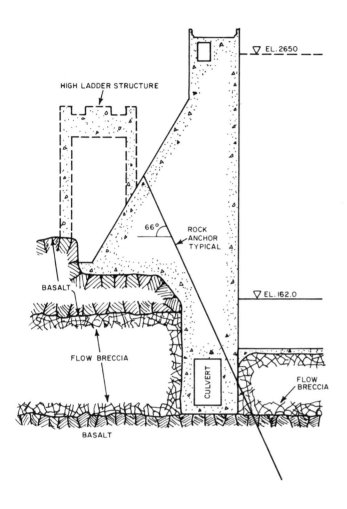

Figure 1. Schematic diagram of John Day Lock and Dam showing placement of post-tensioned tendons.

The remedial repair was successful in reducing the lock wall movement. However, once the system was in service, it was observed that each time the lock was watered, water was forced through the tops of several assemblies. Migration of water up the tendon assemblies could expose the tendons to several types of corrosion. The greatest concern was that EAC of the highly stressed wires could cause failure of some or all of the tendons.

APPROACH - It was noted that the probability of grease flowing out of the tendon assemblies or severe grout cracking increases as water continues to flow over the assemblies during lock watering and dewatering operations. So the likelihood of direct contact of water with the steel increases with the age of the system. When the lock is watered, hydrostatic pressure forces the water out of the assemblies, as the lock is dewatered, water recedes allowing

it to mix with air. This can decrease the pH of the water. Furthermore, previous work has shown that the pH and potential in a pit depends on the concentration and hydrolysis constants of metal cations in solution and and the rate at which hydrogen ions are removed by diffusion or reaction. Often times the pH in a pit is independent of bulk pH through a wide range of bulk pHs.[26] This suggests that the solution in a pit, crevice, or crack can be much more aggressive than that in the bulk.

These factors led to the conclusion that there is a possibility of EAC of the tendons at the John Day Lock. Pits or crevices that were present in the steel when installed or which formed since installation can contain very aggressive environments. If pits form, EAC can result in gradual sharpening of the pits under the applied stress to form a crack. If cracks are initiated (a process that can take years), they can grow under the applied stress until they reach the critical size for wire failure.

An investigation was undertaken to (1) determine the susceptibility of the post-tensioned tendons to corrosion and EAC in environments which could be expected in service, (2) assess whether the repair system might fail by the mechanism of EAC, (3) assess the rate at which a failure might occur and (4) recommend ways to reduce the risk of system failure.[27]

RESULTS OF INVESTIGATION - Pre-cracked samples were loaded to various stress levels in Corrtest proof rings which provide a constant stress to specimens which is linearly proportional to the ring deflection for small specimen strains, and submerged in pH 3 HCl. Estimates of mode 1 stress intensity factors for edge notched specimens were performed.[27] Ten samples failed as they were being loaded. These samples represent the critical stress intensity factor for mode 1 failure (K_{1C}) at t (time) = 0. An initial crack size of 1.3 mm (0.05 in) was assumed for all by the zero-time-to-failure specimens. K_{1C} was estimated at 130 MPa-m$^{1/2}$ (120 ksi-in$^{1/2}$). After the samples were loaded, pH 3 HCl was added to the environmental chamber to provide a corrosive environment. The solution was periodically replenished. The times to failure of samples subjected to long term tests were recorded electronically. Figure 2 shows a plot of the estimated stress intensity as a function of time to failure. Data representing specimens which did not fail during the duration of the experimental programs are indicated by an X with an arrow pointing toward the right. At long times (greater than 100 hr), the value of stress intensity below which failure does not occur was estimated to a between 16 and 33 MPa-m$^{1/2}$ (15 and 30 ksi-in$^{1/2}$). This is a typical value for high strength steels with a yield of about 1400 MPa (200 ksi). For example, K_{1EAC} for D6AC steel with a 1500 MPa (220 ksi) yield stress was estimated to be 40 MPa-m$^{1/2}$ (36 ksi-in$^{1/2}$) in 3 percent NaCl.[28] If it is assumed that K_{1EAC} = 33 MPa-m$^{1/2}$ (30 ksi-in$^{1/2}$) and the working load = 27 kN

(6 kips), the critical crack size (a_{cr}) can be estimated to be about 0.25 mm (0.01 in).

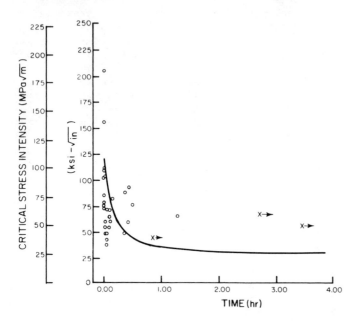

Figure 2. Observed stress intensity versus time to failure observed in laboratory for tendon material used at John Day Lock.

The Damage Tolerant Design Handbook[27] provides an estimate of steady crack growth rate for several high strength steels. The crack growth rates are about 10^{-6} to 10^{-7} cm/sec. These are very rapid rates and are nearly independent of stress intensity for values above K_{IEAC}. For the 0.5mm (0.20 in) diameter wires under a working load of 27 kN (6 kips), failure will occur when the crack reaches about 1.8 mm (0.070 in) (i.e., $K = K_{IC}$ when $a = 1.8$). Thus, the amount of crack growth needed for failure is only 1.5 mm (0.06 in) If the crack grows at 10^{-6} to 10^{-7} cm./sec., it was estimated that an affected wire could fail in 16 to 166 hours.

EFFECTS OF EAC ON REPAIR SYSTEM - The remedial repair system implemented at the John Day Lock and Dam has been in service for more than 4 years with no evidence of EAC related problems. This suggests that there is no steady crack growth in the repair system. Provisions for protecting the steel from aggressive environments were made in the design of the remedial repair system and this is expected to aid in the protection of the tendons from EAC.

The corrosion protection used (grease coating and grouting) appears to be effective. Griess and Naus[28] have investigated these coatings on ASTM-A-416-74 cable in aggressive environments and found them to offer complete protection from corrosion as long as the grease is not flowing out of the system and the grout is not badly cracked (i.e., cracks longer than 1 mm (0.04 in). The John Day Lock cables were greased and covered with a plastic sheath and

then set in secondary grout above the primary grout packer.

Two areas where EAC could occur were identified. Below the primary grout packer, the bare wires were anchored with a primary grout of portland cement. Since the bare wires in the primary grout region are protected only by the primary grout, this area would present problems if the grout became badly cracked. On the other hand, the grout in this region must be in compression to keep the tendon in place, and cracks which do occur should be held closed; thus, EAC problems should be eliminated in this case. The second region of concern is the anchor head area. During normal watering and dewatering cycles, this area is exposed to alternating wet and dry conditions. It was noted that it is unlikely that all of the grease will flow out of the tendon assemblies or that all the grout will crack severely in any one location. As a result, there is no reason to expect an entire tendon (37 cables) to fail at once. It was recommended that all tendon components above the secondary grout should be continuously coated with corrosion-inhibiting grease.

An estimate of the time required for a tendon to fail was made. Figure 3 shows the tendon response to both constant load (assuming the pre-load equals the fully watered load) and constant displacement (lock fully dewatered). For the fully watered or constant load case, the ratio of current stress to original pre-stress versus number of broken strands suggests an unstable response after 11 strands (77 wires) have failed. Since the stage II crack growth rate is nearly independent of the incubation period and not the crack growth rate, at least 77 wires would have to enter stage III crack growth while the lock is fully watered to cause catastrophic failure of a tendon. If each strand (7 wires) entered stage II crack growth sequentially, the estimated time from the first

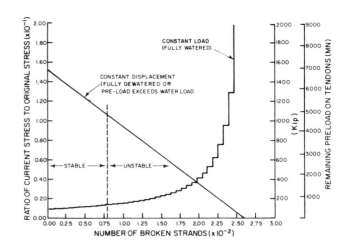

Figure 3. The effect of number of broken strands in tendon on ratio of current stress on tendon to original stress.

sign of failure to complete tendon failure would be about 160 to 1600 hours (i.e., 10 strands times the time to fail per wire). This estimate assumes that all wires in each strand will fail at the same time. It was estimated the time range of 160 to 1600 hours should allow ample opportunity to dewater the lock and begin repairs. Figure 3 also shows the constant displacement response of a tendon which is representative of the lock in the dewatered state or while the pre-load exceeds the watering loads. This plot of remaining load vs. number of strands failed demonstrates that upon watering, the pre-load will drop almost 30 percent before the tendon will fail.

METHODS FOR MITIGATING POTENTIAL EAC PROBLEMS - After a system has been installed, there are actions which can be taken to minimize the effects of EAC. In the case of the John Day Lock and Dam, the following techniques can be used to operate and monitor the in such a way to minimize potential EAC related problems:

1. Maintaining the anchor head areas by insuring that the inhibiting grease coatings are maintained periodically minimizes the chances of EAC

2. Maintaining the design stress on the tendons. The stress on the tendons prevents tendon fatigue during watering and dewatering of the lock because the high stress prevents the opening and closing of the cracks in the lock wall. The temptation to reduce the stress on the tendons to prevent EAC should be avoided, because this could lead to cyclic loading of the tendons. This could ultimately lead to corrosion fatigue, which is a potentially more serious problem than EAC.

3. Monitoring the structure for signs of EAC mechanisms in operation. This may be done by:

a. Noting the rate of shimming required to maintain the desired stress in elongation per year. In this case, a change in compliance because of decrease in net section area would be evident by a change in the amount of shimming required to maintain the desired load.

b. Equipping all tendons which show signs of water intrusion with load transducers to help detect loss of pre-load. It was estimated that significant drops in load due to loss of net section area would be about 180 KN (40 kips) (the load per strand), which is easily detectable. If there is evidence of EAC during the periodic inspections, the tendons should be extracted from the structure and replaced; again, the load should never be reduced to minimize the stress while the remedial repair system is operating.

INVESTIGATION OF FAILURE OF POST-TENSIONED TENDONS ON THE OLD RIVER CONTROL AUXILIARY STRUCTURE

BACKGROUND - In December 1984, a post-tensioned system was installed on the Old River Control Auxiliary Structure in Mississippi. Post-tensioning bars conforming to ASTM-A-722 were initially stressed to 75% of their ultimate tensile strength. Following tensioning, the tendons were left ungrouted for approximately one month and exposed to air and rain. Some relaxation of the load occurred, and the tendons were jacked to achieve the original load. Following grouting, four of the 84 tendons failed during the night and were apparently expelled from their tubes.

RESULTS OF INVESTIGATION - An investigation of the failures showed that it was the result of (environmental assisted) cracking in the presence of hydrogen. A small crack initiated at the base of one of the ridges on the post tensioning bar, and grew to a size of 0.3 to 0.5 mm prior to the brittle fracture of the remaining bar section (Figure 4).

Figure 4. Failed post-tensioned bar showing fracture surface and crack initiation site.

Scanning electron microscopy of the fracture surface revealed growth of the major crack on the fracture plane, with some secondary cracking occurring. Evidence of crack growth and arrest was also found in crack arrest markings on the fracture surface. This suggests that Kishtal and Postnikov[21] hypothesis that load relaxation in post-tensioning systems is an indicator of cracking may be valid for this case. The existence of crack growth and arrest marking on the surface certainly indicates this. The fracture surface condition was not ideal making it impossible to determine exactly when cracking initiated.

Optical microscopy showed that a heavy, adherent mill scale was formed on the surface of the bar during hot rolling and forged into the surface layer of the bar. This mill scale contains numerous laps and cracks which act as stress concentrations, and provide sites for crack propagation into the base materials. Other cracks were found to be growing into the bar from such locations (Figure 5) and it appeared that the crack which resulted in the failure of the bar initiated at such a surface discontinuity.

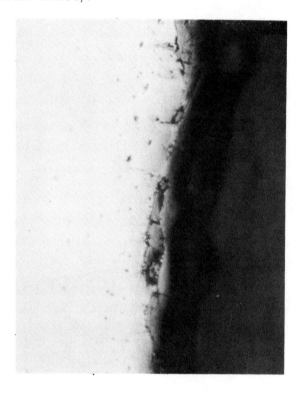

Figure 5. Optical micrograph of failed post tensioned bar showing cracking through mill scale and into post-tensioned bar.

System operators indicated that bars from four heats of steel were used in this project, and that all of the failed bars came from a single heat. This is similar to the behavior noted by McGuinn and Elices[22] in finding variations in EAC susceptibility among different material batches. While the mill certificate for the material involved met the procurement specification (ASTM A-722), with the exception of having a higher than required strength, a lower ductility was found for this heat than the three others.

The EAC problem was aggravated by the fact that an expansive grout was used in the post

tensioning process. The use of aluminum bearing expansive grouts is a current practice for pre-stressed concrete systems.[5] The presence of even minute quantities of hydrogen gas at a crack tip in high strength steel is known to cause crack growth. Because the critical flaw size is so small in strong brittle materials under high stresses, in this case, 1000 MPa (150 ksi), the presence of constituents which contribute to crack growth is unacceptable. K_{IEAC} for this case was determined to be less than 33 MPa-m$^{1/2}$ (30 ksi-in$^{1/2}$), a value similar to that calculated for the John Day Lock and Dam repair tendons.

System designers postulated that hydrogen in the grout may have combined with sulfur present in Portland cement to form H_2S another compound known to cause EAC in some materials. An energy dispersive x-ray analysis (EDAX) of a grout particle found on the fracture surface showed that sulfur was present. This sulfur is a component of the portland cement used in the grout. It is not probable, however, that this sulfur could have combined with the hydrogen gas released by the grout. The presence of grout particles on the fracture surface supports the theory that cracking initiated prior to grouting of the bars.[21]

It was concluded that it was more likely that following the stressing of the bars and during the period when they were left unprotected, cracking occurred on the surface of the bars due to the brittle, as received surface, providing a multitude of highly stressed (plastically strained) regions. When the grouting operation was performed two conditions occurred which may have contributed to the final failure.

1. any grout which was forced into the crack would have increased the stresses at the crack tip as it expanded.

2. more important, hydrogen gas formed by the grout was adsorbed at the crack tip causing the crack to grow under the applied load to a critical size at which brittle fracture occurred.

METHODS TO MITIGATE EAC PROBLEMS - In the case of the post-tensioned system on the Old River Control Auxiliary Structure, standard practice was followed and no efforts were made to protect the steel from the grout environment. Recommended actions to alleviate further potential EAC problems based on the investigation included:

1. Non-Destructive Testing (NDT) of all remaining in-place bars be performed to determine whether any have failed, but are being retained by the grout.

2. Monitoring of remaining bars which have not failed be performed to determine whether further failures occur.

3. A structural analysis of the structure be performed to determine how many additional failures can be tolerated.

4. A study of the stress corrosion cracking behavior of the material be undertaken

to determine what the remaining life of the in-place bars may be.

5. Modification of the procurement specifications materials be made. A material possessing a higher K_{IEAC} would have been less susceptible to failures of the type experienced.

DESIGN OF HIGH-STRENGTH STEEL IN CONCRETE SYSTEMS INCORPORATING MITIGATION OF EAC

The failure of the post-tensioned bars on the Old River Control Auxiliary occurred in the manner predicted in the study on the John Day Lock and Dam and other studies of failures of other post-tensioned systems. Previous experiences with post-tensioned structures also indicate that the potential for failure of tendons as well as catastrophic structural failure exists. It is apparent that environmental assisted attack should be considered in the design of systems incorporating high strength steels in concrete environments.

Although much research on the environment at the site of pits or cracks and the effect of environmental attack on the structural integrity of the components and systems is needed, there are immediate measures which can be taken to avoid the type of failures recently encountered. Specifically, these involve development of specifications for post-tensioned systems which are not susceptible to failure in concrete or hydrogen containing environments at high stresses (over 70% of the UTS).

Specifications are now being developed for materials to be used for future post-tensioned systems and for operation and maintenance of existing post-tensioned systems. These will address: material ductility, strength, and resistance to EAC and hydrogen embrittlement.

Work is being conducted to: assess in service performance of post-tensioned systems, enumerate problems encountered with such structures and evaluate the strength, ductility, and EAC and hydrogen embrittlement resistance of materials used for post-tensioned systems. Existing Corps of Engineers Guide Specifications for post-tensioned repair systems used on civil works structures are being reviewed and revised.

SUMMARY AND CONCLUSIONS

The results of these studies show that environmental assisted attack should be considered in the design of systems incorporating high strength steels in concrete environments.

Studies of high strength stranded steel cable and steel bar showed that K_{IEAC} was 16-30 MPa-m$^{1/2}$ (15 - 30 ksi-in$^{1/2}$); critical flaw sizes were estimated to be 0.25 - 0.5 mm. In one case, the high strength steel failed during installation while the other material is still in service showing no signs of EAC.

The failure of the high strength steel bars during installation of a post-tensioned system supports previous documentation of possible EAC failure modes. The following phenomena associated with the fracture of the high strength steel bars had been previously cited as problems in the literature:

1. Load relaxation prior to fracture
2. Batch to batch variation in high strength steel resistance to EAC
3. The existence of cracking in the presence of hydrogen as a failure mode of high strength steel
4. Reduction in EAC resistance due to inadequate ductility

REFERENCES

1. Schupack, M., J. Am. Concr. Inst., 75 [10], 501-10 (1978)
2. Saglio, R., Puyo, A., and Picaut, J., J. Prestressed Concr. Inst., 16 [1], 48-60 (1971)
3. Balaguru, P. N., Concr. Int., 3 [11], 30-41, (1981)
4. Chung, H.W., J. Am. Concr. Inst., 70 [12], 814-16 (1973)
5. Current Practice Sheets [3PC/10/1 No. 37], Concrete, 63-4 (May 1978)
6. Scott, G. N., Corrosion, 1038-52 (Aug 1965)
7. Hausmann, D.A., J. Am. Concr. Inst., 171-86 (Feb 1964)
8. Cherry, B.W. and Price, Corrosion Sci., 20, 1163-83 (1980)
9. Lewis, D.A. and Copenhagen, W.J., Corrosion, 15 [7] 382t-8t (1959)
10. Finley, H. F., Corrosion, 17, 104t-8t, (1961)
11. Beeby, A.W., Struct. Eng., 56A [3], 77-81 (1978)
12. Treadaway, K., Brit. Corros. J., 6 [3], 66-72 (1971)
13. McGuinn, K.F. and Griffiths, J.R., Tewksbury Symp on Fract, 3rd, Univ. of Melbourne, Aust., pp. 274-285, June 1974
14. Dikii, I.I., Petrivskii, I., Alekseev, S. N. and Krasovskaya, G. M., Translation, Sov. Mater. Sci., 16 [1], 41-4 (1980)
15. Gouda, V.K., Azim, A.A., and El-Sayed, H.A., Corros. Sci. 9 [6], 215-24 (1979)
16. NACA Standard MR-01-75, "Sulfide Stress Cracking Resistant Metallic Material for Oil Field Equipment," National Association of Corrosion Engineers (1975)
17. de Santa Maria, M.S., Verdeja, J.I., and Perosanz, J.A., Microstructural Science, 9 [8], 398-405 (1980)
18. Klodt, D.T., Paper From Proc 25th Conf., Nat. Assoc. Corrosion Eng., NACE. pp. 78-87 (1970)
19. Unz, M., Corrosion, 18 [1], 5t-8t (1962)
20. Gilmour, R.S. and Walker, A.L., University of Melbourne, Australia, 261-73 (1974)
21. Krishtal, M.A. and Postnikov, V.A., Sov. Mater. Sci., 12 [2], 147-8 (1976)
22. McGuinn, K.F. and Elices, M., Br. Corros. J., 16 [3], 132-9 (1981)
23. Ellyin, F. and Matta, R. A., Can. J. Civ. Eng., 8, 416-24 (1981)

24. Isecke, B., Mater. Performance, 36-9 (Dec 1982)
25. Jurell, G., Water Power and Dam Construction, 45-7 (Feb 1985)
26. Turnbull, A., Corrosion Science, $\underline{23}$ [8], 833-70 (1983)
27. Segan, E. G., Socie, D., Morrow, D., Technical Report M-349, (U. S. Army Construction Engineering Research Laboratory) (June 1984)
28. Campbell, J.E., W. E. Berry, and C. E. Feddersen, Damage Tolerant Design Handbook (Metals and Ceramics Information Center, Battelle Columbus Laboratories) (1972)
29. Griess, J. C. and D. J. Naus, pp 32-50, ASTM Special Technical Publications 713, D. E. Tonini, American Hot Dip Galvanizers Assn., Inc., J. M. Gaidis and W. R. Grace & Co., eds., (1980)

THE EMBRITTLEMENT OF MONEL 400
BY HYDROGEN AND MERCURY,
AS A FUNCTION OF TEMPERATURE

C. E. Price, R. K. King*
School of Mechanical and Aerospace Engineering
Stillwater, Oklahoma, USA
*now with Halliburton Services
Duncan, Oklahoma, USA

Abstract

Slow strain rate tensile tests were performed on Monel 400 in the environments of air, liquid mercury, and electrolytic hydrogen over a temperature range of -30°C to + 80°C. Embrittlement did not occur in either environment at -30°C and reached a maximum near room temperature. Embrittlement occurred at 80°C in the presence of mercury but not hydrogen. In general, crack initiation occurred much more readily in hydrogen than in mercury but so did crack blunting from ready plastic deformation. An increase in strain rate reduced the embrittlement level. Differences between hydrogen and mercury embrittlement appeared to be of degree rather than mechanism. The results are discussed in terms of the reduced cohesive strength and enhanced shear models for embrittlement.

ENVIRONMENTALLY INDUCED EMBRITTLEMENT of engineering materials is a substantial problem affecting the cost, reliability and safety of many products and processes. The purpose of this paper is to compare and contrast hydrogen embrittlement (HE) with liquid metal embrittlement (LME) in Monel 400. This is an alloy often used to resist corrosion. Lynch [1-4] has pointed out similarities between HE and LME in nickel single crystals, an Al-6Zn-3Mg alloy, Ti-6Al-4V, and D6-ac steel. He has proposed a model for embrittlement whereby the active environment stimulates dislocation nucleation at the crack tip, with the crack growing subsequently by an alternative slip sequence. Price and Fredell [5], utilizing slow strain rate tensile tests, have recently compared and contrasted the embrittlement of a nickel-copper alloy, Monel 400 (UNS N04400), by electrolytically generated hydrogen and by liquid mercury, at room temperature. They found that the hydrogen caused surface cracks to initiate at low strain levels, but that conspicuous embrittlement, in terms of a loss in tensile strength, did not necessarily ensue, because the cracks blunted easily. The embrittlement decreased with an increase in either strain rate or grain size. It was evident that hydrogen promoted plasticity, which is in accord with the Lynch model. Others have noted previously that hydrogen can enhance the plasticity of nickel itself [6-8].

Price and Fredell [5] found that mercury embrittled Monel 400 to a greater degree than did hydrogen. Cracks did not initiate at such low stress levels in mercury, but they propagated easily once they were initiated. As with hydrogen, the embrittlement decreased with an increase in strain rate. However, unlike with hydrogen, the embrittlement was a maximum at intermediate grain sizes of ~250μm. At finer grain sizes, crack initiation became increasingly difficult, while at coarse grain sizes plastic deformation was more pronounced. A crack propagation sequence of intergranular to transgranular to microvoid coalescence was common in both environments, with the former dominating in the most brittle fractures. Only at the finest grain size studied, ~35μm, was the degree of embrittlement due to hydrogen and mercury similar. In many alloys, LME and HE occur only over a limited range of temperatures, often referred to as a temperature window [9]. The present paper reports the results of an investigation into HE and LME in Monel 400 at temperatures other than room temperature. It is a follow up to the Price and Fredell [5] study and to a complementary study by Traylor and Price [10], whereby the fractographic aspects of HE and LME of Monel 400 and Monel R405 (UNS N04405) at room temperature were compared.

PROCEDURE

The composition of the particular Monel 400 heat was stated by the producer, Inco, to be (weight %) nickel 65.42, copper 31.66, iron 1.5, manganese 1.0, silicon 0.24, carbon 0.14, aluminum 0.041 and sulfur 0.003. Waisted specimens of minimum diameter 6.3mm were machined from 12.7mm diameter rod. The specimens were ground to a 600 grit finish and heat treated in a vacuum of ~15Pa at 700°C for 3.75 hours, followed by furnace cooling over an 8 hour period. The cooling rate can matter because of the segregation of impurities such as phosphorus to the grain boundaries that reduce the degree of embrittlement [11,12]. The specimens were chemically polished subsequently [13]. This procedure gave a grain size of ~35μm. This grain size was chosen because of the similar degree of HE and LME observed at room temperature [5].

The slow strain rate tensile tests were carried out under displacement control on an MTS machine. Strain rates of $1.6 \times 10^{-3} s^{-1}$ and $1.6 \times 10^{-5} s^{-1}$ were utilized. At room temperature, this encompasses the range over which the degree of embrittlement is most sensitive to strain rate [5]. The test environments were air, triple distilled liquid mercury and hydrogen, generated electrolytically. In the latter instance, the specimens formed the cathode of an electrolytic cell that had platinum wire anodes. The electrolyte was dilute sulfuric acid of pH 3.2. For temperatures below 0°C, sufficient methanol was added to prevent the electrolyte freezing. The current density was held constant at 200 Am^{-2}, in accord with the previous studies [5,10]. The specimens were not precharged. This experimental procedure is described in more detail elsewhere [14]. Testing was performed over a temperature range of -30°C to +80°C. It was not found to be necessary to go to lower temperatures, while the upper temperature limit was determined by concern at the mercury vapor pressure, even though a plastic envelope surrounded the specimen and chamber.

RESULTS

a)   AIR ENVIRONMENT - At a strain rate of $1.6 \times 10^{-3} s^{-1}$, the tensile strength decreased from ~750MPa to ~650MPa as the temperature increased from -30°C to 80°C. At a strain rate of $1.6 \times 10^{-5} s^{-1}$, the values were ~50MPa less and showed a similar trend. The fractures were all cup and cone with the reductions in area at fracture decreasing from ~80% to ~70% as the temperature increased.

b)   MERCURY ENVIRONMENT - The mercury did not wet the specimens readily and typically collected as globules on the fracture surface that were easily removed in an ultrasonic cleaner. The variations of tensile strength and reduction in area at fracture with temperature are shown in Figures 1 and 2. The embrittlement is less

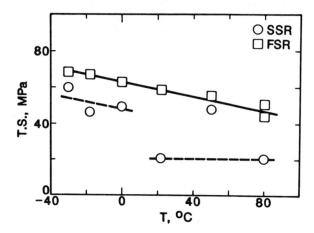

Fig. 1 - The variation of tensile strength with temperature for specimens fractured in mercury at the faster strain rate (FSR) and slower strain rate (SSR).

at the higher strain rate. There is a marked decrease in the embrittlement at the lower temperatures for both strain rates, consistent with the lower temperature end of an embrittlement window. These curves do not show any embrittlement at either strain rate at -30°C, which is above the freezing temperature of mercury. At 80°C there is a hint that the embrittlement is decreasing and there will be an upper end to the embrittlement range, but the data spread preclude a more definite assertion. The reduction in area at necking is of the order of 25%, thus only specimens failing with less than 25% reduction in area showed a loss of tensile strength.

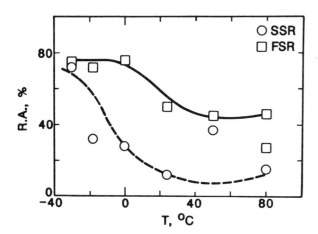

Fig. 2 - The reduction in area of specimens fractured in mercury.

The fracture appearance of the different specimens is sketched in Figure 3. Testing at the 100 fold higher strain

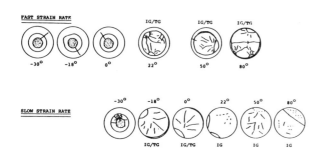

Fig. 3 - The fracture appearance of specimens tested in mercury. The increase in strain rate seems equivalent to a 30-40°C shift in test temperature.

rate is apparently equivalent to testing at a 30 to 40°C lower temperature. This is in accord with the Figures 1 and 2. The sequence begins with ductile cup and cone fractures at the faster strain rate at -30°C, -18°C, and 0°C. The only evidence of embrittlement on the fracture surfaces of these samples was a longitudinal split which extended for 6 to 15mm toward the specimen shoulders, for example, Figure 4. Sometimes, these splits occurred only on one of the fractured halves and sometimes in matching locations on both halves. While one major split per half was usual, occasionally one to three other more minor splits were present. These specimens also exhibited a very small amount of slant

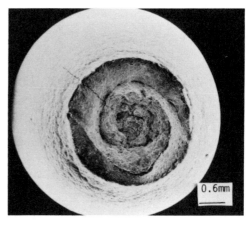

Fig. 4 - A cup and cone fracture with a longitudinal split, in a specimen tested at the FSR at 0°C, in mercury.

surface cracking in the necked region. These cup and cone fractures gave way to brittle, seemingly intergranular, fractures as the test temperature increased. On the fracture surface, secondary longitudinal cracks that usually radiated from one origin were evident, Figure 5. The side

Fig. 5 - A brittle fracture in mercury showing secondary cracks radiating from one origin. This was a FSR test at 50°C.

surfaces of the specimens showed no additional cracks, therefore, the inference was that these fractures were initiation limited. At higher temperatures, multiple crack origins were more usual and the radial secondary cracks were less pronounced, Figure 6. The inference was that crack initiation was no longer the controlling stage and that the embrittlement level was decreasing.

When the fracture surfaces were viewed at higher magnifications in a scanning electron microscope, the features seen were similar to those noted previously for specimens fractured at room temperature in the presence of hydrogen or mercury [10].

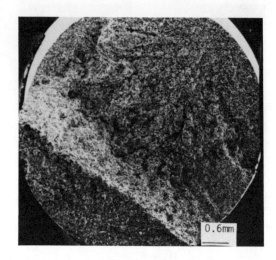

Fig. 6 - A SSR fracture in mercury at 80°C, showing two fracture origins, linked by shear and with less pronounced secondary cracking.

Thus a) intergranular zones always had a minority of transgranular fractures admixed, b) the proportion of intergranular fractures always tended to be highest a little way in from the edge c) the intergranular and transgranular zones were never featureless, showing slip markings to varied degrees, d) if attention was focused on individual grains, there was usually nothing to distinguish whether the fracture had occurred in the presence of hydrogen or mercury. This held for both intergranular and transgranular separations, e) the extent of secondary cracking increased towards the specimen center; this secondary cracking was more conspicuous in specimens tested in mercury than in hydrogen.

c) HYDROGEN ENVIRONMENT - Figures 7 and 8 show the variation of tensile strength and reduction in area at fracture in the presence of hydrogen. In contrast to the behavior in mercury, changing the strain rate was not equivalent to changing the test temperature; instead, an increase in strain rate reduced the degree of embrittlement. The embrittlement was a maximum in the vicinity of room temperature and minimal at both -30°C and +80°C, In nearly all instances the tensile strength in hydrogen was slightly lower than in air, even though the reductions in area at fracture exceeded 25% and necking had begun. Thus, testing in the presence of hydrogen appeared to promote plasticity.

Whether a specimen had fractured in hydrogen or mercury was apparent at a glance because hydrogen caused extensive side cracking throughout the exposed region of the gage length (~18mm) in nearly all specimens, Figure 9. Because of the

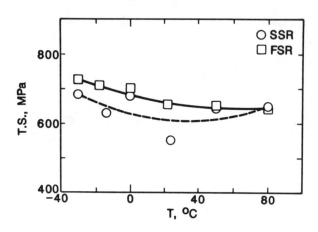

Fig. 7 - The variation of tensile strength with temperature for specimens fractured in the presence of hydrogen.

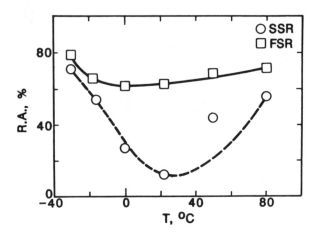

Fig. 8 - The reduction in area at fracture of specimens tested in the presence of hydrogen.

waisted geometry, there is a factor of more than two difference in the cross sectional area over this exposed zone, hence there is at least a factor of two difference between the stress at crack initiation and the stress at fracture in such cases. The only specimens without side cracking were some tested at -30°C. The extent of side cracking was also reduced in some specimens broken at 80°C. The side cracking was overwhelmingly intergranular for specimens tested between -18°C and 50°C inclusive. Slant transgranular cracks co-existed in the necked zones of the more ductile specimens and two specimens tested at 80°C showed slant cracking only.

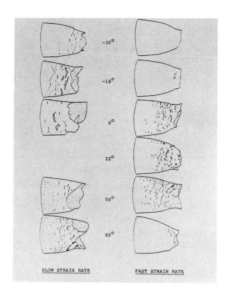

Fig. 9 - A schematic of side cracking in specimens fractured in the presence of hydrogen.

With reference to the fracture surfaces, specimens not embrittled by hydrogen had cup and cone fractures and failed by microvoid coalescence. The embrittled specimens had fractures with multiple surface origins. These surface cracks were predominantly intergranular. At varied distances in from the edge, the cracking became predominantly transgranular, with a central zone of microvoid coalescence. Typically, intergranular cracks on different levels, on different sides of the specimens, linked by shear, giving a stepped or eared appearance to the fracture. The more brittle the fracture, the greater the extent of the predominantly intergranular zone. For the lower strain rate series, it was evident that crack initiation began at lower strain levels as the temperature increased, but crack blunting also became easier. Figures 10 and 11 are side views of a specimen tested at 80°C. Seemingly, every grain boundary has separated, yet the fracture surface only had an intergranular zone 4 to 10 grains deep. Thus, as the temperature increased, a recovery from embrittlement occurred because the ready crack blunting did not allow the cracks to propagate.

From Figure 9, tests at the faster strain rate showed similar trends, but the embrittlement temperature range was compressed. Up to 80°C, increasing the test temperature resulted in crack initiation at lower stress levels and readier crack blunting subsequently. The

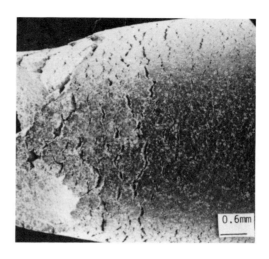

Fig. 10 - Side cracking due to hydrogen in a SSR specimen tested at 80°C.

Fig. 11 - A more detailed side view of the specimen of Fig. 10 - All grain boundaries have parted and many cracks gape.

test at 80°C did not fit this trend in that the fracture was cup and cone and the only evidence of embrittlement was a small amount of 45° cracking in the neck. A like result was observed when the test was repeated.

DISCUSSION

The thrust of this investigation has been the variation of HE and LME by mercury with temperature. The pertinent findings requiring discussion are a) the onset of both LME and HE above -30°C, b) the recovery from HE at ~ 80°C and the beginnings of a recovery from LME, c) the greater severity of LME and d) the much more extensive cracking and crack blunting due to hydrogen. The study has also substantiated the findings of the earlier room temperature study, of Monel 400 that neither LME nor HE produce any unique fractographic features and that intergranular, transgranular, and microvoid coalescence fracture modes can be got in varied proportions in either environment [5, 10].

The lack of embrittlement at low temperatures can be explained by lack of adsorption of hydrogen and lack of wettability by mercury. Wedler [15] has described the results of diverse methods of investigating the adsorption of hydrogen onto nickel, including determining the entropy and enthalpy of adsorption, isotope exchange measurements, field emission microscopy, and low energy electron diffraction. A number of different adsorption states are reported including one physiadsorption and at least two chemisorption states. Of particular relevance is that the chemisorbed hydrogen is immobile below about 195-240°K. It is reasonable that mobile hydrogen would be a prerequisite for embrittlement, hence the existence of a lower temperature boundary. Wedler also reports that hydrogen does not adsorb onto copper. Copper is embrittled by mercury but not by hydrogen [16, 17]

Wetting is a prerequisite for LME [18, 19]. Mercury does not wet Monel well, witness the existence of the mercury as globules on the fracture surface. This indicates that dewetting has occurred after fracture. Lynch [20] observed mercury globules on fractured surfaces of zinc and D6-ac steel and suggested that the adsorption of mercury at a crack tip may be promoted by strain. This idea of strain activated adsorption was advanced earlier by Westwood and Kamdar [21]. They suggested that the role of the elastic strain may be to either supply an activation energy (a chemical effect) or to produce the appropriate atomic spacing for the electronic interaction needed for chemisorption (a geometrical effect). An increase in temperature would also increase the lattice spacing, and chemisorption typically increases with increasing temperature [22]. The LME embrittlement window can now be explained on the basis of Figure 12. At low temperatures, the specimens fracture in the usual ductile manner before the strain can get high enough to promote wetting. At intermediate temperatures, wetting first becomes possible locally at sites associated with high strains such as grain boundaries and

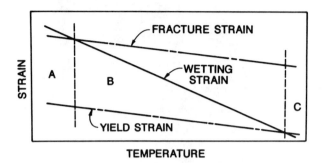

Fig. 12 - The LME window. The suggestion is that in region B there is localized strain induced wetting.

slip bands. This localized wetting leads to embrittlement. At higher temperatures, general wetting occurs without the necessity for plastic deformation. The general wetting is not a localized affect and embrittlement would not be expected. Liquid and solid metal couples with significant solubilities and good wetting do not typically undergo LME [23]. For example, Breyer and Johnson [24] reported that the LME of 4145 steel by lead was more severe in the presence of antimony which did not wet the steel, than in the presence of tin which wetted well.

The clearest indication that LME by mercury is decreasing with increasing temperature is provided by the changing nature of the secondary longitudinal cracks. These cracks develop in the interior at the crack tip under conditions of plastic constraint. The observed transition with increasing temperature is of increasing plasticity, with a change from deep cracks running many mm to blunter shallow cracks. A further point about the secondary cracks is that they must arise without any (or a very small) incubation time. For example, when the cup and cone failures occurred at the lower temperatures, the primary fracture spread from the interior. Only when this crack broke through to the surface in the final throes of the separation could mercury ingress occur to give the observed longitudinal splitting. Price and Good [25] obtained similar cup and cone fractures and longitudinal splitting due to mercury in low cycle fatigue tests at room temperature on cold drawn Monel 400. Hence, with load control, the incubation time must have been essentially zero and the crack propagation rate many mms^{-1}.

The HE window can be explained with reference to Figure 13, which is in accord

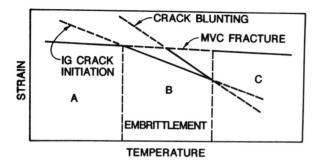

Fig. 13 - The HE window. In region A, a microvoid coalescence (MVC) fracture occurs before intergranular (IG) cracks can be initiated. In region C, plastic deformation occurs too easily.

with the observations. The low temperature cut off has been discussed already. At intermediate temperatures the adsorbed hydrogen is mobile and stimulates crack initiation at regions of strain concentration. This is a localized effect. However, the hydrogen also promotes plastic deformation and, as the cracks propagate and the stress (and strain) intensity increases, or the applied stress increases, so crack blunting occurs. With increasing temperature, the crack blunting process becomes easier, until a limiting embrittlement temperature is reached beyond which the necessary strain concentration to initiate brittle cracking cannot develop.

In view of the foregoing, it is not immediately obvious why the embrittlement due to both hydrogen and mercury should be less at a faster strain rate at all temperatures. An answer for the case of mercury embrittlement is that crack initiation at an external surface is under plane stress conditions and that a finite time is needed for a crack to grow through this zone and develop the necessary stress intensity for unstable propagation. In a rapid strain rate test, more plastic deformation has been imposed by the external system before the plane stress zone is penetrated. Price and Fredell [5] found that rapid loading to near the fracture stress in mercury followed by a hold period under load did not give a delayed fracture, such as should occur if fracture was governed by a time dependent adsorption, or time dependent penetration of mercury along grain boundaries or slip

bands.

A similar line of reasoning can apply to HE. The observed difference between HE and LME is that hydrogen is much more effective than mercury at both initiating and blunting cracks. The latter follows if hydrogen, unlike mercury, can diffuse a little way ahead of the crack tip and thereby act over a larger volume of metal than can mercury. Beacham [26] has suggested that dissolved hydrogen ahead of a crack tip may promote dislocation generation or motion or both. Vehoff and Rothe [27] estimated, from their tests of nickel single crystals in gaseous hydrogen, that the embrittlement is controlled by the hydrogen concentration in the fracture process zone, less than 100nm ahead of the crack tip. At temperatures of 20-100°C, they found that the HE was not dependent on the crack growth rate. At lower temperatures, HE decreased with increasing crack growth rate which they attributed to a limitation in hydrogen transport; at higher temperatures they found the HE was limited by excess plasticity.

The final question is whether the same fracture mechanisms apply to the HE and LME by mercury of Monel 400. The most widely accepted model of LME is that the adsorbed atoms lower the tensile cohesive stress [21, 28]. Stoloff [9] has commented that adsorbed atoms are likely to reduce the shear strength, also. In the present study intergranular cracking, transgranular cracking and crack blunting have been noted, much as was previously observed at room temperature [5, 10]. Therefore it is likely that, under conditions least condusive to plasticity, both hydrogen and mercury cause embrittlement by reducing the strength of atomic bonds. Under less restrictive conditions, mercury and especially hydrogen facilitate dislocation nucleation, as proposed by Lynch [1], Beacham [26] and also by Clum [29]. Between these two extremes, transgranular cracking may occur because of the shear strength reduction by the active environment.

SUMMARY

The embrittlement of Monel 400 by both hydrogen and mercury is most severe near room temperature. Embrittlement does not occur at low temperatures probably because of limited wettability and mobility of mercury and adsorbed hydrogen respectively. At elevated temperatures, the embrittlement is curtailed by the ease of plastic deformation. Hydrogen is much more effective than mercury both at initiating cracks and at promoting plasticity subsequently. Accordingly, cracks initiate at low strain levels in the presence of hydrogen but become blunted. When cracks are initiated by mercury, they do not blunt so early, hence the resultant embrittlement is usually more severe in mercury. The embrittlement is less at a higher strain rate in both environments at all temperatures. It is reasoned that this is due to difficulties of crack propagation through the surface zone, where plane stress conditions prevail. It is likely that both hydrogen and mercury act to reduce both the tensile and shear cohesive strengths, and also promote dislocation nucleation under conditions more favorable for plasticity. The greater potency of hydrogen may be due to an ability to diffuse a little way ahead of the crack tip, thereby having a larger zone of influence than mercury.

REFERENCES

1. Lynch, S. P., Scripta Met. 13, 1051-1056 (1979)
2. Lynch, S. P., "Hydrogen in Metals", pp. 863-871, Met. Soc. AIME, New York (1981)
3. Lynch, S. P., Acta Met. 29, 325-340 (1981)
4. Lynch, S. P., Acta Met. 32, 79-90 (1984)
5. Price, C. E. and R. S. Fredell, Met. Trans., accepted for publication (1985)
6. Windle, A. H. and G. C. Smith, Met. Sci. J. 2, 187-191 (1968)
7. Latanision, R. M. and R. W. Staehle, Scripta Met., 2, 667-672 (1968)
8. Eastman, J., T. Matsumoto, N. Narita, F. Heubaum and H. K. Birnbaum, "Hydrogen Effects in Metals", I. M. Bernstein and A. W. Thompson (eds.), pp. 397-409, Met. Soc. AIME, New York (1981)
9. Stoloff, N. S., "Atomistics of Fracture", R. M. Latanision and J. R. Pickens (eds.), NATO Conf. Series VI, 5, 921-953, Plenum Press, New York (1983)
10. Traylor, L. B. and C. E. Price, J. Eng. Matls. and Tech., accepted for publication (1985)
11. Costas, L. P., Corrosion 31, 91-96 (1975)
12. Funkenbusch, A. W., L. D. Heldt and D. F. Stein, Met. Trans. 13A, 611-618 (1982)
13. Tegart, W. J. M., "The Electrolytic and Chemical Polishing of Metals", p. 102, Pergamon Press, Oxford, England, 2nd Edn. (1959)
14. King, R. K. Ph. D. Thesis, Oklahoma State University, Stillwater, Oklahoma (1985)
15. Wedler, G., Chemisorption: An Experimental Approach, Butterworths, Boston (1976)
16. Louthan, M. R., Jr., G. R. Caskey, Jr., J. A. Donovan and D. E. Rawl, Jr., Mat. Sci. and Eng. 10 357-368 (1972)
17. Rosenberg R. and I. Cadoff, "Fracture of Solids", D. C. Drucker and J. J. Gilman (eds.), pp. 697-636, Wiley-Interscience, New York (1963)
18. Kamdar, M. H., "Progress in Mat. Sci.", B. Chalmers, J. W. Christian and T. B. Massalski (eds.), 15, 289-374, Pergamon Press, Oxford, England (1973)
19. Nicholas, M. G., "A Survey of Literature on Liquid Metal Embrittlement of Metals and Alloys", Embrittlement by Liquid and Solid Metals", 27-50, TMS-AIME, Warrendale, Pennsylvania (1984)
20. Lynch, S. P., "Mechanism of Fracture in Liquid-Metal Environments", Embrittlement by Liquid and Solid Metals, 3-26, TMS-AIME, Warrendale, Pennsylvania (1984)
21. Westwood, A. R. C. and M. H. Kamdar, Phil. Mag. 8, 787-804 (1963)
22. Castellan, G. W., "Physical Chemistry", second edn., p 437, Addison-Wesley, Reading, Massachusetts (1971)
23. Nicholas, M. G. and C. F. Old, J. of Mat. Sci. 14, 1-18 (1979)
24. Breyer, N. N. and K. L. Johnson, J. Testing and Evaluation, ASTM, 2, 471-477 (1974)
25. Price, C. E. and J. K. Good, J. Eng. Matls. and Tech. 106, 178-183 (1984)
26. Beacham, C. D. Met. Trans. 3, 437-451 (1972)
27. Vehoff, H. and W. Rothe, Acta Met. 11, 1781-1793 (1983)
28. Stoloff, N. S. and T. L. Johnston, Acta Met. 11, 251-256 (1963)
29. Clum, J. A., Scripta Met., 9, 51-58 (1975)

SOME UNIQUE CASE STUDIES ON HYDROGEN EMBRITTLEMENT FAILURES IN COMPONENTS USED IN AERONAUTICAL INDUSTRY

A. K. Das, B. M. Thippeswamy, J. Prasad
Hindustan Aeronautics Ltd.
Bangalore, India

THE PURPOSE OF THIS PAPER is to highlight some unique case studies on hydrogen embrittlement failure encountered in aeronautical components, more particularly, during production/fabrication stage. The problems of hydrogen embrittlement assume greater criticality when the parts are vital and highly stressed and as such, pose a serious concern to the designers/engineers as far as safety of the aircraft is concerned. This paper also emphasises the fact that failures due to hydrogen embrittlement are mostly unpredictable in nature, sometimes occuring prior to usage. One dangerous aspect of the failure is that unlike other propagative type of failures, viz. fatigue, stress-corrosion, this may suddenly erupt and disrupt the sustenance capability of the metallic life during service.

BREAKAGE OF MAIN UNDERCARRIAGE AXLE OF LANDING GEAR SYSTEM OF AN AIRCRAFT UNDER SIMULATED TESTING

One main undercarriage axle made of high strength alloy steel (C 0.33%, Cr 1.15%, Ni 1.50%, Mn 1.05%, Si 0.96%) was subjected to simulated fatigue test for 6000 hours of service. But only after 300 hours it prematurely broke into two along the sharp radius (Fig.1).

On fractographic examination, fracture showed a distinct (bright) crescent shaped zone with fracture lines emanating from several zones of 'ratchet' marks in the outer periphery of crescent envelope (Fig.2).

On close examination, fracture revealed coarse, irregular and brittle surface before final fracture by thick angular shearlip zone (Fig.2).

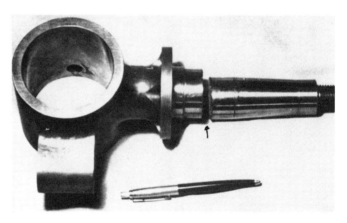

Fig.1 Line of breakage along the sharp radius indicated by the arrow, at x 0.25

Fig.2 Crescent shaped envelope with coarse brittle fracture and 'ratchet' marks at x 2

Presence of smooth layer zones of characteristic 'ratchet' type marks in the periphery of the bright envelope indicated the crack initiated by fatigue (Fig.2). SEM examination at higher magnification showed intergranular fracture with deep secondary cracks (i.e. yawning grain boundaries) between the grains in the zone close to ratchet marks. The fracture examined at different locations within the crescent shaped envelope presented distinctly two fracture modes intermingled with each other i.e., brittle cleavage (facets) and ductile dimple zones of crack propagation (Fig.3), followed by the predominance of dimples with further progress in the crack propagation.

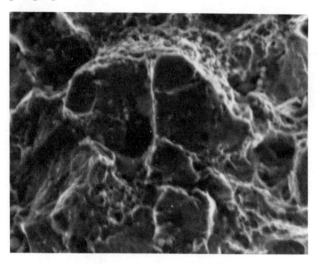

Fig.3   Fracture mode of brittle cleavage and ductile dimple characteristics under SEM examination at x1000

Presence of micropores in the cleavage facets as well as at the grain boundaries and hairline type crack indications under SEM examination were all suggestive of hydrogen embrittlement. Further a few elongated rod shaped manganese sulphide inclusions (a preferred site for hydrogen absorption) along with a hosts of globular oxide inclusions were observed in the fracture.

Microexamination of a section across the fracture in the initiation zone revealed steel material very unclean due to presence of non-metallic inclusions, mostly of globular oxides (D4 thin series) and long stringers of oxides in several places along with large inclusions in the form of irregular and flat slag (Fig.4) near the initiation point.

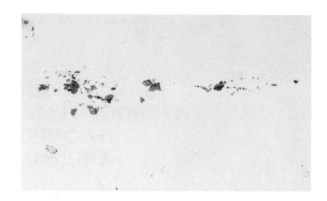

Fig.4   Large irregular and flat slag inclusions (as polished), at x100

Chemical composition, metallurgical structure and hardness of the axle material were found meeting the specification requirements besides satisfactory flow pattern, tensile strength and impact properties.

SUMMARY AND CONCLUSIONS: During simulated fatigue test, the high strength (Cr-Ni-Mn-Si alloy) Main Undercarriage Axle part broke prematurely along the sharp radius. Detailed investigation was carried out to determine the cause and mode of failure using scanning electron microscope and optical microscope. On the basis of investigation results and observations, it was concluded that the transverse breakage of the axle had occured intergranularly in a brittle manner, possibly, initiated by a shallow zone of fatigue along the sharp radius acting as stress riser.

In this case, the following facts were suggestive of occurence of a failure phenomenon similar to the one induced by hydrogen embrittlement: (i) Crack propagation path was intergranular throughout the breakage right from the initiation zone to the fracture at the bore edge; (ii) Fracture phenomenon in the initial zone of failure (i.e.bright crescent shaped envelope) consisted of two distinct modes(1) intermingled with each other as a mixed mode of crack propagation;; and (iii) Axle material was found very unclean due to presence of numerous non-metallic oxide inclusions as well as slag inclusions. As is well known(2), inclusion sites act as origin for hydrogen absorption.

CRACKING OF HIGH STRENGTH STEEL PISTON ROD DURING CHROME PLATING

A few Cr-Mo (C 0.3%, Cr 3.0%, Mo 0.4%) steel piston rods from different production batches were found identically cracked in the eye end near the radius (Fig.5) after chrome plating and baking treatment. Two of them cracked in the plating stage

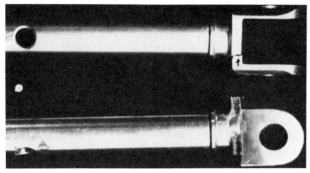

Fig.5   Cracking of the eye end portion near the radius, at x 0.75

itself instantly broke on slight tapping. Cracking was found to have initiated from the outer base surface of the forked eye end (i.e. from the plated side, the inner surface of the eye ends being masked). It was ascertained during investigation that 40 mm. diameter forged piston rods were subjected to plating after carrying out heavy machining on the part without any stress-relieving treatment. Also, time lapses between plating and baking were noted to vary from 3 to 11 hours.

Macroexamination revealed numerous tiny blisters all over the bright chrome plated piston rod surface. On examination under stereoscopic microscope, fracture showed identically coarse, irregular crystalline (brittle) surface features, typical of intercrystalline failures (Fig.6).

Fig.6   Coarse, irregular crystalline surface features, at x 2.

Microexamination of a section across the fractures revealed intergranular cracking with lifting up of crystals in places, characteristics of a sudden intergranular mode of failure (Fig.7).

Scanning electron micropic examination of the same fracture confirmed cracking of intergranular mode along with transcrystalline fracture structure of micro-quasi-cleavage type, as indicated by the presence of 'botonical appearance'(3) resembling curved leaves (Fig.8) - a feature suggestive of failure under hydrogen

embrittlement.

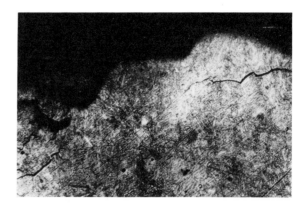

Fig.7   Lifting up of crystals in the microstructure, at x 100

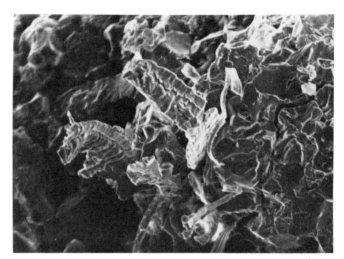

Fig.8   Fracture with botanical appearance resembling curved leaves under SEM examination, at x 450

Chemical composition and microstructure of the piston rod material were found satisfactory as per specification. However, strengthwise the piston rods that failed during plating and after baking showed slightly higher hardness values compared to the drawing requirement. The failed piston rod was also subjected to instrumental gas analysis and the hydrogen gas content was found to be of the order of 18 ppm.

SUMMARY AND CONCLUSIONS: A number of Cr-Mo steel piston rods got cracked during chrome plating and some after baking treatment. A detailed investigation was undertaken to pinpoint the source of cracking with the aid of optical as well as scanning electron microscope including gas analysis. Based on the investigation results and observations, the brittle cracking along forked eye-end radius portion was attributed to hydrogen embrittlement occuring during chrome plating process. In this connection, the following factors were believed to be primarily responsible for enhancement of hydrogen gas absorption leading to instant cracking in extreme cases, (i) presence of high residual stresses in the part, particularly in the location of complex shaped thin forged eye-end radius zone due to removal of heavy mass by machining in the heat-treated condition; (ii) presence of numerous tiny gas bubbles/blisters all over the surface of the affected parts noticed prior to baking indicating higher degree of gas absorption; (iii) time lapse between plating and baking for different batches facilitating chances of hydrogen embrittlement; and (iv) Hydrogen content of the order of 18 ppm being sufficient enough to cause hydrogen embrittlement cracking without any external stress.

CRACKING OF MACHINED END FRAME STEEL FORGINGS AFTER HEAT TREATMENT

A number of machined end frame steel forgings made of Cr-Si-Mn alloy were showing tiny cracks during magnetic particle inspection after heat treatment. The cracks were mostly confined to base edges and fillet radius (Fig.9).

Fig.9 Tiny cracks at the fillet radius and base edges, at x 0.75

No significant abnormality was observed in chemical composition and microstructure. Initially, the cracks appeared to be due to faulty heat-treatment. Subsequent batches also showed similar cracking, although heat-treatment parameters were carefully checked and controlled. Further investigation was conducted to pinpoint the cause of cracking. From the raw stock of the same batch (annealed condition), two discs were machined to a 50 mm. diameter hole in the centre similar to the end frame. Both the machined parts were subjected to magnetic particle inspection. Tiny cracks in 50 mm. diameter bore were observed similar to the end frame (Fig.10).

Fig.10 Indication of cracking on magnetic particle inspection in the 50 mm diameter machined bore surface in annealed condition, at x 1.

This suggested the fact that cracks noticed in the finished forgings were existing prior to heat-treatment. Heat-treatment could only aggravate the crack situation.

Microexamination of the sections taken across the discontinuities of both annealed and heat-treated parts, revealed sub-surface discontinuous hairline cracks close to the bore surface (Fig.11) which were intergranular in

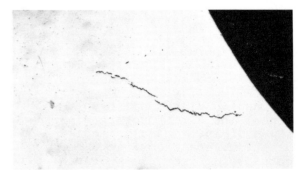

Fig.11 Sub-surface discontinuous hairline cracks close to the bore surface, (as polished) at x50

nature without any evidence of branching effect. A few sulphide and oxide type inclusions were seen along the path of cracking in places.

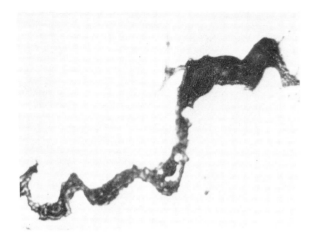

Fig.12   Intergranular crack in the as-polished condition, at x 400

Fig.13   Discontinuous intergranular cracks with Mn-sulphide inclusions along the path of cracking (2% Nital etched), at x 400

SEM examination of the fracture (on opening out the tiny crack) revealed dimpled structure associated with brittle facets, ductile hairline cracks, micropores (Fig.14) and rod-shaped sulphide (Fig.15) inclusions indicating the characteristic features of hydrogen embrttlement.

The end frame material was comparatively found cleaner as regards non-metallic inclusions. Chemical analysis confirmed the material used was alloy steel (0.34%C, 1%Cr, 1.9%Mn, 1%Si). Chapry impact properties determined in the locations both near and away from the defective zone in the longitudinal and transverse directions were found generally low (i.e. 4.5 & 3.5 kg.M/cm^2 restpectively against the specification requirement of 5 kg.M/cm^2).

Based on the above, the cracked end frame was subjected to instrumental gas analysis and hydrogen gas content was found to be of the order of 22 ppm.

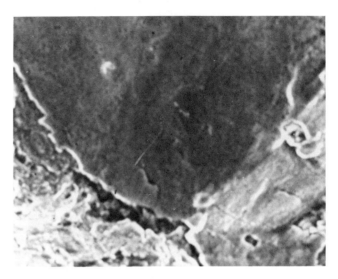

Fig.14   Brittle facets, dimples, hairline cracks and micropores under SEM examination, at x 450

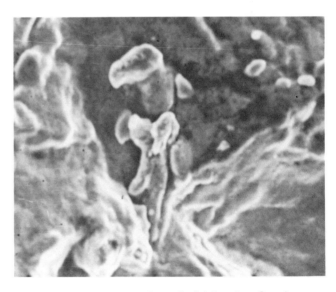

Fig.15   Rod-shaped sulphide inclusions under SEM examination, at x 450

SUMMARY AND CONCLUSIONS: A number of Cr-Si-Mn alloy machined end frame steel forgings developed random tiny cracks, mostly in the fillet radius portion on heat-treatment. It was suspected to be due to either to improper heat-treatment or wrong material. As both the chemistry and microstructure of the failed parts were found satisfactory with regard to heat-treatment parameters, further detailed probe was made into the reasons for the development of cracks with the help of

scanning electron microscope, optical
microscope and gas analysis. Based
on the investigation observations and
findings it was concluded that (i) the
sub-surface discontinuous cracks noticed
at the bore edges as well as in the
fillet radius of the heat-treated end
frame component in the final machined
form had occured due to hydrogen
embrittlement and definitely not due to
faulty heat-treatment, not due to
drastic quenching as appeared to be in
the initial stage of investigation from
the radial pattern of cracking etc.,
which is well supported by the presence
of crack like indications in machined
bore surface of the annealed part; and
(ii) the fact that heavy machining
involving a large diameter hole (as done
in the case under investigation to
develop to develop sub-surface crack)
can induce severe residual stresses
resulting in the increase of hydrogen
concentration level for failure to
occur by hydrogen embrittlement phenomenon (4).   However, presence of high
level concentration of hydrogen gas
(22 ppm) in the annealed part as
measured by instrumental analysis was
indicative of a potentially dangerous
situation even prior to service.

ACKNOWLEDGEMENT

The authors wish to record their
deep gratitude to Dr.C.G.Krishnadas Nair,
General Manager, Foundry & Forge Division
Hindustan Aeronautics Ltd., Bangalore
for his keen interest and kind permission
to present and publish this paper in the
International Conference organised by
American Society for Metals, at Salt Lake
City. The authors also take this
opportunity to thank the organiser for
accepting the paper for presentation
and publication.

Reference:

1. Metals Hand Book 8th edition Vol.9 p.186,
American Society for Metals,Ohio (1974)

2-4 Lothar Engle and Hermann Klesigele
"An Atlas of Metal Damage"(English version),
p.121, 122 & 123, Wolfe Publishing Ltd.,
Conway Street, London (1981)

A SLIP DISSOLUTION-REPASSIVATION MODEL AND A PREDICTION OF CRACK PROPAGATION RATE OF 304 STAINLESS STEEL IN 42% MgCl$_2$

Michinori Takano
The Research Institute of Iron, Steel and Other Metals
Tohuku University
Sendai, Japan

Takenori Nakayama
Graduate School, Tohoku University
Present Address:
Central Research Laboratory
Kobe Steel Ltd.
Kobe, Japan

Observation of crack tip of Type 304 stainless steel in a boiling 42% MgCl$_2$ by a high voltage electron microscope revealed that transgranular stress corrosion cracking (SCC) always proceeds along {111} slip planes. SCC behavior of the steel in the same solution has been investigated using a slow strain rate technique in the range of crosshead speed (CHS) from 6×10^{-5} to 1.5 mm/min at different potentials. The highest susceptibility to SCC appeared at a slow CHS of 1.5×10^{-4} mm/min. With increasing the susceptibility to SCC increased the ratio of transgranular cracking on a fracture surface. Dissolution repassivation behavior has been investigated using straining electrode method at different potentials. Dissolution current i(t) is expressed as $i(t) = i^0 \cdot \exp(-\beta t)$ where, i is initial dissolution current density of slip steps, β is decay constant and t is time. The results obtained will be discussed on the basis of the dissolution-repassivation model. Velocity of transgranular cracking has been expressed by the equation;
$$da/dt = 2.02 \cdot 10^{-6} \cdot i^0/\beta \cdot \dot{n}_s \cdot [1 - \exp(-\beta/\dot{n}_s)]$$
where, $\dot{n}_s$ is formation rate of slip steps. The equation has been confirmed by observed values. Further, potential (environmental factor)-$\dot{n}_s$ (stress factor) - da/dt(SCC susceptibility) diagram in this system will be represented.

INTRODUCTION

The mechanism of stress corrosion cracking (SCC) of austenitic stainless steels has been widely studied and discussed by numerous investigators, but no successful mechanism for crack propagation has been proposed.

Behavior of SCC proceeded by active path corrosion depends largely upon the formation rate of slip steps under stress and the dissolution rate of alloys on the slip steps[1,2]. Materials show general corrosion, transgranular or intergranular cracking and/or ductile failure depending upon above two factors, even though in the same material/solution[2]. The former factor can be controlled generally using slow strain rate test (SSRT) method and the latter by solution temperature and/or applied potential.

In this paper mechanism of transgranular cracking of Type 304 stainless steel in 42% MgCl$_2$ solution at 143°C using SSRT will be discussed and the crack propagation rate will be calculated based on the dissolution/repassivation model. Finally environmental factor - stress factor - SCC susceptibility diagram in this system will be represented.

EXPERIMENTAL PROCEDURES

Commercial Type 304 stainless steel rod and plate were used as specimen and their chemical compositions are given in Table 1. Each specimen was machined to dimensions of 6ϕx140 mm and 14x8x200 mm, respectively. The specimens for SCC testing were given a round notch on their centers to initiate cracks. After machining, they were all solution annealed at 1050°C for 30min in Argon, and lightly electropolished in H$_2$SO$_4$(2 vol) + H$_3$PO$_4$(3 vol) at 80°C. A boiling 42% MgCl$_2$ solution at 143°C was used as a corrosive.

For the observations of SCC tips of the plate specimens using a high voltage electron microscope (HVEM: 1000 KV), the specimens were prepared

TABLE 1   Chemical compositions of Type 304 stainless steels

SPECIMEN	C	Si	Mn	P	S	Ni	Cr	Mo	Cu
				(wt %)					
ROD	0.05	0.33	0.99	0.034	0.009	9.11	18.43	0.24	0.20
PLATE	0.06	0.05	1.29	0.028	0.002	8.47	18.20	0.08	0.20

as follows: Stress corrosion tests were stopped at the maximum load in each load-elongation curve under SSRT, where numerous cracks occurred. After then cracked specimens were thinned by electropolishing and/or by an ion milling technique. SCC test has been carried out for the rod specimens using SSRT in the range of crosshead speed (CHS) from 6×10^{-5} to 5 mm/min at different potentials.

Rod specimens, which were protected from the test solution by teflon tape except for the central part (1.885 cm^2), were also used for the straining electrode test. In this test, they were deformed plastically in above solution at different potentials at a CHS of 2 mm/min by using a Tensilon tensile test machine, and the decaying anodic current was measured by x-t recorder. The increment of surface area rusulting from straining was calcurated from eq.(1), which is based on the assumption that volume is constant before and after straining for a small strain.

$$S = 1/2 \cdot S^o \cdot \Delta l / l^o \qquad (1)$$

where, S^o is original surface area, l^o is original specimen length (11 cm), Δl is increment of the specimen length by straining. In the present study the amount of strain ($\Delta l / l^o$) was 1 to 2 %.

In the plate specimens, crack length was directly measured by a travelling microscope to obtain the crack growth rate.

Fracture surfaces of specimens after SCC test were examined by the scanning electron microscope.

RESULTS

OBSERVATION OF CRACK TIPS BY A HIGH VOLTAGE ELECTRON MICROSCOPE   Typical example of crack tip obtained from the specimen tested at CHS of 1.5×10^{-2} mm/min is shown in Figure 1. The cracks are branched before crossing the grain boundary and there exists a deformation structure containing high densed dislocations and stacking faults in the vicinity of the crack tips. The slip planes $(1\bar{1}\bar{1})$, $(\bar{1}1\bar{1})$ and $(\bar{1}11)$ in the photograph were determined by analyzing diffraction patterns and direction of the slip planes appearing in the photograph. The cracks exist in the slip planes $(1\bar{1}\bar{1})$ and $(\bar{1}1\bar{1})$.

Transgranular cracking in the solution always proceeded along {111} slip planes. From these observations, it was found that the cracks propagated by a preferential dissolution of active slip planes containing moving dislications under stress.

SCC TEST BY A SLOW STRAIN RATE TECHNIQUE
Figure 2 shows the effect of CHS on maximum stress at different potentials. It is found that susceptibility to SCC at anodic potential (E_a) was higher than that at corrosion potential (E_{corr}, -0.34V) for all CHS and that at cathodic potential (E_c) the susceptibility decreased and the maximum susceptibility shifted to the high CHS region.

FIGURE 1. Transmission electron micrograph showing stress corrosion crack tip of Type 304 stainless steel in 42% $MgCl_2$ at 143°C.

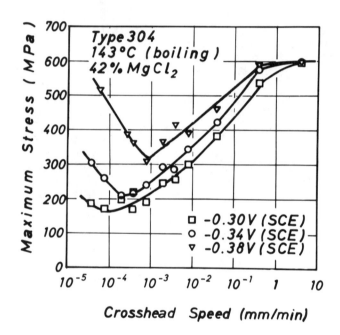

FIGURE 2. Effect of crosshead speed on maximum stress of Type 304 stainless steel in 42% $MgCl_2$ at 143°C at different potentials.

It has been already shown[1-3] that in SCC proceeded by active path corrosion, film rupture due to slip steps, and dissolution-repassivation of fresh slip steps are important factors, and that these factors are related to the formation rate of slip steps under stress and corrosivity of environment.

Figure 3 shows a schematic diagram of a relationship between susceptibility to SCC and CHS in Figure 2. As seen in Figure 3 the curve is classified from region I where SCC does not occur at high CHS to region V where no SCC takes place at very slow CHS. SCC region is $(CHS)_{IV/V} < CHS < (CHS)_{I/II}$.

Considering the behavior of one slip step

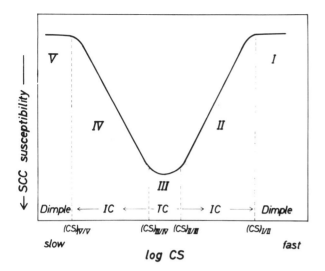

FIGURE 3. Schematic illustration showing the effect of crosshead speed (CS) on SCC susceptibility and fracture mode. TC; transgranular cracking, IC; intergranular cracking.

regarding above two factors, film repair (repassivation) begins after immediately slip step creates. At this time, dissolution current density (i) is expressed as following equation from the result[4] of strain electrode study.

$$i(t)=i^o \cdot \exp(-\beta t) \qquad (1)$$

where, i^o is initial dissolution current density on slip step. β is decay constant. In SSRT, CHS corresponds to formation rate of slip steps ($\dot{n}_s$). If slip steps nucleate every $t(1/\dot{n}_s)$ then the behavior of dissolution and repassivation of alloy on slip steps could be shown by II, III, IV and V in Figure 4, respectively, depending on the amount of t. In region II and III, it is difficult for repassivation to occur because of small t (large $\dot{n}_s$), and high crack propagation rate could be expected. With increasing CHS (nearing region I), however, most part of the fracture surface becomes to show dimple pattern rather than SCC, and susceptibility to SCC decreases.

The results in Figure 2 are explained as follows by the dissolution and repassivation model above mentioned. Dissolution-repassivation behavior in eq(1) is shown schematically in Figure 5 (a) at three applied potentials. Average dissolution current density (i), which corresponds to velocity of SCC, is also shown in Figure 5 (b) together with the relation of CHS and SCC susceptibility. In this case following relation can be established for all CHS, $i_a > i_{corr} > i_c$. High dissolution current density at E_a, therefore, results high susceptibility and low dissolution current at E_c results low susceptibility. Further, it is found from Figure 5 that under anodic condition, repassivation at crack tip occurs not so easily even though in slow CHS (small $\dot{n}_s$) and that at cathodic potential dissolution of alloy on slip steps is suppressed and repassivation occurs easily. The CHS correspond-

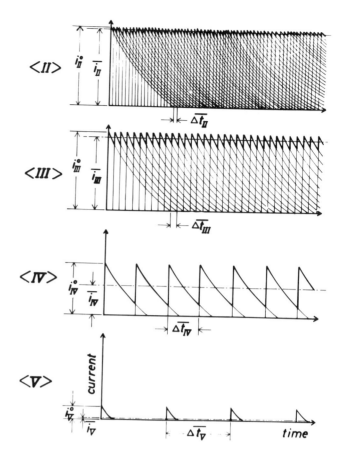

FIGURE 4. Schematic figures of dissolution-repassivation current density of slip step emerged at crack tip at different crosshead speed. i^o; initial dissolution current density, $\bar{i}$; average dissolution current density. $\overline{\Delta t}$; average interval of slip step emergence at crack tip ($1/\dot{n}_s$).

ing to the maximum sysceptibility, therefore, shows smaller value at E_a than at E_c.

Thus, the mechanism of SCC of Type 304 stainless steel in 42% $MgCl_2$ could be explained by film rupture and dissolution-repassivation model.

CRACK PROPAGATION RATE   Details of dissolution-repassivation behavior of the specimen in the solution was investigated using a strain electrode technique at different potentials.

Figure 6 shows a relationship between current density i(t) and time. It is found from the measured points that the decaying current density followed the exponential curve given by eq.(1). The β at different potentials obtained from eq. (1) is shown in Figure 7, in which i^o is also shown. It is clear from Figure 7 that the i^o increases and the β decreases with potential. The solid curves in Figure 6 are obtained from eq.(1).

Crack growth rate (da/dt) of specimens in the solution under SSRT was measured in the range of CHS from 4×10^{-5} to 4×10^{-1} mm/min. Figure 8

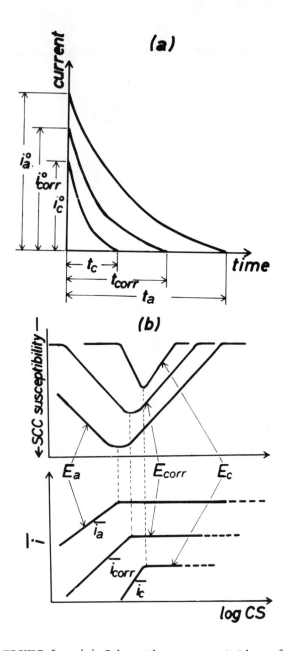

FIGURE 5. (a) Schematic representation of dissolution-repassivation current density on slip step at different potentials. (b) Schematic representation showing the relationship between SCC susceptibility and average dissolution current density on slip steps at different potentials.

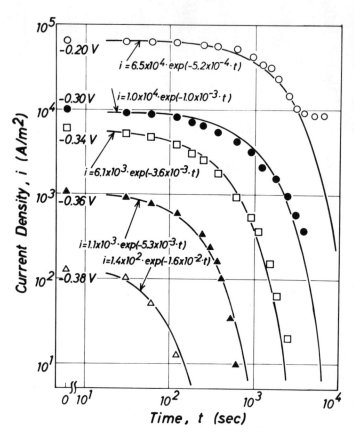

FIGURE 6. Dependence of dissolution current density of Type 304 stainless steel upon time at different potentials tested in 42% $MgCl_2$.

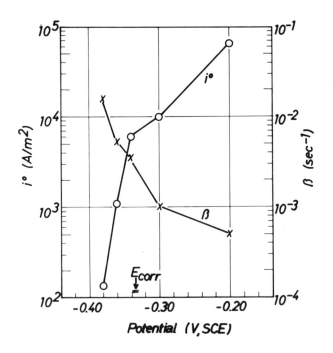

FIGURE 7. Dependence of dissolution current density (i^o) and decay constant (β) in strain electrode test on potential.

shows the relationship between da/dt and crack length (a). In the range of CHS$\geq$4x10^{-4} mm/min, da/dt increased with increasing a or CHS. At a CHS of 4x10^{-5} mm/min, however, da/dt was constant in the range of a from 3 to 5 mm. Intergranular cracking was observed in the upper region of the broken line (da/dt=1.5x10^{-2} mm/min), and transgranular cracking was in the lower region. The crack length resulting in marco branching increased generally with decreasing CHS. At this CHS, da/dt showed two different behaviors against a (in Figure 8). One is region I where da/dt

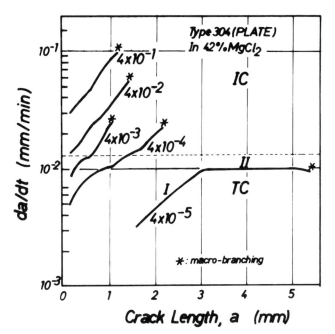

FIGURE 8. Relationship between crack propagation rate and crack length of Type 304 stainless steel in 42% MgCl$_2$ at 143°C at different crosshead speeds.

depends upon a, and another is region II where da/dt is independent of a.

DISCUSSION

CRACK PROPAGATION MODEL   It was already shown that the dissolution-repassivation behavior of alloy in the fresh slip step are given by eq.(1). In actual process of SCC, however, it is considered that successive dissolution of the slip planes would occur, because the film rupthre due to slip steps under straining continuously takes place.

Dissolution current density $i_c(t)$ in time t at crack tip for one slip step is given by eq.(2) according to eq.(1).

$$i_c(t)=i^O \cdot \exp(-\beta t) \qquad (2)$$

Further, if the film rupture takes place for every slip step formation, then the fresh slip steps would be emerged at an interval of $1/\dot{n}_s$ (see Figure 4). Average dissolution current density $\bar{i}_c$ is, therefore, given by eq.(3).

$$\bar{i}_c = \int_0^{1/\dot{n}_s} \cdot i_c(t) \cdot dt/(1/\dot{n}_s)$$

$$=i^O/\beta \cdot \dot{n}_s \cdot [1-\exp(-\beta/\dot{n}_s)] \qquad (3)$$

As eq.(3) is a function of a monotonic increment, the following relation can be obtained.

$$\left.\begin{array}{ll} \bar{i}_c=i^O & \text{for } \beta/\dot{n}_s \ll 1 \\[2mm] \bar{i}_c=i^O/\beta \cdot \dot{n}_s & \text{for } \beta/\dot{n}_s \gg 1 \end{array}\right\} \quad (4)$$

da/dt is given by eq.(5) according to Faraday law.

$$da/dt=\bar{i}_c \cdot M/(z \cdot F \cdot \rho) \qquad (5)$$

where, M is atomic weight, z is the number of moles of electrons which are taking part in the process, F is Faraday constant (96500 C/mol), ρ is the density of material. The crack propagation rate for Type 304 stainless steel is, therefore, given as follows;

$$da/dt=2.02 \cdot 10^{-6} \cdot i^O$$

$$=2.02 \cdot 10^{-6} \cdot i^O/\beta \cdot \dot{n}_s \cdot [1-\exp(-\beta/\dot{n}_s)] \qquad (6)$$

CORRESPONDANCE OF THEORETICAL VALUES DEDUCED STRAIN ELECTRODE METHOD AND OBSERVED VALUES Figure 9 shows the calculated relationship between da/dt and $\dot{n}_s$, where da/dt can be obtained by substituting the values of i^O and β at each specimen in strain electrode technique for eq. (6). It is clear from Figure 9 that (1) two regions exist at all potentials; one is region I where da/dt depends upon $\dot{n}_s$ and another is region II where da/dt is independent of $\dot{n}_s$, (2) at corrosion potential of -0.34 V, the value of da/dt agrees with the results shown in Figure 8, and (3) with increasing applied potential da/dt increase and the values of $\dot{n}_s$ corresponding to the transition point from region I to region II lower.

Figure 10 is the potential-$\dot{n}_s$-da/dt diagram, which can be deduced from Figure 9. Broken line aa' in Figure 10 indicates corrosion potential of the specimen in the solution. This diagram represents the values of da/dt for both factors of stress ($\dot{n}_s$) and environment (potential), and offers the valuable informations for the crack propagation behavior of Type 304 stainless steel in boiling 42% MgCl$_2$.

It is found in Figures 9 and 10 that under a constant $\dot{n}_s$ (stress condition is constant), da/dt increases with applied potential. This agrees with the results in Figure 2.

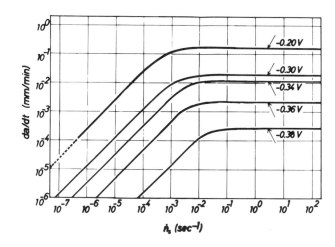

FIGURE 9. Relationship between calculated da/dt and $\dot{n}_s$ at different potentials.

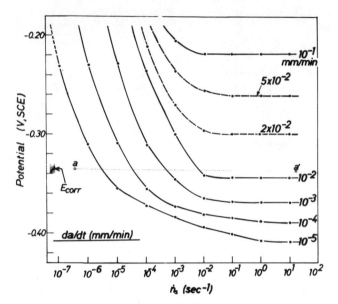

FIGURE 10. Potential-$\dot{n}_s$-da/dt diagram obtained from Fig. 9.

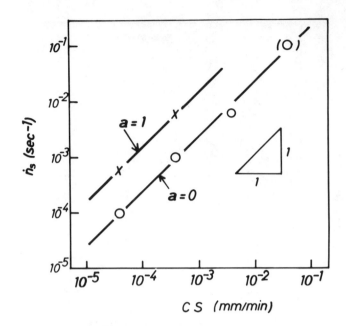

FIGURE 11. Correlation between $\dot{n}_s$ and crosshead speeds.

Considering that n_s is a function of crosshead speed, K value in fracture mechanics or crack length, it is said that the region I and region II appeared in the case of transgranular cracking in Figure 8 are substantially the same as those in Figure 9.

(a) Region II   The observed values of da/dt in region II in Figure 8 were approximately 10^{-2} mm/min, and this agrees well with the calculated value at -0.34 V shown in Figure 9 and 10.

As $\beta/\dot{n}_s \ll 1$ in region II, the value of da/dt can be given by following equation from eq.(6).

$$da/dt = 2.02 \cdot 10^{-6} \cdot i^o \qquad (7)$$

It is found from eq.(7) that the dissolution rate of alloy in fresh slip step would be rate controlling factor in the region II where the formation rate of slip steps is high.

Sgibata et al[5] determined i^o of Type 304 stainless steel in 45% $MgCl_2$ at 154°C at -0.35 V (SCE) by a strain electrode technique and obtained the value of 3×10^4 A/m^2. Substituting the i^o for eq.(7), da/dt can be given as 6×10^{-2} mm/min. Kitagawa et al[6] also measured da/dt of the same material in the same environmental condition but at corrosion potential, and showed that da/dt was 2×10^{-1} mm/min. Considering that the former's potential was a little cathodic, these values of da/dt are reasonable. These results also show that the eq.(7) in the present study is generally acceptable.

(b) Region I   It is more general to express $\dot{n}_s$ as CHS in SCC test under SSRT. In order to establish the correlation between $\dot{n}_s$ and CHS in region I, therefore, Figure 11 was made from the relation between the values of da/dt which are obtained by extrapolation to a=0 and a=1 at each

CHS in Figure 8, and $\dot{n}_s$-da/dt relation at corrosion potential of -0.34 V in Figure 9.

It is found that the slope of the straight lines in Figure 11 gives 1, Further, following equation can be obtained for a=0.

$$\dot{n}_s (sec^{-1}) = 114 \cdot CHS (mm/sec) \qquad (8)$$

where, CHS can be expressed by following equation.

$$CHS = N \cdot \dot{n}_s \cdot n \cdot b \cdot 1/\ 2^{1/2} \qquad (9)$$

where, N is a number of active slip planes, n is a number of dislocations participating formation of slip steps, b is Burger's vector (2.5×10^{-7} mm).

da/dt for a=0 corresponds to that of crack starting, plastic deformation under this condition, therefore, takes place uniformly within notch width (0.25 mm) of the specimen. Assuming that a distance between active slip planes is 2.5 micron meter, which was obtained from the observation of dislocations in 5% strained specimen by transmission electron microscope, then N becomes 100. As a hight of slip band is generally considered to be 20 to 1000 atomic diameters for fcc materials[8], assuming n=500, eq.(9) gives rise eq. (8).

Thus, eq.(9) is also confirmed theoretically. It is obvious, therefore, that eq.(6) is also applicable for region I.

The values of da/dt in the range of CHS =4x 10^{-4} mm/min in Figure 8 did not show region I and II. This is due to that intergranular cracking occurred easily in the specimen at higher CHS seen in Figure 2 (and Figure 3).

Thus, the observed values of da/dt for transgranular cracking in region I and region II agree generally with the calculated values obtained on the basis of film rupture model (dis-

solution-repassivation model). This indicates
that transgranular cracking of Type 304 stainless
steel in boiling 42% $MgCl_2$ proceeded with the
mechanism mentioned above.

Further, if the values of i^o and β in other
material/environment system are obtained, then
da/dt could be calculated. It is possible,
therefore, that a predictable SCC diagram of the
system is made.

SUMMARY

Using a high voltage electron microscope, observations of crack tips of Type 304 stainless steel
in boiling 42% $MgCl_2$ were carried out. Behavior
of SCC of the specimen in the solution was also
studied using slow strain rate technique and
strain electrode method. The following summary
can be given.

1. Transgranular cracking always propagates by
   a preferential dissolution of active slip
   planes containing moving dislocations.
2. Mechanism of SCC in the solution can be explained by considering the formation rate of
   slip steps and their characteristics of slip
   dissolution-repassivation.
3. Propagation rate for transgranular cracking
   has been expressed by the following equation.

$$da/dt = 2.02 \cdot 10^{-6} \cdot i^o / \beta \cdot \dot{n}_s \cdot [1 - \exp(-\beta/\dot{n}_s)]$$

   where i^o is initial dissolution current density of slip steps, β is decay constant, $\dot{n}_s$ is
   formation rate of slip steps. The equation
   has been confirmed by observed values of crack
   propagation rate.
4. Potential (environmental factor)-$\dot{n}_s$ (stress
   factor)-da/dt (SCC susceptibility) diagram has
   been presented for SCC prediction in this
   system.

REFERENCES

1. T. Nakayama and M. Takano. Boshoku Gijutsu,
   28, 540-47 (1979).
2. T. Nakayama and M. Takano. Corrosion, 37,
   226-31 (1981).
3. T. Murata and R. W. Staehle. 5th Int. Metallic Corrosion. NACE, p. 513-18 (1974).
4. M. Takano and S. Shimodaira. Trans. JIM, 9,
   294-301 (1968).
5. T. Shibata and T. Takeyama. Boshoku Gijutsu,
   23, 379-83 (1979).
6. H. Kitagawa and H. Ohira. Kikoron, 740-11,
   263- (1974).
7. T. Nakayama and M. Takano. Corrosion, 41,
   592- (1985).
8. H. Ohtsuka and A. Ikushima. "Translation from
   Mechanical Behavior" (The Structure and
   Properties of Materials. Vol. III) by W.
   Hayden, W. G. Moffatt and J. Wulff, John Wiley
   & Sons, N.Y. (1965), p. 92, Iwanami (1967).

ANALYSIS OF SUB-CRITICAL CRACKING IN A Ti-5Al-2.5Sn LIQUID HYDROGEN CONTROL VALVE

D. A. Meyn, R. A. Bayles
Code 6312, Naval Research Lab.
Washington, DC, USA

A liquid hydrogen main fuel control valve for a rocket engine failed by fracture of the Ti-5Al-2.5Sn body during the last of a series of static engine test firings. Fractographic, metallurgical, and stress analyses determined that a combination of fatigue and unexpected aqueous stress-corrosion cracking initiated and propagated the crack which caused failure. The failure analysis approach and its results are described to illustrate how fractography and fracture mechanics, together with a knowledge of the crack initiation and propagation mechanisms of the valve material under various stress states and environments, helped investigators to trace the cause of failure.

THE FRACTURE OF A COMPONENT which is, so far as can be determined by material design data and careful stress analysis, operating completely within design limits, is a matter of great concern to project managers. Such a failure may be an indication that the design data are wrong, the stress analysis is wrong, or that something happened which was not expected and went unnoticed during testing or operation of the component. Such an event may be detected after the fact by careful analysis of the mechanical, metallurgical and environmental features of the failure event and of the component itself. An analytical technique of great power combines careful study of the fracture surface micro-features (fractography) and correlation of those features through fracture mechanics with the expected crack propagation characteristics of the alloy under conditions of stress and environment which the component might have experienced before failure.

This paper shows how fracture mechanics was used to correlate the environment and stress history with microscopic fracture surface features in a titanium alloy liquid hydrogen control valve fracture; and how by comparing the fractographic evidence with knowledge of crack propagation mechanisms in the alloy, investigators were able to point to a probable cause of the failure and to minimize the probability of a recurrence.

DESCRIPTION OF THE FAILURE

The failed part was the body or housing of a liquid hydrogen main fuel control valve, which ruptured during testing of the rocket engine of which it was a part. The area where the crack initiated comprised a cut-out in the internal surface to accomodate a cam follower bearing which was part of the valve operating mechanism. A schematic of the longitudinal cross-section of the valve is shown in Fig. 1(a) indicating the location of the cut-out. Figure 1(b) shows a transverse cross-section through the cut-out, which resembles a "U" shaped longitudinal trough in the bottom of the valve body. The location of the crack which caused the failure is indicated in Fig. 1(b). The valve was carrying liquid hydrogen at a pressure of 41.3 MPa when it ruptured, and the temperature of the metal at the site of crack initiation on the internal surface was approximately 77 K. The theoretical stress concentration (K_t) at that point is 1.7, resulting in a calculated stress of 1345 MPa, which is approximately equal to the yield stress of the alloy at 77K. The stresses during the test were essentially static, with the superposition of an alternating component of approximately 13 MPa peak-to-peak caused by mechanical and hydraulic vibration. This alternating component amounted to about 1% of the static stress.

When the ruptured valve body was examined, several crack initiation sites were found along the edge of the fracture within the cut-out area. Figure 2 is a sketch of the slow or stable crack propagation areas of the fracture surface, and adjacent portions of the overload fracture areas. The main portion of stable cracking comprises areas A, B and C. Area C is the visible portion of an originally somewhat larger crack parallel to and partly overlaying area B, and part of it was apparently retained by the missing mating fracture remnant. Areas D and E are not significant. Areas A and B appeared to be the most significant and fruitful for fractographic analysis. At first glance areas A and B appear to have little organization, being complex mixtures of cleavage and fatigue. Much study of these areas convinced the author that there are four major episodes of stable crack growth represented. Cracking in area A will be discussed in some detail, but area B is essentially similar.

Cracking initiated in area A at the arrow, apparently by transgranular cleavage, Fig. 3. A line perpendicular to the fracture edge passes through four zones of stable crack growth before terminating at the onset of overload fracture. These zones are(referring to Fig. 2b): C1, the initial cleavage area which extends in to a depth of 0.635 mm (Fig. 4); F1, the following band of fatigue, which is 0.165 mm across, with striations of 1-4 μm spacing (Fig. 5); C2, the second cleavage band which is 0.533 mm across (Fig. 6); and F2, the final fatigue band which is .152 μm across, with striations of 5-10 μm spacing (Fig. 7). These four zones are duplicated in the smaller area B, along a line bearing toward 10 0'clock from the origin of area B. Cleavage zones C1 and C2 have no detectable fatigue striations, and the cleavage facets often terminate with flutes (12). The appearance is very similar to the cleavage plus flutes topography caused by aqueous SCC in Ti-5Al-2.5 Sn (3). The fatigue bands F1 and F2 have scattered fields of striations together with other flat, characteristic fatigue markings. Other areas marked "F" are similar.

Striations in zones F1 and F2 of area A were studied to determine how many load cycles were necessary to complete each zone. At no point in zone F1 were striations even approximately continuous across the zone, so the number of load cycles was estimated by dividing the width of the zone by the average spacing of stiations found in the zone. This produced a value of 50-100 cycles for zone F1. In two parts of zone F2 striations were nearly continuous across the width of the zone, and a reasonably accurate estimate of load cycles was made. There were 25 to 30 load cycles accounted for by zone F2. Essential agreement was obtained by a similar analysis of zones F1 and F2 in area B.

These fractographic results suggested a four-stage crack propagation history: (1) some initial period of sustained stress, or of high mean stress low amplitude fatigue, together with an environmental factor to account for the un-striated cleavage-plus-flutes increment; (2) a period of low cycle, high stress fatigue comprising 50-100 load cycles; (3) another period similar to the first one; (4) and finally another period of low cycle, high stress fatigue comprising 25-30 load cycles, culminating in overload fracture when the combination of maximum load and crack depth exceeded the fracture toughness of the material. At this point, it is necessary to examine the processing, proof testing and operating history of the valve to correlate the stresses and environments with the fractographic observations.

MATERIAL AND PROCESSING

The valve body was produced from Ti-5Al-2.5Sn ELI (extra low interstitial) and had 0.095% oxygen and 21 ppm hydrogen (by weight), and conformed to the specified composition limits. The material was cross-forged at a starting temperature of 1227 K, which is in the middle of the alpha-beta two-phase field (beta transus = 1294K). After forging, the part was vacuum-annealed at 1033 K, air cooled, surface-machined to remove scale, and inspected for cracks; none were found. Forging prolongations were mechanically tested at room temperature and at 20K; yield strengths were 730-760 MPa and 1345-1445 MPa respectively. The microstructure was according to specifications. The valve body was then machined to rough shape, pressure tested at room temperature, partly finish-machined, inspected for dimensions, finish machined, dye penetrant-inspected for cracks, cleaned (non-cholorinated solvent), and inserts and studs were assembled to the body. Then followed a series of exterior cleaning and coating procedures (orifices sealed), after which the valve body was stress-relieved in vacuum at 922K, air cooled, coated with foam insulation, nickel-plated, thermal shock tested (immersed in liquid nitrogen then allowed to warm up to ambient), then thoroughly flush-cleaned with Freon TF ($C_2F_3Cl_3$) and dried. The operating mechanisms of the valve were then installed, and the valve was ready for proof testing and subsequent use on an engine.

OPERATING AND PROOF TEST HISTORY

Figure 8 is a chart of the pressurization history of the valve. The only significant pressurization not included is the initial pressure test after rough machining and before finish machining. This pressure test might have initiated cracks, but the appearance of the origins of the various cracks suggests that no subsequent machining was performed after initiation. The first series of tests was performed at room temperature, after which followed a series of cryogenic proof tests. Then followed several engine tests, for each of which the valve was pressurized once for time varying from a few seconds to several dozen seconds. These engine tests were interspersed with a few additional proof tests.

One very significant episode is represented by the heavy vertical line. At this point (during the test just before this line) the valve was involved in a fire, which did not apparently damage the well-insulated valve. It was disassembled, the body was stripped of insulation, cleaned, and carefully inspected. No damage was found, so it was re-processed, beginning with the dye-penetrant inspection and including the stress relief anneal at 922K in vacuum. After two room temperature and one cryogenic proof tests, a series of engine tests followed, during the last of which failure occured by rupture of the valve body.

This division of the valve pressurization history into two major parts suggested that the series of pressurizations before the fire might correspond to zones Cl and Fl, and those after the fire to C2 and F2. A comparison of fatigue striation spacings and crack propagation distances with predictions based on fracture mechanics principles, materials properties, and the loading history of the valve, was performed.

FRACTURE MECHANICS ANALYSIS AND DISCUSSION

The concepts of fracture mechanics and the significance of the crack tip stress intensity factor, K, are explained in several standard references, for example reference 4, and those needing further background should consult that reference. The application of fracture mechanics to fatigue is based upon the observation made many years ago by Paris (5) that the propagation rate of a fatigue crack per load cycle is a function of the cyclic amplitude or range of the stress intensity factor. For many materials over a considerable range of crack growth rates this function takes the form (5):

$$da/dn = A (\Delta K)^n \qquad (1)$$

where a = crack depth, N= number = of load cycles, A and n are experimentally determined constants, and ΔK is the stress intensity factor range. The term "range" here refers to the difference between the minimum and maximum value of load during each cycle. This is equivalent to the peak-to-peak double amplitude for a sinusoidal load-vs-time waveform. When the log-log plot of crack growth rate versus ΔK is reasonably linear over the growth rate range of interest, the expression in Eq. 1 can be used to predict growth rates for any given value of ΔK, or conversely to translate striation spacings into values of ΔK and thus, given the crack depth, estimate the applied stress values causing cracking. If the log-log plot is excessively curved, then information can be taken directly from the plot, which is not difficult for the number of load cycles experienced by this valve.

A chronological listing of all pressurization cycles is presented in Table I, with the crack depth and calculated ΔK and da/dN values. For room temperature cycles, values of A = 3.9 x 10^{-5} and n=2.9 (from data of Frost and Denton (6)) were substituted into Eq. 1. For cryogenic cycles values of da/dN were read from graphical data supplied by the valve manufacturer (7).

The stress intensity values were calculated from the following relationship:

$$\Delta K = 0.025 \Delta \sigma \sqrt{\pi a} \quad (MPa\sqrt{m}) \qquad (2)$$

where 0.025 includes a correction factor accounting for the shape of the semi-elliptical surface crack (8), $\Delta \sigma$ is the cyclic stress range in MPa produced by the applied pressure, and a is the crack depth in mm. The stress at the crack location is related to the applied pressure, p, by:

$$\Delta \sigma = 32.5p \quad (MPa) \qquad (3)$$

There are several entries in Table I whose da/dN values are enclosed in parentheses; these are for pressurizations at very low pressure following high level pressurizations. There is much evidence to suggest that crack growth would be severely retarded if not completely suppressed for such cycles (9). These are accordingly not counted in making the correlation with fractographic observations.

One can assume that some event, not yet specified, caused the initial cleavage crack (zone Cl) early in the valve's test history; that the first series of proof and engine test cycles caused Fl; that after the fire and re-processing of the valve another cleavage crack increment occurred (zone C2); and that the last series of proof and engine tests caused F2. Accordingly, Table I predicts for zone Fl probably 55 striations of spacings from 1 to 6 μm, possibly several other much smaller increments, for a total estimated propagaion of 0.151-0.158 mm. The actually measured width of Fl is 0.165 mm, and fractographic measurements indicate 50-100 cycles of loading with striation spacings of 1 to 4 μm. Similarly Table I predicts for zone F2 28-30 striations of 2-8 μm spacing and a total propagation of 0.115-0.120 mm. This compares with a measured width of 0.152 mm and a fairly accurate fractographic estimate of 25-30 load cycles with striations of 5-10 μm spacing. The agreement with respect to striations spacings and total propagation per zone can be regarded as somewhat fortuitous, because for such small cracks, about 1mm deep, the relationship between ΔK and da/dN may not be accurately predicted by standard crack propagation data, which are obtained from specimens with much larger cracks (10). Furthermore, the effects of crack growth rate retardation or enhancement caused by prior high or low loads, respectively (9), have not been systematically accounted for. Nevertheless, the degree of agreement satisfactorily supports the overall hypothesis that Cl and Fl occurred before the fire and C2 and F2 occurred after the fire. Furthermore, it is evident that Cl and C2 in particular are not the product of engine test cycles, nor even of any proof tests except 1 or 2 at the start of each of the 2 segments of the valve history. The crack Cl + Fl should have been found during re-processing after the fire, but could well have been missed because of its small size.

The final piece of information required is the cause of Cl and C2. Initially, it was conjectured that high frequency vibrations caused high mean stress low amplitude fatigue, which can cause striationless cleavage-like cracking (2). This conjecture seems unlikely because such vibrations occurred only during engine test, not during proof test. If high mean stress low amplitude fatigue occurred during any engine test, it should occur during all of them, and its effects would then be uniformly interspersed amongst the coarse striations. Also, the vilbratory stress range, 13 MPa, was extremely low, and the calculated stress intensity factor range was less than 0.8 MPa m; too low for fatigue crack initiation or propagation, respectively.

It was evident that the cause of C1 and C2 must be looked for in the processing and initial proof testing of the valve body, and that the cause must be a static load cleavage mechanism. Hot salt stress corrosion (HSSC) (11) seemed a possibility, because accidental contamination of the surface by salt is always possible, and after each of the two valve body processing episodes a stress relief anneal was performed at a temperature and for a time sufficient to cause cracking. However, HSSC experiments performed by the author on specimens cut from samples of identical processing always showed the cracking mechanism to be mixed cleavage and intergranular cracking, and no instance of pure fluted cleavage has been reported for HSSC of titanium alloys.

Another possibility was that during an initial engine test following each of the two processing episodes, hydrogen embrittlement occurred. This seems unlikely for two reasons. First, the valves are only pressurized while in contact with hydrogen at or below 77K. At this temperature any embrittlement reaction is extremely unlikely for thermodynanic reasons and efforts to produce embrittlement effects by hydrogen contact in this temperature range have been unsuccessful (12). Second, there is no reason to suppose succeeding engine tests would not cause similar episodes of cracking, and no evidence of such cracking was observed. There was the possibility that the valve might have warmed up to room temperature in contact with ambient pressure hydrogen gas, abosrbing hdyrogen to a great enough depth to become embrittled and suffer cleavage cracking on subsequent tests. This was most likely to happen after a "cluster" test, in which three engines are connected and fired as a unit. The diagram in Fig. 8 shows such tests took place only after the fire and second processing, and could not have created the Cl-F1-C2-F2 sequence observed.

The most likely cause of the observed pattern of Cl and C2, both as to fractographic character and the two-part pattern (Cl-F1 then C2-F2), was that a proof test after each processing episode was conducted using an aqueous or organic pressurizing fluid capable of causing SCC, or conducted while traces of such a fluid remained inside the valve body.

After much investigation by the valve manufacturer, it was finally determined that an initial room temperature proof test had been conducted using tap water to pressurize the valve, both after primary manufacture, and after re-processing following the fire. Aqueous solutions are known to cause SCC of the fractographic character observed in zones Cl and C2 (3), and experiments conducted by the author during this investigation demonstrated even distilled water caused such cracking. Figure 9 shows the typical appearance. It is conjectured that a microcrack opened in a thin oxide layer formed during annealing, allowing an SCC crack to initiate and propagate during water pressurization to form Cl. Probably a precrack was not necessary at the high stresses experienced in proof tests; SCC cracks can initiate in this alloy in aqueous solution under straining at and above the yield stress without precracks (3). C2 could of course initiate from the already existing fatigue crack front (F1). There is another difficulty with zone Cl besides initiation of a crack. SCC in water usually requires a threshold value of stress intensity factor, K_{ISCC}, of approximately 20 MPa$\sqrt{m}$. At a maximum pressure of 38 MPa, the local stress would be 1235 MPa, and solving Eq. 2 for crack depth:

$$a_{min} = K^2_{ISCC}/(.025^2 \Delta\sigma^2 \pi) = 0.53mm.$$

This implies that SCC growth should not occur until over halfway through zone Cl, despite the fractographic observations. One can only conclude that very small cracks do not obey fracture mechanics criteria very rigorously under SCC conditions, and that for small cracks aqueous SCC growth is possible even though the nominal stress intensity factor might be well below K_{ICC}.

SUMMARY AND CONCLUSIONS

The foregoing results have shown that fracture mechanics can be used with fractographic techniques, a knowledge of the circumstances of failure and of the mechanical and metallurgical characteristics of the failed part to determine the probable cause of failure.

1. The fracture surface of the valve was examined, and the fracture micromechanisms present carefully mapped, the initiation sites identified, and quantitative fatigue data obtained by counting striations and estimating their average spacing and measuring the widths of fatigue propagation bands.

2. The known fatigue crack propagation rates as a function of stress intensity factor range, ΔK, were used to correlate the fractographic fatigue data with the proof test and operation history of the valve.

3. It was possible thereby to conclude that all of the engine tests and most of the proof tests caused only ordinary fatigue cracking in two separate bands (F1 and F2), and that crack initiation (Cl) and most of the propagation (Cl

and C2) were probably SCC caused by the use of water to initially proof test the valve body after primary manufacture and again after secondary processing following a fire during one of the engine tests.

4. The results of these findings induced the manufacturer to stop the use of water for proof testing the valve body, to redesign the cutout for a larger fillet radius thus reducing the local stress, and as a precaution to ensure that pure hydrogen did not remain in contact with the valve as it warmed up to room temperature. This latter step was taken on the reasonable grounds that it is best to avoid any possibility of hydrogen embrittlement, and it can be achieved by either contaminating the hydrogen with small amounts of water vapor or carbon dioxide, or by flushing out the hydrogen with an inert gas while the system is still cold.

ACKNOWLEDGEMENT

The investigation on which this paper is based was the joint effort of many individuals. However, the fracture analysis and the conclusions drawn therefrom are the author's, and are not necessarily agreed to in all respects by the valve manufacturer nor by NASA. R.P. Jewett (Rocketdyne Division) coordinated the investigation and conducted metallurgical investigations which provided much background information, and with others at Rocketdyne made available technical data (stresses, processing history, etc) necessary to the proper interpretation of the failure. The author is indebted to the other consultants on the failure analysis team for useful information, ideas and discussions H. G. Nelson, NASA-Ames, Moffat Field, CA; H. R. Gray, NASA-Lewis, Cleveland, OH; E. N. Pugh, Nat'l Bureau of Standards, Gaithersburg, MD; H. Gilmore, Marshall Space Flight Center, Huntsville, AL; J. C. Williams, Carnegie-Mellon University, Pittsburgh, PA; N. E. Paton and J. C. Chestnutt, Rockwell International Science Center, Thousand Oaks, CA; and R. R. Boyer, Boeing Commercial Airplane Div., Renton, WA. The Naval Air Systems Command provided partial support to the author for research on which metallurgical aspects of this analysis were based.

REFERENCES

1. Aitchison, I. and Cox, B., Corrosion, 28, 83-87 (1972)

2. Meyn, D. A., "Fractography and Materials Science", ASTM STP 733, 5-31, American Society for Testing and Materials (1981)

3. Powell, D. T. and Scully, J. C., Corrosion, 24, 151-158 (1968).

4. Tetelman, A. S. and McEvily, A. J., Jr., "Fracture of Structural Materials", Ch. 2 and 3, John Wiley and Sons, New York (1967)

5. Paris, P. C. and Erdogan, F., J. Basic Engineering, Trans. ASME, Series D, 85, 528-534 (1963)

6. Frost, N. E. and Denton, K., NEL Report No. 76, National Engineering Laboratory, East Killbride, Glasgow, Scotland (1963)

7. Unpublished Data, Rocketdyne Div., Rockwell International (1979)

8. Smith, H. R. and Piper, D. E., "Stress-Corrosion Cracking in High Strength Steel and in Titanium and Aluminum Alloys", 48-49, Naval Research Laboratory (1972)

9. von Euw, E. F. J., Hertzberg, R. W. and Roberts, R., "Stress Analysis and Growth of Cracks", ASTM STP 513, 230-259, Amer. Soc. for Testing and Materials (1972)

10. Hudak, S. J., Jr., Trans. ASME, J. Engineering Materials and Technology, 103, 26-35 (1981)

11. Sinigaglia, D., Taccani, G. and Vicentini, B., Corrosion Science, 18, 781-796 (1978)

12. Williams, D. P. and Nelson, H. G. Met. Trans., 3, 2107-2113 (1972). Also: H. G. Nelson, Private Communication, Canoga Park, CA (Oct 1979)

Table 1 - Crack Propagation Predictions

Type of Test	p MPa	σ MPa	a (mm)	ΔK (MPa√m̄)	da/dN (μm)	Cycles
Proof @ 298K	33.1	1075	0.71	40.7	1.8	2
Proof @ 298K	38.6	1254	0.71	47.4	2.9	14
Proof @ 77K	34.4	1121	0.71	42.4	1.52	1
Proof @ 77K	57.2	1859	0.71	70.3	5.8	5
Engine @ 77K	21.4	694	0.71	26.3	(.3)	(1)
Engine @ 77K	36.5	1187	0.71	44.9	1.8	2
Engine @ 77K	30.3	985	0.71	37.3	1.02	2
Engine @ 77K	21.4	694	0.71	26.3	(.30)	(5)
Proof @ 298K	34.4	1120	0.71	42.4	2.1	1
Proof @ 298K	38.6	1254	0.71	47.4	2.9	5
Proof @ 77K	34.4	1120	0.71	42.4	1.52	1
Engine @ 77K	34.4	1120	0.71	42.4	1.52	3
Engine @ 77K	21.4	694	0.71	26.3	(.30)	(2)
Proof @ 298K	37.9	1232	0.71	46.6	2.7	1
Engine @ 77K	30.3	985	0.71	37.3	1.02	2
Engine @ 77K	22.7	739	0.71	27.9	(.38)	(1)
Engine @ 77K	30.3	985	0.71	37.3	1.02	1
Engine @ 77K	36.5	1187	0.71	44.9	1.83	1
Engine @ 77K	25.5	829	0.71	31.4	(.57)	(2)
Engine @ 77K	41.3	1344	0.71	50.8	2.49	1
Engine @ 77K	37.9	1232	0.71	46.6	2.01	1
Engine @ 77K	43.4	1411	0.71	53.4	2.84	2
Engine @ 77K	22.7	739	0.71	27.9	(.38)	(1)
Engine @ 77K	43.4	1411	0.71	53.4	2.84	1
Engine @ 77K	26.2	851	0.71	32.2	(.62)	(1)
Engine @ 77K	51.0	1657	0.71	62.7	4.1	1
Engine @ 77K	26.2	851	0.71	32.2	(.62)	(1)
Engine @ 77K	37.9	1232	0.71	46.6	2.01	1
Engine @ 77K	47.5	1545	0.71	58.4	3.4	1
Engine @ 77K	42.7	1388	0.71	52.5	2.8	1
Engine @ 77K	26.9	873	0.71	33.0	(.7)	(1)
Engine @ 77K	45.5	1478	0.71	55.9	3.1	1
Engine @ 77K	38.0	1233	0.71	46.6	2.01	1
Engine @ 77K	27.6	896	0.71	33.9	(.75)	(1)
Engine @ 77K	44.8	1456	0.71	55.0	3.05	2
Engine @ 77K	48.9	1590	0.71	60.2	3.68	1
				Totals	151 (7.0)	55 (16)

Fire & Re-Processing, then:

Type of Test	p MPa	σ MPa	a (mm)	ΔK (MPa√m̄)	da/dN (μm)	Cycles
Proof @ 298K	34.4	1120	1.40	59.4	5.5	1
Proof @ 298K	37.9	1232	1.40	65.3	7.2	1
Proof @ 77K	34.4	1120	1.40	59.4	3.6	1
Engine @ 77K	35.8	1164	1.40	61.8	4.0	2
Engine @ 77K	41.3	1344	1.40	71.3	6.2	2
Engine @ 77K	44.1	1433	1.40	76.0	8.3	1
Engine @ 77K	41.3	1344	1.40	76.0	6.2	8
Engine @ 77K	36.5	1187	1.40	62.9	4.2	1
Engine @ 77K	29.6	963	1.40	51.0	(2.5)	(1)
Engine @ 77K	41.3	1344	1.40	71.3	6.2	1
Engine @ 77K	28.9	940	1.40	49.9	(2.4)	(1)
Engine @ 77K	36.6	1188	1.40	62.9	4.2	2
Engine @ 77K	41.3	1344	1.40	71.3	6.2	8
				Totals	115 (4.9)	28 (2)

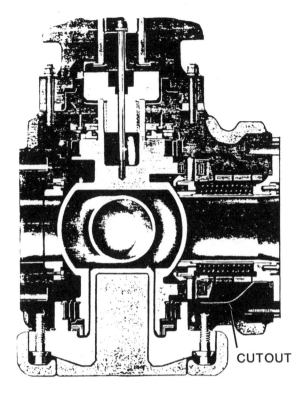

CUTOUT

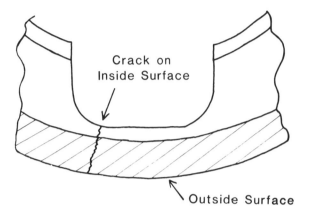

MAIN FUEL VALVE HOUSING

CAM FOLLOWER BEARING CUTOUT

Crack on
Inside Surface

Outside Surface

1(a). Longitudiual cutaway of valve assembly. The cam follower bearing cutout is under the right-hand bore of the valve (courtesy Rocketdyne Div., Rockwell International). Approx. 1/5X.

1(b). Transverse cross section through the cut-out showing crack location. The crack is parallel to the bore. Approx. 1/2X.

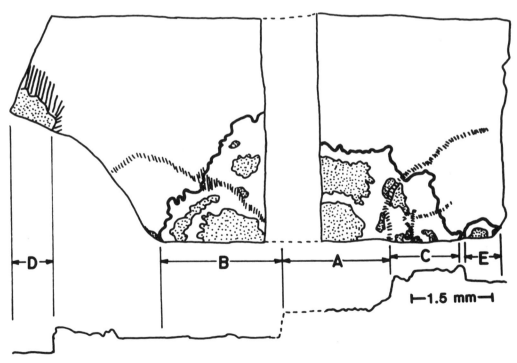

D   B   A   C   E

⊢1.5 mm⊣

2(a). Sketch of the fracture surface as received by the author. Part of the center section was removed by a saw cut. Cleavage zones are dotted, the rest of slow growth area (inside heavy lines) is fatigue. The shaded lines represent steps between parallel cracks at different elevations. The bottom edge corresponds to the inside surface of the valve. The vertical profile of the fracture edge is below. 20X.

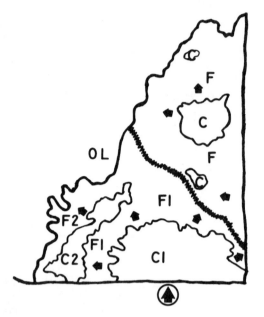 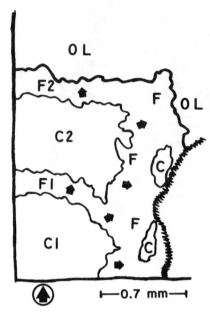

2(b). Enlargement of areas A and B. Initiation sites are indicated by large arrows. Local crack growth directions are indicated by smaller arrows on the fracture. See text for interpretation of other symbols. 50X.

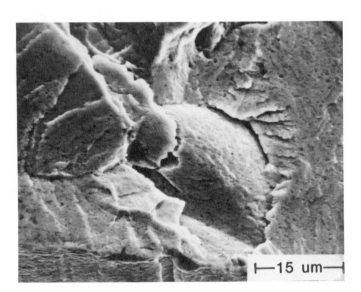

3. SEM fractograph showing cleavage at initiation site of area A, zone Cl (see text). Fracture edge is at bottom. 2000X.

4. SEM fractograph showing cleavage plus flutes in zone Cl of area A. 1500X.

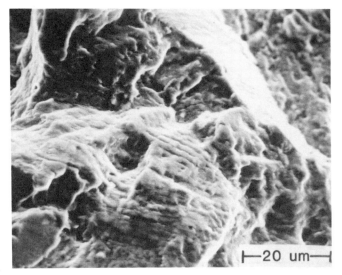

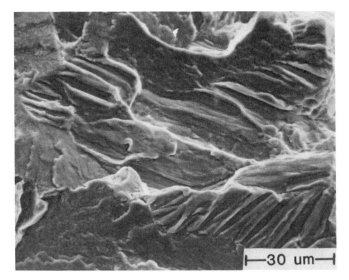

5. SEM fractograph showing fatigue striations in zone F1, area A. 1500X.

6. SEM fractograph showing cleavage and flutes in zone C2, area A. 1000X.

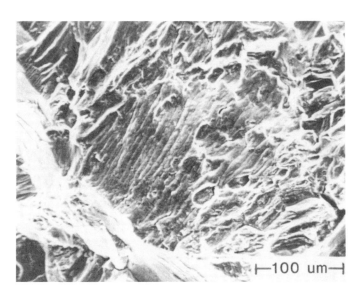

7. SEM fractograph showing fatigue striations in zone F2, area A. 300X.

H2 VALVE TEST HISTORY

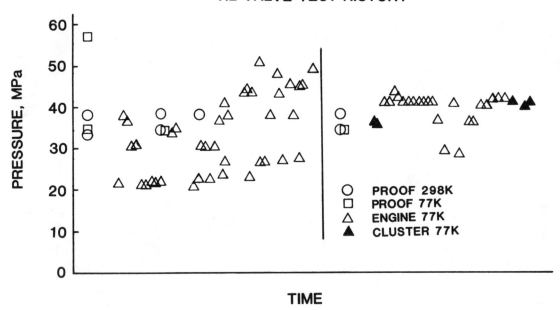

8. Chart of proof test and engine test pressurization history of the valve body. Each point represents an excursion from zero to indicated pressure and back to zero.

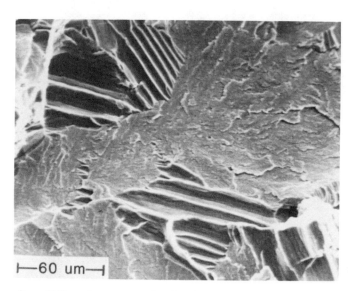

9. SEM fractograph showing typical cleavage and flutes caused by SCC in water in Ti-5Al-2.5Sn from similar valve body forging. 500X.

EFFECT OF SEA WATER ON THE FATIGUE CRACK PROPAGATION CHARACTERISTICS OF WELDS FOR OFFSHORE STRUCTURES

D. K. Matlock, G. R. Edwards
D. L. Olson, S. Ibarra

Center for Welding Research
Colorado School of Mines
Golden, Colorado, USA

Amoco Research Center
Naperville, Illinois, USA

ABSTRACT

The fatigue crack propagation characteristics of ASTM A36 steel weldments produced by both conventional surface welding operations and underwater repair welding techniques were evaluated in both air and seawater. The resulting fatigue crack growth rates were shown to depend sensitively on pore density and pore distribution, factors which significantly varied with welding procedure and environment. Variations in growth rate were correlated with scanning electron fractographs of the stable fatigue crack zone.

The results of this study show that underwater wet welding procedures produce fatigue resistant weld metal which is adequate for use at low applied stresses in offshore structures.

INTRODUCTION

Offshore structures of sufficient structural integrity are normally fabricated with weldments designed to provide both adequate static strength and good fatigue resistance. Such structures often must be repaired, and, if structural integrity is to be maintained, repair welds must be designed and installed with the same rigor as that applied to initial construction. These welds most often must be made under water.

Two main types of underwater welding are in common use: "dry" habitat and "wet". Dry habitat welding involves the construction and evacuation of an enclosure around the submerged portion of the structure to be welded. Air or other suitable gases at ambient pressures controlled by the depth of the operation, then make up the welding atmosphere. Wet welding is similar to normal shielded manual metal arc welding, except that the work environment is water. As a

result, there are many advantages to wet welding. For example, the welder can work on portions of structures which, because of the complexity, cannot be welded by another method. In comparison to dry habitat welding, wet welding can be performed faster and at a lower cost, since the construction of an enclosure is not required.

As a result of the expansion in offshore construction in recent years, interest in underwater welding has grown, resulting in several publications dealing primarily with the welding process and design applications (1-3). Limited systematic data on mechanical properties has been obtained to date. However, it has been recognized that underwater manual metal-arc (MMA) welds have mechanical properties which are generally inferior to those of surface welds made in air. The reported degree by which the properties are modified varies, but the average degradation appears to be a 20% reduction in tensile properties and a 50% reduction of ductility (1). Also, the conditions present during underwater welding are particularly conducive to hydrogen absorption and cracking, both in the weld metal and the heat-affected zone (HAZ). Limitations on parent plate composition to prevent hydrogen cracking have been reported elsewhere (2,3). Most of the problems associated with the underwater wet welding are removed by the use of dry habitats.

An earlier study by the authors of this paper systematically compared the mechanical properties of underwater wet welds with welds produced in dry habitats or by normal surface welding procedures (4,5). In the previous study, the weld metal fatigue and fracture properties of two series of weldments for offshore structures were presented. Specifically, the room temperature fatigue crack growth and fracture behavior in dry air for surface, habitat, and underwater wet welds fabricated by two commercial offshore welding companies (referred to as supplier L and supplier M) were compared (4,5). In a more recent study, the previous program was expanded to consider the effects of sea water on the fatigue and fracture characteristics. This paper reviews data from the previous study, and compares these data to the recent data concerning the effects of seawater on fatigue and fracture of underwater welds.

EXPERIMENTAL PROCEDURE

FATIGUE CRACK GROWTH IN AIR – In the previous study (4,5) nine experimental welds representative of standard surface, habitat, and underwater wet welding procedures were fabricated according to commercial practice by two offshore welding companies. The nine 0.56m-long multiple pass butt weldments, fabricated by joining two 0.3 x 0.6m plate sections of ASTM A-36 (6) steel, were divided into 3 groups. Groups I and II consisted of surface, habitat, and underwater wet welds from suppliers L and M respectively; Group III consisted of three underwater wet welds, each fabricated by supplier M with E6013 electrodes having different coatings.

The fatigue crack growth characteristics of the welded plates were evaluated with 12.7mm-thick compact tension specimens machined and tested according to ASTM E-647-81T (7). The samples were oriented so that the fatigue crack would propagate along the centerline of the weld bead. The 12.7mm specimen thickness, which is larger than that normally used in fatigue studies, was chosen to ensure that a representative distribution of weld defects was present in the crack propagation plane. All testing was performed according to ASTM E-647-81T (7). Multiple samples of each weldment were tested.

The fatigue tests were performed in load control on a 89kN capacity servohydraulic test system. All tests were performed in air at room temperature with a 30Hz tension/tension sinusoidal load cycle of 0.4 to 7.1kN. Samples were removed from the test system for visual crack length measurement with a traveling microscope.

FATIGUE CRACK GROWTH IN SEA WATER – In the later study, three welds from the previous study were chosen for evaluation in sea water. The welds included a fully dense weld (weld #L1), a weld containing a fine distribution of pores (weld #M6), and a weld containing coarse pores (weld #M4).

Fatigue crack growth data were obtained with samples submerged in simulated sea water (8) contained in a corrosion test cell mounted on the hydraulic actuator. The load line, consisting of standard fracture mechanics grips and pull rods, was electrically insulated from the test frame with non-conductive epoxy bushings. Dual 5-gpm Diatom filters continuously circulated

the water through the chamber, keeping corrosion particulates from depositing on the test specimen. The solution pH was continuously monitored and controlled with additions of 0.1 molar NaOH and HCl as necessary. With this system, samples were tested in a free corroding condition without the application of an imposed potential.

Crack lengths were measured with a modified compliance technique (9) in which an extensometer extension, along with a standard commercial fracture mechanics clip gage, was used to measure mouth opening displacement. The extensometer extension was machined from titanium and electrically insulated from both the samples and the clip gage. The mouth opening displacement measurements were correlated with crack lengths through compliance measurements (10) and the resulting crack lengths were shown to be equivalent to the lengths determined by means of the optical technique used previously (4,5). An overview of the corrosion cell with clip gage assembly is presented in Figure 1.

The fatigue data in simulated sea water were obtained at 32°C and for loading frequencies of 30Hz and 0.3Hz. Three samples of each of the three experimental weldments were tested at 0.3Hz, and two samples were tested at 30Hz.

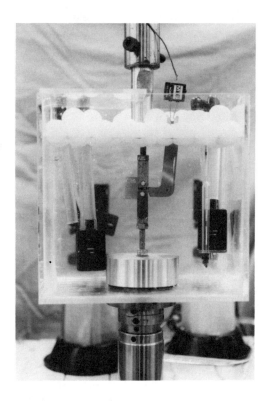

Fig. 1. Experimental apparatus for the seawater corrosion fatigue study, showing the plexiglass tank and sample with clip gage extender.

In both the seawater corrosion fatigue study and the initial fatigue study in air (4,5), after the stable fatigue crack growth data were recorded, the samples were overloaded at an approximately constant displacement rate, and the peak loads at failure were recorded as a function of fatigue crack length. Fracture surfaces were then examined by techniques of both optical and scanning electron microscopy.

RESULTS

In this section the initial results obtained in air (4,5) are briefly reviewed and then used as a basis for evaluating the effects of sea water on the fatigue behavior of underwater welds. The nine experimental welds of the initial study exhibited similar chemical compositions as summarized in reference 4 except that the underwater wet welds had lower manganese and higher oxygen contents.

Photomacrographs representative of the nine initial welds are presented in Fig. 2 and summarize several important microstructural differences observed. First, the number of passes used for the wet welds was significantly greater than for the surface and habitat welds. Furthermore, the surface and habitat welds from both suppliers were essentially free of porosity, while the wet welds contained a high density of uniformly distributed pores. The pore density in the wet welds was further shown (4,5) to vary with supplier and welding consumable. Analysis by optical microscopy

indicated that the primary microstructural constituents in all the welds were similar and typical of microstructures in A-36 steel weld metal. The average weld metal centerline and heat affected zone (HAZ) hardness values were also determined and found to be HRB 82-92 for all welds, indicating that the strength of all the welds was similar.

Photomicrographs of the three weldments chosen for the seawater corrosion fatigue study are presented in Fig. 3. Weld L1 is a surface weld which represents the fully dense surface and habitat welds. Weld M6 is an underwater wet weld with low porosity, approximately 3%. Weld M4 is an underwater wet weld with high porosity, approximately 12%.

FATIGUE CRACK GROWTH BEHAVIOR IN AIR (4,5) – The fatigue crack growth rate data, da/dN, were correlated with the imposed stress intensity factor range, ΔK, and analyzed according to the Paris crack growth expression (11),

$$\frac{da}{dN} = A \ (\Delta K)^n \qquad (1)$$

where for a constant stress ratio R, a given environment, material, and set of test conditions, A and n are constants.

The fatigue crack growth data for the nine welds in the three groups of samples tested in air (4,5) are summarized in Figs. 4, 5, and 6. As shown previously (4),

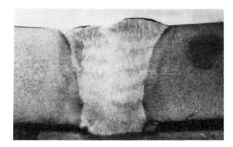

SURFACE

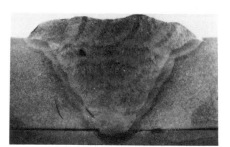

HABITAT

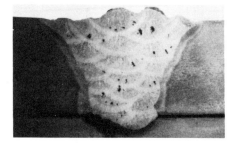

UNDERWATER WET

10 mm

Fig. 2. Representative macrographs of surface, habitat and underwater wet welds. Etched in 2% Nital (4,5).

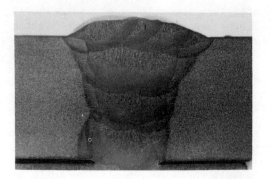

SURFACE

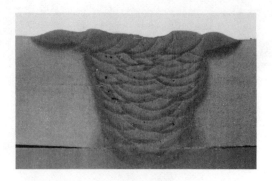

WET – LOW POROSITY

WET – HIGH POROSITY

Fig. 3. Macrographs of welds used for the corrosion fatigue study in simulated seawater.

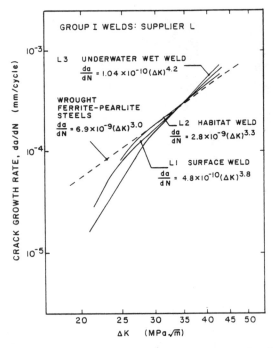

Fig. 4. Fatigue crack growth characteristics of surface, habitat, and underwater wet welds fabricated by supplier L.

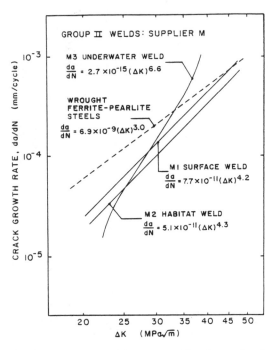

Fig. 5. Fatigue crack growth characteristics of surface, habitat, and underwater wet welds fabricated by supplier M.

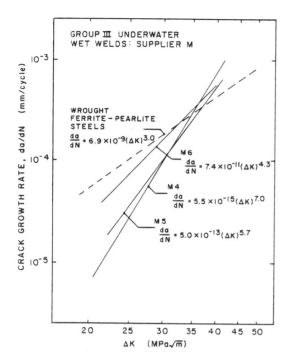

GROUP III UNDERWATER
WET WELDS: SUPPLIER M

WROUGHT
FERRITE-PEARLITE
STEELS
$\frac{da}{dN} = 6.9 \times 10^{-9} (\Delta K)^{3.0}$

M6
$\frac{da}{dN} = 7.4 \times 10^{-11} (\Delta K)^{4.3}$

M4
$\frac{da}{dN} = 5.5 \times 10^{-15} (\Delta K)^{7.0}$

M5
$\frac{da}{dN} = 5.0 \times 10^{-13} (\Delta K)^{5.7}$

Fig. 6. Fatigue crack growth characteristics of three underwater wet welds fabricated with different welding rods by supplier M.

reproducibility was excellent and the data points are omitted here for clarity. The appropriate crack growth equations for the high ΔK linear data are included along with data from the literature (12) for wrought ferrite-pearlite steels.

Several important observations are obtained from an analysis of the fatigue crack growth rate data presented in Figs 4-6. First, in comparison to normal wrought ferrite-pearlite steels, all nine experimental welds exhibited lower growth rates for low values of ΔK. This observation is consistent with previous studies on fatigue in welds (13). Also, all of the welds exhibited a higher sensitivity of the fatigue crack growth rate to ΔK, as evidenced by the magnitude of the exponent n in Eq. 1 (n = 3.0 for wrought ferrite-pearlite steel, while n = 3.3 to 7.0 for the nine welds).

The differences between the nine welds were shown (4) to result primarily from a difference in weld porosity, which varied with supplier, welding consumables, and welding procedure. The effect of porosity on fracture surface appearance is illustrated in the optical fractographs in Fig. 7. The stable fatigue crack growth zones of the surface and habitat welds were characterized by relatively smooth, flat fracture

essentially free of voids or inclusions. The stable fatigue crack growth zones of the underwater wet welds from supplier M (shown in Fig. 7) were significantly different from the zones of the surface and habitat welds, and were characterized as topologically rough, with a high uniform porosity.

The pore volume fractions in both the fatigue and overload zones of each weld were measured using standard point counting techniques. Typical results are included with Fig. 7. The higher values obtained in the overload zone reflect the effects of the interconnecting of voids which were out of the primary fatigue fracture plane. Similar data concerning pore densities were obtained for the welds represented by the data of Figs. 4-6. Specifically, the surface welds L1 and M1 and the habitat welds L2 and M2 were found to be fully dense. The underwater wet welds M6 and L3 contained a low (3%) porosity while welds M3, M4, and M5 contained a high (up to 12%) pore density.

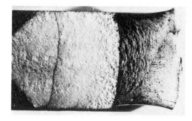

SURFACE

HABITAT

|10 mm|

WET

Fig. 7. Representative fractographs of fractured compact tension fatigue specimens. Pore volume fractions in the underwater wet welds were approximately 0.10 and 0.19 in the fatigue zone and overload zone, respectively. The surface and habitat welds were fully dense.

An evaluation of the fracture surfaces of the welds showed that for low values of crack extension (and thus low values of ΔK) the pores effectively pinned the crack front and retarded crack growth. Thus, for low values of ΔK, the observed crack growth rate decreased with an increase in pore density. However, for higher values of ΔK, the pores decreased the total load bearing area, and increased the local stress at the crack tip. Correspondingly, the growth rates at high ΔK were greater in porous welds.

Pores which retard crack growth at low ΔK but enhance crack growth at high ΔK result in a value of n in Eq. 1 which increases with porosity. This is verified in the comparison of the fully dense surface weld M1 (n = 4.2) with the high porosity underwater wet weld M3 (n = 6.6) in Fig. 5.

The fatigue crack growth rate data in Figs. 4-6 also support the interpretation presented above on the effects of porosity on fatigue. First, all three welds from supplier L were low porosity welds and correspondingly exhibited similar fatigue properties as shown in Fig. 4. Second, the fatigue crack growth characteristics of the three underwater wet welds (Sample Group III) presented in Fig. 6 directly reflect the pore structure, which was shown to vary with welding rod. In particular, a weld with smaller pores (weld M6) exhibited fatigue crack growth characteristics similar to those of the low porosity welds of supplier L and to the surface and habitat welds of supplier M. Underwater wet welds M4 and M5 with high porosity volume fractions exhibited fatigue crack growth characteristics wherein the pores retarded fatigue crack growth rates at low values of the imposed stress intensity factor range.

FATIGUE CRACK GROWTH BEHAVIOR IN SEA WATER – The fatigue crack growth behavior of the three welds selected to represent the broad variations in porosity (identified in the earlier work) are summarized in Figs. 8, 9, and 10. In each Figure, representative data for individual samples tested in seawater at 0.3 and 30 Hz are shown, along with the corresponding data obtained in air (indicated by the dotted lines). Data on multiple samples exhibited excellent reproducibility, equivalent to the results discussed above in air, and thus the data obtained on other samples at the indicated loading conditions are omitted for clarity.

Figure 8 presents the data for the fully dense surface weld L1. Consistent with previous studies of corrosion-fatigue (14), a decrease in loading frequency resulted in an increase in growth rate for all values of ΔK. However, testing in sea water at 30 Hz did not result in the same crack propagation rate as testing in air at 30 Hz. At low ΔK and high (30 Hz) frequency, crack propagation rates were substantially less in sea water

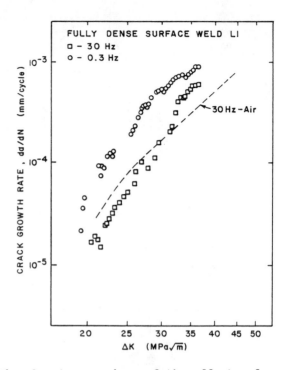

Fig. 8. A comparison of the effects of frequency and environment on the fatigue crack growth behavior of surface weld L1. Also shown by the dotted line are the results of tests performed in air at 30 Hz and presented in Figure 4.

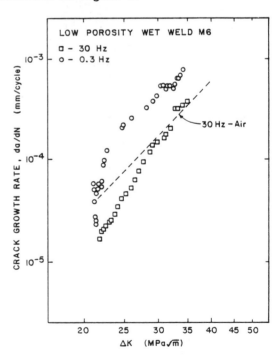

Fig. 9. A comparison of the effects of frequency and environment on the fatigue crack growth behavior of the low porosity underwater wet weld M6. Also shown by the dotted line are the results of tests performed in air at 30 Hz and presented in Figure 6.

than in air, implying that corrosive crack blunting may retard growth if the applied stress intensity is sufficiently low. At high ΔK and 30 Hz, crack propagation rates in sea water were greater than corresponding rates in air.

The fatigue crack growth data for the low porosity underwater wet weld M6 are presented in Fig. 9. Based on a point count of the fracture surface, this weld was shown (4) to contain approximately 3% porosity on the fatigue fracture surface. The results are analogous to the fully dense weld shown in Fig. 8. Again, crack growth rates at low ΔK and high (30 Hz) frequency in sea water were less than those in air, while crack growth rates at 0.3 Hz in sea water were significantly higher than corresponding rates at 30 Hz.

The results for the high porosity underwater wet weld M4, which contained approximately 12% porosity on the fatigue fracture surface, are presented in Fig. 10. In this case, results at 30 Hz for both air and sea water were similar, implying that the crack retarding effect of sea water at low ΔK is somehow negated by the presence of pores in the crack path. Again, crack growth rates at 0.3 Hz in sea water were observed to be significantly greater than those at 30 Hz.

Figure 11 is a summary presentation of the effects of both frequency and porosity on the sea water corrosion fatigue behavior of underwater welds in which the results of Figs. 8-10 are presented without data points. As shown in Fig. 11, fully dense welds, fatigued at 30 Hz, as well as welds containing both high and low area fractions of porosity, exhibited similar crack growth characteristics. However, at 0.3 Hz, both the fully dense (L1) and the low porosity (M6) welds exhibited crack growth rates significantly greater than those at 30 Hz. High weld porosity reduced the frequency effect as shown in Fig. 10. Although the 0.3 Hz crack growth rates for the high porosity weld M4 were greater than the corresponding rates at 30 Hz, the difference in growth rates attributed to frequency was substantially less than the same growth rate difference for the fully dense and low porosity welds.

LIMIT LOAD ANALYSIS – After completion of stable fatigue crack growth, the samples from both the initial study and the seawater corrosion fatigue study were loaded at a constant displacement rate, and the peak loads at failure were recorded as a function of fatigue crack length. Data from samples tested in air are presented in Fig. 12, a plot of peak load at failure versus fatigue crack length. Corresponding data from this study for samples loaded to failure in sea water are presented in Fig. 13. In both figures the data naturally divide into two bands, depending on porosity. In each

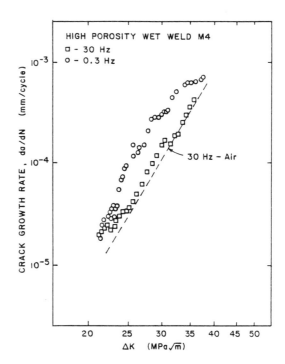

Fig. 10. A comparison of the effect of frequency and environment on the fatigue crack growth behavior of the high porosity underwater wet weld M4. Also shown by the dotted line are the results of tests performed in air at 30 Hz and presented in Figure 4.

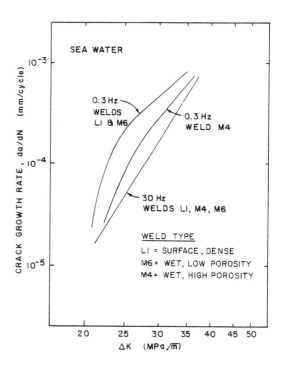

Fig. 11. A comparison of the effects of porosity and frequency on the fatigue crack growth behavior of welds tested in seawater.

figure, data in the upper band were obtained from the essentially fully dense samples, while data in the lower band were obtained from the porous underwater wet welds.

If it is assumed that the primary effect of the porosity is to decrease the cross sectional area of the weldment, then the peak load at failure simply is a measure of the limiting load for ductile rupture in the remaining ligament. This fact can be mathematically described (15) as follows:

$$P_L = \frac{0.35\sigma_o \, B \, b^2}{(a + b/2)} \qquad (2)$$

where: P_L = limit load, σ_o = average flow stress, usually taken as the average of the yield stress and ultimate tensile strength, B = specimen thickness, a = crack length, and b = ligament length. In Eq. 2, the specimen thickness B can be viewed as a porosity dependent thickness, B_{eff}, where:

$$B_{eff} = B \, (1 - f_p) \qquad (3)$$

and f_p is the volume percent porosity associated with the fracture plane. Then Eq. 3 is rewritten as:

$$P_L = \frac{0.35\sigma_o \, B \, (1 - f_p) \, b^2}{(a + b/2)} \qquad (4)$$

The data in Figs. 12 and 13 indicate the anticipated (15) decrease in fracture load with an increase in crack length. Furthermore, in both figures, the predictions of Eq. 4 with f_p = 0.0 (i.e. fully dense) and f_p = 0.2 are also shown. For the prediction of Eq. 4, σ_o was taken as 70 ksi (based on hardness data all weldments were estimated to have a yield strength of 62 ksi and an ultimate tensile strength of 78 ksi.). A comparison of the experimental data with the predicted crack length dependence of the limit load in Fig. 12 indicates that the limit load analysis for B_{eff} = B (i.e. no porosity) directly describes the behavior of the pore-free surface and habitat welds tested in air.

The underwater wet weld data are also adequately described by Eq. 4 if a porosity of 20% is assumed. As shown in Fig. 13, the limit load analysis for samples tested in air (solid lines) is equally valid for samples tested in sea water. While the porosity effect displayed in Figs. 12 and 13 is not completely explained by Eq. 4, a comparison of Fig. 12 with Fig. 13 clearly indicates that sea water exposure did not alter bulk properties which control overload failure processes. This implies that sea water only modified the local plasticity and fracture properties within the plastic zone at the tip of a propagating fatigue crack.

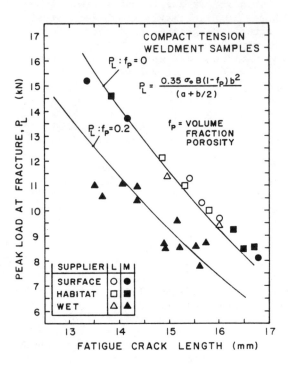

Fig. 12. The variation in peak load at failure with fatigue crack length for samples tested previously in air (4,5). The solid lines represent predictions of Equation 4 with porosity volume fractions of 0 and 0.2.

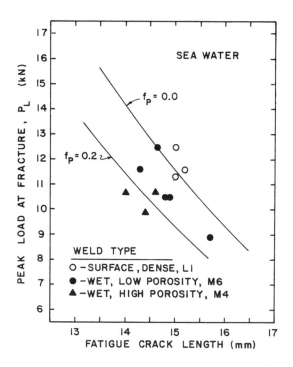

Fig. 13. The variation in peak load at failure with fatigue crack length for samples tested in seawater. The solid lines represent predictions of Equation 4 with porosity volume fractions of 0 and 0.2.

120

DISCUSSION

The fatigue crack propagation of underwater wet welds in sea water exhibits characteristics which are similar to those observed for wrought steel in sea water. Specifically, a decrease in frequency results in an increased growth rate and the appearance of two distinct regions in ΔK in which the slope of the da/dn vs ΔK curve differs significantly from the corresponding data for samples tested either in air or in sea water at high loading frequencies. This change in growth rate behavior with ΔK has been interpreted previously to reflect the relationship between K_{ISCC} (i.e. the fracture toughness in a stress corrosion cracking environment) and the maximum K imposed in a fatigue test (16). Through a comparison of da/dn vs ΔK curves with K_{ISCC} values, Wei and Landes (17) divided the observed behavior into three classes and developed a generalized superposition model for describing fatigue crack growth data. Based on their analysis, the total crack growth rate $(da/dN)_t$, is the sum of that due to pure fatigue, $(da/dN)_f$, the component due to stress corrosion cracking $(da/dN)_{scc}$, and the corrosion fatigue component of cyclic crack growth, $(da/dN)_{CF}$ as shown in the following generalized equation:

$$\left(\frac{da}{dN}\right)_t = \left(\frac{da}{dN}\right)_f + \left(\frac{da}{dN}\right)_{scc} + \left(\frac{da}{dN}\right)_{CF} \qquad (5)$$

To evaluate the applicability of this interpretation to the underwater weldments, the fatigue fracture surfaces at a crack length which corresponded to an imposed ΔK of 27.5 MPa m were investigated with scanning electron microscopy, because contributions due to stress corrosion cracking should be observable as changes in the fracture mode. The sample of the fully dense weld (L1) and low porosity weld (L6) exhibited a distinct increase in the density of grains with a flat faceted fracture surface appearance when tested in sea water. Typical SEM fractographs are presented in Fig. 14 for the dense surface weld for samples tested at 30 Hz in air and 0.3 Hz in sea water. The appearance of the flat faceted grains in Fig. 14b are interpreted as evidence of stress corrosion cracking due to the sea water. The appearance of faceted grains are direct evidence (18) for rapid crack front advancements during corrosion-fatigue. It is interesting to note that in the high porosity weld, M4, the flat faceted features were not observed, and the corresponding effects of sea water were not as pronounced (see Fig. 11).

Another interesting observation from the data in Figs. 8-11 is that when weld metal specimens were tested at 30 Hz in sea water, the effects of porosity observed for tests made in air disappeared. Furthermore, at low ΔK the effect of the sea water was to retard crack growth in both the fully dense and low porosity welds in much the same way as porosity did in the high porosity welds. As yet this observation has not been adequately explained.

SUMMARY

The results of this study have further clarified several of the factors which control the mechanical properties of welds for offshore structures. Specific new results are presented for samples tested in sea water. These data show that at low loading frequencies, sea water enhances fatigue crack growth in low-porosity welds, and the observed enhancement is similar to that observed in wrought steels. Seawater was further shown to alter only the fatigue crack growth characteristics while leaving unaffected the overload fracture properties. The fatigue resistance of high porosity wet welds was shown to be less sensitive to the presence of sea water than the fatigue resistance of low porosity welds.

ACKNOWLEDGEMENTS

The authors acknowledge the support of the Center for Welding Research at the Colorado School of Mines. The assistance of Mr. Craig Dallam in performing the fatigue tests is greatly appreciated.

REFERENCES

1. Cotton, H. C., Welding in the World, 21, 2-16 (1983).

2. Grubbs, C.E. and Seth, O.W., "Multipass all Position Wet Welding - A New Underwater Tool", 4th Annual Offshore Technology Conference, Houston, TX, April (1972).

3. Helburn, S., Welding Design and Fab., 52, 53-9, July (1979).

4. Matlock, D.K., Edwards, G.R., Olson, D.L., and Ibarra, S., "Proceedings of the Second International Conference on Offshore Welded Structures", H.C. Cotton, Ed., The Welding Institute, Cambridge, England, 15-1 to 15-10, (1982).

5. Matlock, D.K., Edwards, G.R., Olson, D.L., and Ibarra, S., in "Underwater Welding", IIW, Pergamon Press, 303-10 (1983).

6.   ASTM Standard Specification for
     Structural Steels, A36-81a.

7.   ASTM Constant Load-Amplitude Fatigue
     Crack Growth Rates Above 10^{-8}m/cycle,
     E647-81.

8.   ASTM Standard Specification for
     Substitute Ocean Water, D1141-75.

9.   Hammon, D.L., "An Analysis of the
     Effects of a Liquid-Lithium Environment
     on Fatigue Crack Propagation", Ph.D.
     Thesis, T-2688, Colorado School of
     Mines, (1983).

10. McHenry, H.I., Jour. of Mat., JMSLA, 6,
     862-73, (1971).

11. Paris, P.C. and Erdogan, F., J. Basic
     Engr., 85, 528-34 (1963).

12. Rolfe, S.T. and Barsom, J.M., "Fracture
     and Fatigue Control in Structures",
     Prentice Hall, New Jersey, 239 (1977).

13. Rolfe, S.T. and Barsom, J.M., "Fracture
     and Fatigue Control in Structures", New
     Jersey, Prentice-Hall, 252 (1977).

14. Jaske, C.E., Payer, J.H., and Balint,
     V.S., "Corrosion Fatigue of Metals in
     Marin Environments", Springer-Verlag,
     New York, (1981).

15. Paris, P.C., Tanada, H., Zahoor, A., and
     Ernst, H., in "Elastic-Plastic
     Fracture", ASTM STP668, American Society
     for Testing and Materials, Philadelphia,
     Pennsylvania, 5-36 (1979).

16. McEvily, A.J. and Wei, R.P., in
     "Corrosion Fatigue: Chemistry Mechanics
     and Microstructure", NACE-2, 381-95
     (1975).

17. Wei, R.P. and Landes, J.D., Materials
     Research and Standards, 9, (7), 25
     (1969).

18. Spenser, R.E., Matlock, D.K., and
     Olson, D.L., J. of Materials for Energy
     Systems, 4, (4), 187-94 (1983).

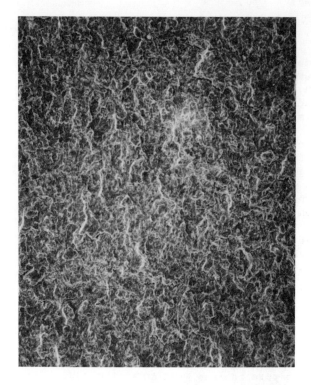

a.

$\overline{100\mu m}$

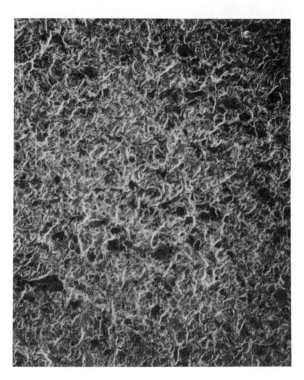

b.

Fig. 14. SEM fractographs of fatigue
fracture surfaces of the fully dense surface
weld L1. The fractographs were obtained at a
crack length which corresponds to a ΔK of
27.5 MPa m. (a) 30 Hz in air. (b) 0.3 Hz in
sea water.

CORROSION FATIGUE AND STRESS CORROSION CRACKING OF 7475-T7351 ALUMINUM ALLOY

Eun U. Lee
Naval Air Development Center
Wayminster, Pennsylvania, USA

ABSTRACT

This study was undertaken to characterize the corrosion fatigue and stress corrosion cracking behavior of 7475-T7351 aluminum alloy. The fatigue tests were performed under constant and variable amplitude loadings in a controlled laboratory atmosphere and in a 3.5% NaCl aqueous solution. The stress corrosion cracking tests were conducted under sustained tension in a 3.5% NaCl aqueous solution.

The fatigue crack initiation life, N_i, and the fatigue fracture life, N_f, in both environments can be related to the applied stress range, $\Delta\sigma$, by equations of the form $\log N = A - B\Delta\sigma$, where A and B are constants. The reductions of fatigue crack initiation life and fatigue fracture life by exposure to a 3.5% NaCl aqueous solution are represented by $\Delta(\log N_i) = 2.22 - 0.08 \Delta\sigma$ and $\Delta(\log N_f) = 1.87 - 0.06 \Delta\sigma$, respectively. The fatigue crack growth rate in a 3.5% NaCl aqueous solution is greater than that in a controlled laboratory atmosphere by a factor of $6.20 (\Delta K)^{-0.11}$. There is no stress corrosion cracking detectable in any crack plane orientation, indicating absence of stress corrosion cracking contribution to the corrosion fatigue of the employed specimens. The crack retardation effect of overloading is reduced by exposure to a 3.5% aqueous solution.

CORROSION FATIGUE, attributed to the combined action of a repetitive dynamic loading and a corrosive environment [1-7], has been a principal cause of metallic structure failures. However, the complexity of the phenomenon has hampered its quantitative characterization and understanding, essential to the service life prediction and the failure control of metallic structures.

A new aluminum alloy, 7475-T7351, has begun to be used for aircraft structural parts. However, its corrosion fatigue behavior has not been understood well. The objective of this study is to characterize the corrosion fatigue behavior of this alloy.

The corrosion fatigue crack initiation and growth, corrosion fatigue fracture, and stress corrosion cracking in a 7475-T7351 aluminum alloy have been investigated, and the corresponding empirical equations have been formulated. Furthermore, the effect of corrosive environment on the overload-induced crack retardation has also been studied.

EXPERIMENTAL PROCEDURE

MATERIAL AND SPECIMEN PREPARATION - As the specimen material, a 6.35 mm (1/4 in.) thick plate of 7475-T7351 aluminum alloy was selected. Its chemical composition is shown in Table 1. From the plate, open-hole dogbone specimens were machined, Fig. 1. The longitudinal axis of the specimen was in the original plate rolling direction.

Table 1 - Chemical Composition of 7475 Aluminum Alloy

Element	Min (%)	Max (%)
Cu	1.2	1.9
Mg	1.9	2.6
Mn	–	0.06
Fe	–	0.12
Si	–	0.10
Zn	5.2	6.2
Cr	0.18	0.25
Ti	–	0.06
Al	Balance	

FATIGUE TEST - To get the baseline data, the fatigue test was performed in a controlled laboratory atmosphere of 75°F and 45% relative humidity, employing a 88.96 KN (20 kips) closed-loop servo-hydraulic Materials Testing System machine. The corrosion fatigue test was carried out using a salt

water container of Plexi-glas, and the specimen hole was **immersed** in a 3.5% NaCl aqueous
solution throughout the fatigue loading
period. In all tests, the fatigue crack was
initiated and propagated under tension-to-
tension haversine waveform loading with a
stress ratio of 0.05 and a frequency of 10
Hz.

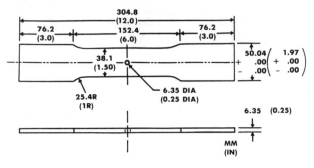

Fig. 1 – Specimen

At various intervals, fatigue loading
was stopped, and the specimen hole was inspected with a MAGNAFLUX HT-100 Eddy Current
Hole Scanner for possible crack initiation.
Any crack presence was indicated by a group
of eddy current signal spikes whose heights
were proportional to the depths at several
crack tip points. The greatest height was
converted to the corresponding crack depth
using a calibration curve. The calibration
curve was drawn with the spike heights of
eddy current signal from Electrical Discharge
Machining notches of known depths, 0.203 mm
(0.008 in), **0.381 mm** (0.015 in), 0.762 mm
(0.030 in), and 1.524 mm (0.060 in), in a
geometrically similar calibration standard of
the same alloy. The determined crack depth
was plotted against the corresponding number
of loading cycles, and a crack growth curve
was established. From this curve, a particular number of loading cycles to produce a
0.254 mm (0.01 in.) deep crack was taken and
defined as the fatigue crack initiation life
in this study. (Though the Eddy Current Hole
Scanner is capable of finding smaller cracks,
e.g., 0.127 mm (0.005 in.) deep crack, it can
detect 0.254 mm (0.01 in.) deep cracks more
accurately and reliably.)

When the crack emanating from the fastener hole became visible, the fatigue loading was stopped, and the crack length was
measured visually using a traveling microscope. This measurement was repeated at
shorter loading intervals as the crack grew
until its tip reached an edge of the specimen
or the specimen was fractured. The rate of
crack growth, da/dN, was then determined.
The corresponding stress intensity factor
range, ΔK, was calculated, employing the following equation [8, 9].

$$\Delta K = \Delta\sigma \sqrt{\pi a} \cdot F(a/r) \cdot$$
$$\sqrt{\sec\{\pi(a+2r)/2W\}} \qquad (1)$$

where

  Δσ: stress range

  a: length of crack emanating from
     open-hole

  F(a/r): Bowie factor

$\sqrt{\sec\{\pi(a+2r)/2W\}}$: correction
factor for specimen width

  r: radius of open-hole

  W: specimen width

FRACTOGRAPHIC EXAMINATION – The fractographic examination of fatigue-fractured
specimens was carried out using an Advanced
Metals Research (AMR) 1000 Scanning Electron
Microscope, operated at an accelerating potential of 20 kV. For each specimen, the
fracture surface morphologies of fatigue
crack initiation and growth regions were
studied.

STRESS CORROSION CRACKING TEST – This
test was done with double cantilever beam
specimens, Fig. 2, under sustained tension
at room temperature. Except for the loading-
bolt portion, most part of the specimen, including the notch tip, was immersed in a 3.5%

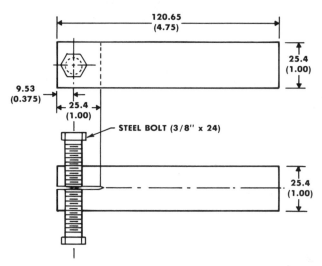

Fig. 2 – Double cantilever beam specimen

NaCl aqueous solution throughout the loading
period. The loading lasted 3 weeks to 8
months, and its directions were longitudinal,
transverse, and short transverse. The stress
intensity factors, corresponding to the sustained tensions applied, ranged from 840 to
1,800 MPa$\sqrt{\text{mm}}$ (24.2 to 51.8 ksi$\sqrt{\text{in}}$).

CRACK RETARDATION TEST – In order to
investigate the crack retardation phenomenon,
fatigue tests were performed, using open-hole
dogbone specimens, under variable amplitude

loading in a controlled laboratory atmosphere and in a 3.5% NaCl aqueous solution. The load waveform contains periodic tensile over-load spikes, as shown in Fig. 3, and the loading frequency was 10 Hz.

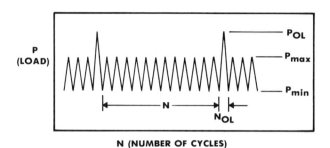

$P_{OL}/P_{max} = 1.35$, $P_{min}/P_{max} = 0.05$
$N = 8,000$, $N_{OL} = 1$

Fig. 3 - Load waveform of crack retardation test

RESULTS AND DISCUSSION

The results and discussion consist of six parts: fatigue crack initiation life, fatigue fracture life, fatigue crack growth, fracture surface morphology, stress corrosion cracking, and crack retardation.

FATIGUE CRACK INITIATION LIFE - Under fatigue loading, mostly a single crack was initiated at an edge of the specimen hole in a controlled laboratory atmosphere and in a 3.5% NaCl aqueous solution. Based on the results of fatigue tests in both environments and crack inspection, the applied stress range, $\Delta\sigma$, is plotted against the fatigue crack initiation life, N_i, in a semi-log scale, Fig. 4.

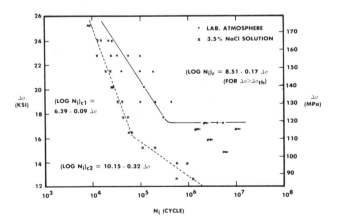

Fig. 4 - Variation of fatigue crack initiation life, N_i, with applied stress range, $\Delta\sigma$

The plot for the fatigue test in a controlled laboratory atmosphere has the feature of a typical S-N curve with a knee (or the threshold stress range for fatigue crack initiation, $\Delta\sigma_{th}$) at $\Delta\sigma_{th} = 118.59$ MPa (17.2 ksi). For $\Delta\sigma > \Delta\sigma_{th}$, the plot is a straight line, defined by the equation.

$$(\log N_i)_r = 8.51 - 0.17 \Delta\sigma \qquad (2)$$

where the unit of $\Delta\sigma$ is ksi. For $\Delta\sigma < \Delta\sigma_{th}$, no detectable crack was initiated within the limit numbers of loading cycles employed.

In the case of corrosion fatigue test, the plot of $\Delta\sigma$ vs. log N_i has a bend and consists of two straight line portions, defined by the following equations, respectively.
(Portion 1) $(\log N_i)_{C1} = 6.29 - 0.09 \Delta\sigma \qquad (3)$
(Portion 2) $(\log N_i)_{C2} = 10.15 - 0.32 \Delta\sigma \qquad (4)$
where the unit of $\Delta\sigma$ is ksi.

It is interesting to notice the change in the form of $\Delta\sigma$ vs. log N_i plot accompanying the change of test environment from a controlled laboratory atmosphere to a 3.5% NaCl aqueous solution. In the case of fatigue test in a controlled laboratory atmosphere, the plot has a knee or a threshold stress range for fatigue crack initiation, $\Delta\sigma_{th}$, below which no fatigue crack initiation occurs within the limit numbers of loading cycles employed. On the other hand, in the case of corrosion fatigue test in a 3.5% NaCl aqueous solution, the plot has a bend, not a knee, and fatigue cracks are initiated below the bend-level too. Similar changes in the form of S-N plot were also observed by Gough and Sopwith [10] in their fatigue tests of 0.15% C steel, 15% Cr steel, and 17/7 Cr-Ni steel in air and brine.

Besides the knee and bend, the plots of $\Delta\sigma$ vs. log N_i have differences in the intercept and slope too. As shown by Eq. (2) and (3), both of the intercept and slope for the portion of $\Delta\sigma > \Delta\sigma_{th} = 118.59$ MPa (17.2 ksi) are reduced by changing the fatigue test environment from a controlled laboratory atmosphere to a 3.5% NaCl aqueous solution. The reduction must be the quantitative measure of corrosive environment effect on fatigue crack initiation life. From Eq. (2) and (3), the reduction of fatigue crack initiation life by exposure to a 3.5% NaCl aqueous solution can be given by

$$\Delta(\log N_i) = (\log N_i)_r - (\log N_i)_{C1}$$
$$= 2.22 - 0.08 \Delta\sigma \qquad (5)$$

This equation indicates that the larger the applied stress range, the smaller is the life reduction effect of corrosive environment.

FATIGUE FRACTURE LIFE - The plot of stress range, $\Delta\sigma$, vs. logarithm of fatigue fracture life (or total fatigue life), log N_f, in a controlled laboratory atmosphere has also the feature of a typical S-N curve with a knee at $\Delta\sigma = 118.59$ MPa (17.2 ksi), Fig. 5. For $\Delta\sigma > 118.59$ MPa (17.2 ksi), the plot is a straight line, defined by the equation.

$$(\log N_f)_r = 8.19 - 0.15 \Delta\sigma \qquad (6)$$

where the unit of $\Delta\sigma$ is ksi.

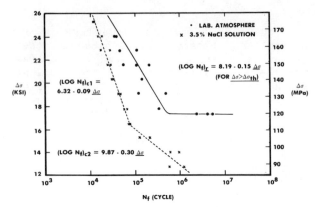

Fig. 5 – Variation of fatigue fracture life, N_f, with applied stress range, $\Delta\sigma$

The corrosion fatigue test plot of $\Delta\sigma$ vs. log N_f has a bend and consists of two straight line portions, defined by the following equations, respectively.

(Portion 1) $(\log N_f)_{C1} = 6.32 - 0.09\ \Delta\sigma$ (7)
(Portion 2) $(\log N_f)_{C2} = 9.87 - 0.30\ \Delta\sigma$ (8)
where the unit of $\Delta\sigma$ is ksi.

The change in the form of $\Delta\sigma$ vs. log N_f plot, accompanying the change of fatigue test environment, is similar to that of fatigue crack initiation life plot. In those plots of fatigue crack initiation and fatigue fracture lives, the knees and the bends are at $\Delta\sigma = 118.59$ MPa (17.2 ksi) and 113.77 MPa (16.5 ksi), respectively. In the $\Delta\sigma > 118.59$ MPa (17.2 ksi) portion of $\Delta\sigma$ vs. log N_f plot, th intercept and slope are also reduced by changing the fatigue test environment from a controlled laboratory atmosphere to a 3.5% NaCl aqueous solution. From Eq. (6) and (7), the reduction of fatigue fracture life by exposure to a 3.5% NaCl aqueous solution can be given by

$$\Delta(\log N_f) = (\log N_f)_r - (\log N_f)_{C1}$$
$$= 1.87 - 0.06\ \Delta\sigma \qquad (9)$$

As in the case of fatigue crack initiation life, the larger the applied stress range, the smaller is the fatigue fracture life reduction effect of corrosive environment.

FATIGUE CRACK GROWTH – The initiated single crack grew mostly alone. Occasionally, before the crack reached an edge of the specimen, the second crack was initiated at the other edge of the specimen hole and began to grow. However, for the calculation of the stress intensity factor range, ΔK, it was assumed that only a single crack was initiated and grown. (This assumption is reasonably valid as long as the second crack is very small compared to the first one.) The fatigue crack growth rates, da/dN, in a controlled laboratory atmosphere and in a 3.5% NaCl aqueous solution are plotted against the stress intensity factor range, ΔK, in a log-log scale, Fig. 6. For the fatigue test in a

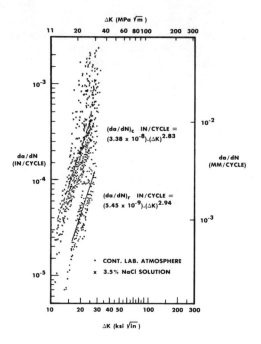

Fig. 6 – Variation of fatigue crack growth rate, da/dN, with stress intensity factor range, ΔK

controlled laboratory atmosphere, the log-linear portion of the curve takes a form of Paris' equation [11].

$$(da/dN)_r = (5.45 \times 10^{-9}) \cdot (\Delta K)^{2.94} \qquad (10)$$

For the fatigue test in a 3.5% NaCl aqueous solution,

$$(da/dN)_c = (3.38 \times 10^{-8}) \cdot (\Delta K)^{2.83} \qquad (11)$$

From Eq. (10) and (11),

$$(da/dN)_c = [6.20\ (\Delta K)^{-0.11}] \cdot (da/dN)_r \qquad (12)$$

(In Eq. (10), (11), and (12), the units of da/dN and ΔK are in/cycle and ksi $\sqrt{in.}$) This equation indicates that the exposure to a 3.5% NaCl aqueous solution increases the fatigue crack growth rate by a factor of $[6.20\ (\Delta K)^{-0.11}]$.

From Eq. (12), the general form of corrosion fatigue crack growth rate equation can be written as

$$(da/dN)_c = [\alpha \cdot \Delta k^{-\beta}] \cdot (da/dN)_r \qquad (13)$$

where α and β are constants for a given condition of corrosion fatigue test. $\alpha=1$ and $\beta=0$ for the fatigue test in the controlled laboratory atmosphere of 75°F and 45% relative humidity employed. Eq. (13) indicates that the greater the magnitude of ΔK, the less is the crack growth rate enhancement effect of corrosive environment.

FRACTURE SURFACE MORPHOLOGY – The fracture surfaces of the specimens, fatigue-tested in a controlled laboratory atmosphere, have long, narrow, and smooth **plateaus** with well-developed ductile fatigue striations, Fig. 7(a). On the other hand, those, fatigue-tested in a 3.5% NaCl aqueous solution, have cleavage facets with river patterns and brittle fatigue striations, Fig. 7(b). Such a morphology is a characteristics

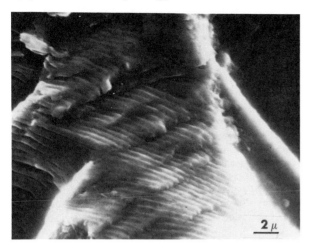

(a)

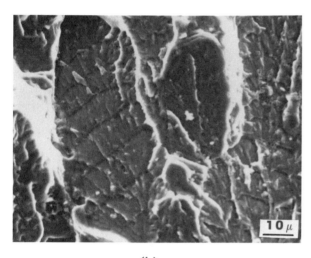

(b)

Fig. 7 - Scanning electron microscope fractographs of specimens fatigued-tested: (a) in a controlled laboratory atmosphere and (b) in a 3.5% NaCl aqueous solution

and an indication of corrosion fatigue crack growth [12]. No fractographic evidence of intergranular cracking was observable in specimens fatigue-tested in either environment.

STRESS CORROSION CRACKING - There was no detectable evidence of stress corrosion cracking in any of the L-T, T-L, and S-L crack plane orientations under the sustained tensions applied within the employed test periods. This indicates that stress corrosion cracking is not a contributing factor to the corrosion fatigue behavior of those specimens used in this study.

CRACK RETARDATION - As a plot of crack length, a, vs. number of fatigue loading cycles, N, shows, Fig. 8, the crack growth is

retarded by a single spike overload in a 3.5% NaCl aqueous solution too. The retardation behavior is also observable in fatigue crack

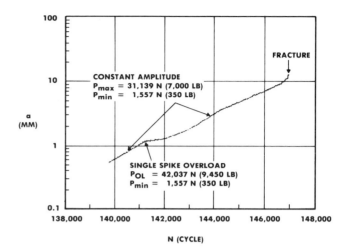

Fig. 8 - Variation of crack length, a, with number of fatigue loading cycles, N, under variable amplitude loading in a 3.5% NaCl aqueous solution

initiation. In this study, the measure of retardation is defined as the ratio of the life extension due to overloading and the life under constant amplitude loading, $\Delta N/(N)_{CA}$. Table 2 shows the overload-induced retardations of fatigue crack **initiation and growth and fatigue fracture in a controlled laboratory atmosphere and in a 3.5% NaCl aqueous solution. Those retardations or life extensions**

Table 2 - Overload-Induced Retardation

Atmosphere	$\dfrac{\Delta N_i}{(N_i)_{CA}}$	$\dfrac{\Delta N_p}{(N_p)_{CA}}$	$\dfrac{\Delta N_f}{(N_f)_{CA}}$
Controlled Lab. Atmosphere	0.81	0.51	0.77
3.5% NaCl Aqueous Solution	0.17	0.34	0.20

ΔN: life extension due to overloading
$(N)_{CA}$: life under constant amplitude loading
Subscripts, i, p, and f indicate the crack initiation, crack growth, and fracture, respectively.

are much less in a 3.5% NaCl aqueous solution than in a controlled laboratory atmosphere. A similar observation was also made by Chanani [13] during his investigation of saltwater effects on retardation behavior of 2024 and 7075 aluminum alloys.

SUMMARY AND CONCLUSIONS

1. The fatigue crack initiation and fracture lives are reduced by increasing the applied stress range, $\Delta\sigma$, in a controlled laboratory atmosphere and a 3.5% NaCl aqueous solution. (In a controlled laboratory atmosphere, this is true only for $\Delta\sigma > \Delta\sigma_{th}$, where $\Delta\sigma_{th}$ is the threshold stress range for fatigue crack initiation.) The relationship is defined by the equations of the form $\log N = A - B\Delta\sigma$, where A and B are constants.

2. The reductions of fatigue crack initiation and fracture lives, N_i and N_f, by exposure to a 3.5% NaCl aqueous solution are indicated by $\Delta(\log N_i) = 2.22 - 0.08 \cdot \Delta\sigma$ and $\Delta(\log N_f) = 1.87 - 0.06\,\Delta\sigma$.

3. The fatigue crack growth rate is greater in a 3.5% NaCl aqueous solution than in a controlled laboratory atmosphere by a factor of $[6.20 \cdot (\Delta k)^{-0.11}]$. That is, $(da/dN)_c = [6.20 \cdot (\Delta k)^{-0.11}] \cdot (da/dN)_r$.

4. The corrosion fatigue behavior is also evidenced by a fractographic feature: brittle fatigue striations.

5. Stress corrosion cracking does not occur in the employed specimens, and it is not a contributing factor to their corrosion fatigue behavior in a 3.5% NaCl aqueous solution.

6. Exposure of specimens to a 3.5% NaCl aqueous solution reduces the retardation effect of overloading on fatigue crack initiation and growth.

ACKNOWLEDGEMENT

The author gratefully acknowledges the provision of specimen material, 7475-T7351 aluminum alloy, for the stress corrosion cracking test by Reynold Metal Company.

REFERENCES

1. Achter, M.R., "Fatigue Crack Propagation," p. 181, ASTM STP 415, American Society for Testing and Materials (1967)

2. Hudson, C.M. and K.N. Raju, NASA TN D-5702, National Aeronautics and Space Administration (1970)

3. Wei, R.P., Engineering Fracture Mechanics, 1, 633 (1970)

4. McEvily, A.J. and R.P. Wei, "Corrosion Fatigue: Chemistry, Mechanics and Microstructure," p. 381, NACE-2, National Association of Corrosion Engineers (1972)

5. Wei, R.P. and M.O. Speidel, "Corrosion Fatigue: Chemistry, Mechanics and Microstructure," p. 379, NACE-2, National Association of Corrosion Engineers (1972)

6. Gallagher, J.P. and R.P. Wei, "Corrosion Fatigue: Chemistry, Mechanics and Microstructure," p. 409, NACE-2, National Association of Corrosion Engineers (1972)

7. Wei, R.P, "Fatigue Mechanisms," p. 816, ASTM STP 675, American Society for Testing and Materials (1979)

8. Bowie, O.L., Journal of Mathematics and Physics, 35, 60 (1956)

9. Isida, M., Journal of Applied Mechanics, 33, 674 (1966)

10. Gough, H.J. and D.G. Sopwith, Engineering, 136, 75 (1933)

11. Paris, P. and F. Erdogan, Journal of Basic Engineering, 85, 528 (1968)

12. Engel, L. and H. Klingele, "An Atlas of Metal Damage," p. 98, Prentice-Hall, Inc., Engelwood Cliffs, NJ (1981)

13. Chanani, G.R., "Corrosion Fatigue Technology," p. 51, ASTM STP 642, American Society for Testing and Materials (1978)

SIMULATION OF CRACK INITIATION AND GROWTH IN CORROSION FATIGUE OF CARBON STEELS IN SEAWATER

Masanori Kawahara, Takahiro Fujita, Hirosuke Inagaki, Toshio Iwasaki, Akihiko Katoh
Nippon Kokan K.K.
Kawasaki, Japan

Abstract

Corrosion fatigue process of carbon steels in seawater consists of the multiple steps; 1) formation and growth of corrosion pits, 2) initiation of cracks at the pit edges, 3) coalescence and growth of distributed short cracks, and 4) propagation of long cracks.

The present paper aims to formulate the phenomena in each stage by referring a series of fatigue tests on two 500 MPa class carbon steels. Total of these formulation enables to simulate numerically the process of corrosion fatigue and to estimate fatigue life in seawater. A size factor common to both pits and cracks is used to describe the progress of corrosion fatigue. Simulations were performed by use of four parameters; pit dissolution constant, number of pits or cracks per unit area, and crack propagation constants, C and m. Discussion is directed to the influences upon corrosion fatigue features of some metallurgical factors. The difference in fatigue life between two steels with different pearlite grain sizes are interpreted as the difference in the pit dissolution constant.

INTRODUCTION

Corrosion fatigue life estimation is one of the most important subjects in the design of offshore structures. Accumulation of experimental data is of the first importance for this purpose. However, the experiments are extremely time consuming to simulate the sea wave loadings in offshore structures: it takes, for example, 140 days to obtain the data of $N=2 \times 10^6$ to perform the tests in sea wave loading rate 10 cpm. As the number of testing machines and testing periods are limited for economical reason, numerical simulation is necessary to estimate alternative results and to supplement the data to know their scatters which are difficult to obtain in practical tests. Extensive studies have recently been performed on numerical simulation of growth and coalescence of distributed short cracks (ref. 1-4).

The present paper deals with the case of corrosion fatigue of carbon steels in seawater, where the pit formation process acts as equally important role as the crack growth behavior. In order to apply in the simulation of distributed pits and cracks, formulations are to be simplified. Critical points are how to joint the pits and cracks, and how to characterize each stage in view of the whole process.

PROCESS OF CORROSION FATIGUE

Corrosion fatigue process of carbon steels in seawater consists of the following steps:
(i)   Formation and growth of pits
(ii) Initiation of crack at pit edges
(iii) Growth and coalescence of distributed short cracks
(iv) Growth of long cracks

Figures 1 and 2 show examples of specimen surfaces of a carbon steel subjected to corrosion fatigue test in seawater (ref. 5). Many pits are initiated. Pits are circular when they are small and become elliptical as they grow. Pits tend to transform to cracks. Longer cracks are formed by growth and coalescence of cracks.

CRITICAL POINTS FOR FORMULATION

1. Formation of pits

It is well recognized that corrosion pits are formed by preferential dissolution at the sites of accumulated slip bands, micro segregation, inclusions, or second phase particles. Figure 3 shows an example of corrosion pits in a carbon steel. The sizes of the smallest pits are order of grain size. The location and population of pits are dependent on metallurgical factors such as grain sizes or second phase distribution, as well as various local chemical or electro-chemical conditions.

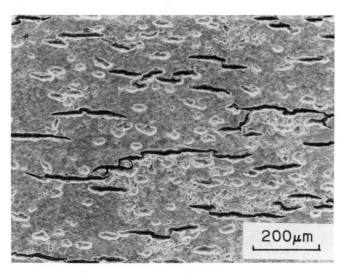

Fig. 1   Surface of a carbon steel specimen subjected to corrosion fatigue test in seawater

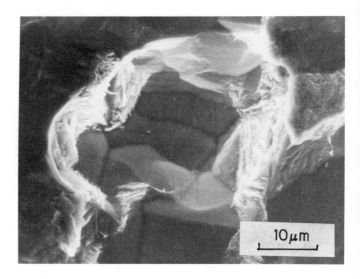

Fig. 3   Enlarged view of a small circular pit

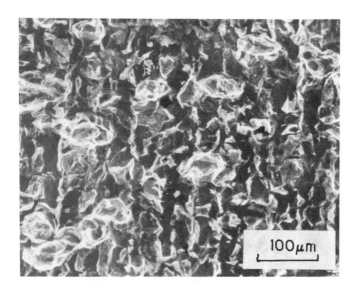

Fig. 2   Pits on a specimen surface of carbon steel subjected to corrosion fatigue test in seawater

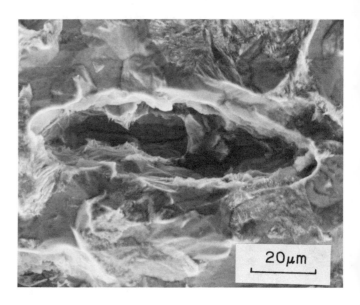

Fig. 4   An example of elliptical pit

2. Growth of pits

Pits are approximately circular when they are initiated, and become elliptical as they grow in size. Figure 4 shows an example of elliptical pits. The major axis of the ellipse is perpendicular to the principal stress direction. The minor diameter of the ellipse or the width of pits are mostly 20 to 50 micron, a few number times grain size.

As for the growth rates of pits, data are not yet sufficiently accumulated. A long term investigation on the corrosion of structural ferrous materials (ref. 6) shows that pit growth rates follows approximately to a parabolic law.

In case of carbon steels in sea water, no protection films are formed on specimen surface. By assuming a constant bulk dissolution rate, B, the relationship between the pit sizes and time, t, can be expressed as follows; for the semi-spherical pit of radius, a,

$$(2/3) \pi a^3 = B t \qquad (1)$$

$$da/dt = B / (2 \pi a^2) \qquad (1a)$$

and for the elliptical shape pit of major radius, a, and minor radius, b (considered to be fixed),

$$(2/3) \pi a^2 b = B t \qquad (2)$$

$$da/dt = 3B / (4 \pi a b) \qquad (2a)$$

In the present paper, it was assumed that the pit growth follows to Eq. (1) for a < (grain size), and to Eq. (2) for a > (grain size). As the dependency of coefficient B upon applied stress are not known, it is assumed to have a form;

$$B = Const \ \sigma^n \qquad (3)$$

where the value of n is chosen empirically 1 or 2.

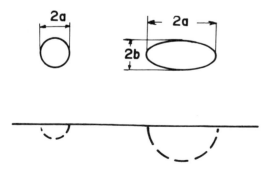

Fig. 5 Models of pits

3. Initiation of crack at pit edges

The curvature radius of the pit edge contour is apt to be very small depending upon the states of local dissolution of grains. In well grown elliptical pits, cracks can easily be initiated at such steep edged sites. The condition of crack initiation is not yet well established for such cases.

As far as confirmed in several experimental observations, the minimum length of such micro-cracks across the pits is approximately 100-200 microM for the specimens subjected to lower stress levels 180-220 MPa. A rough estimation gives the threshold condition as;

$$\Delta K_{th} = 1.7-2.3 \ MPa \ m \qquad (4)$$

However, as it is not sure that ΔK_{th} can be defined in such a small crack around a pit, the following procedure was adopted to determined the transition.

Pits always include some intrusions or intrusion-like cracks at their edge corners, and these intrusions of cracks are always ready to grow as a fatigue crack. However, when the dissolution is stronger, crack can not have the time to propagate and there occurs only the pit growth. As the pit increases in its size, the growth rates go down if they follows to the Eqs. (1) or (2), and finally the fatigue crack growth will overwhelm the pit growth.

Therefore, the crack initiation conditions are given by a criterion as follows;

(Pit dissolution rate)
< (latent crack growth rate)   (5)

4. Growth of cracks

An experimental analysis (ref. 7) shows that growth of a surface crack under cyclic tensile stresses can be expressed by the following simplified formulas:

(i) Increase in half length, a

$$da/dN = C(\Delta\sigma \sqrt{\pi a})^m \qquad (6)$$

(ii) Change in shape

$$dc/da = (a/c)^{m/2} \qquad (7)$$

Equation (9) is exact for elliptical crack of major radius a and minor radius c in a infinite body. Cracks are initially regarded as semi-circular. Semi-elliptical cracks are formed by crack coalescence.

5. Coalescence of adjacent cracks

Crack coalescence is considered as a discrete increase in crack length. Each crack has its territory of influence. Interactions occur when other cracks enter into this territory or the territories of adjacent cracks overlap. Coalescence is a result of strong

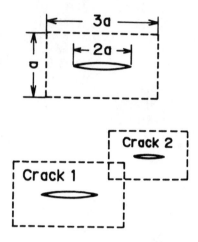

Fig.6   Conditions of crack coalescence

interaction between two cracks. The conditions of coalescence can be defined by the the difinition of territories around a crack.
    Coalescence occurs in the following two modes;
(i) Tip-to-tip coalescence,
(ii) Tip-to-flank coalescence.

    Tip-to-tip coalescence occurs when two cracks situate each other at the tip front of the other. Kitagawa et al. (ref.2) proposed a definition of crack territory for this type of coalescence by the analysis on stress intensity factor variation.
    Another type of coalescence occurs when two cracks situate at parallel positions. Crack tips approaches each other to the flank of other's.
    In the present analysis, noting these two modes of coalescence, the territory of influence for a crack of length 2a was defined as a rectangle with sizes 3a x a of which the longer side is in the crack direction.
    Probability of crack coalescence is in this case calculated as the probability of the overlapping of crack territories. The critical condition due to crack coalescence can be given as follows

$$3 \, a_{rms}^2 * n_p * F = 1 \qquad (8)$$

where a_{rms} is root mean square of crack length, n_p number of cracks in unit area, and F a coeffecent order of 1 depending upon the crack length distribution.

RESULTS OF SIMULATION

1. S-N curves of two carbon steels

    Numerical simulation was performed on the growth of pits and cracks by using the following formulation.

(i)   The size parameter, a, was used as a common parameter between pit and crack: the radius of a spherical pit, the major radius of elliptical pits, or the half length of surface cracks.

(ii)   The rate of increase in the parameter, a, is determined by the following equation.

$$da/dt = max[\; C_e \, \sigma^n / a \quad , \; C(\Delta\sigma \sqrt{\pi a})^m f] \qquad (9)$$

(iii)   The location of pits (or cracks) are determined by random number sampling. The number of pits in unit area, n, was considered as a given constant. In future analyses, n is to be treated as the function of time and location coordinates.

(iv)   The territory of influence was defined for each pit or crack as a rectrangle 3a x a, as discussed in the previous chapter. Adjacent cracks are considered as coalesced when the territories overlap each other.

The simulation was applied to interpret a series of experimental results of two carbon steels shown in Table 1.

Table 1   Two Carbon Steels Tested (ref. 5)

	Steel A	Steel B
Production	Thermo-Mech. Proc. Cooling (Total Reduction 62% below 870 C, Finish Roll Temp. 750 C)	Conventional Normalising (finish Roll Temp. 950 C)
Chem. Comp.(%)		
C	0.10	0.14
Si	0.36	0.31
Mn	1.45	1.48
P	0.012	0.021
S	0.003	0.007
others	-	Cu 0.15, Ni 0.02 Cr 0.15
Mech. Prop.		
Y.S.(MPa)	373	353
T.S.(MPa)	510	530
El. (%)	36	33
Grain Sizes		
ferrite	8.9	14.2
pearlite	4.7	10.0
Volume Fraction		
ferrite	83.5	75.1
pearlite	16.5	24.9

Fatigue test results are shown in Fig. 7. Fatigue lives of Steel A are longer than those of Steel B, especially at lower stress levels. Two materials have nearly the same tensile properties. The difference in chemical compositions are not so spectacular: Steel B is slightly more alloyed than Steel A.

Distinct differences are noted in grain sizes. Ferrite grain size in Steel B is about 1.6 times larger than that in Steel A, and pealite grain size in Steel B is 2 times larger than that in Steel A.

The surface roughness of specimens subjected to corrosion fatigue tests in seawater was significantly different between two steels. A measurement showed average roughness $Ra=2$ micron for Steel A and 6 micron for Steel B.

Taking account of these differences, numerical simulations were conducted for two steels by assuming a difference in pit dissolution constant, C_e. Table 2 shows the values of constants used in the simulation. The results of simulations are denoted as the increase in representative sizes of pits or cracks (the maximum value in a sample surface). The failure was defined as the point where the size factor, a, reached to 5 mm. A set of obtained curves are shown in Appendix. Table 3 shows the comparison of fatigue lives between experiments and simulations. Coincidence was fairly good for both Steel A and Steel B. The results of simulations are denoted as the increase in the size factor, a, representative

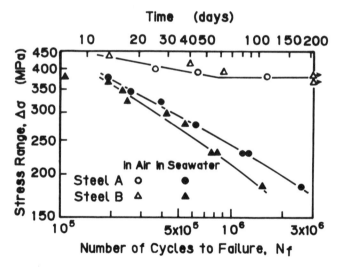

Fig. 7 Corrosion fatigue results for two carbon steels

DISCUSSION

As shown in Table 3, the present model can simulate fairly well the features of corrosion fatigue in carbon steels. The difference in fatigue life between Steel A and Steel B was interpreted as a difference in pit dissolution constant C_e.

Table 2   Constants Used in the Simulations

	Steel A	Steel B
Pits		
dissolu. coef. C_e	8×10^{-8}	1.6×10^{-7}
num. per $1mm^2$	250	250
Cracks		
propagation const.	$C = 1.952\times10^{-10}$	
(mm/cycle vs MPa m)	$m = 4.0$	

Table 3   Comparison of Fatigue Lives between Experiments and Simulations

Steel	Stress (MPa)	Fatigue lives ($\times10^3$ cycles) Experiment	Simultion		
A	179.3	2,530 1,971	1,807 1,471	2,277 2,801	2,261
	224.4	1,261 1,154	940 1,176	1,207 814	1,181
	269.5	622	781 732	821 684	732
	314.6	390	480 552	353 458	458
	370.4	188	302 312	331	268
B	179.3	1,504	1,235 1,145	1,286 1,732	1,411
	224.4	764 825	648 554	840 756	821
	269.5	541	384 307	502 446	499
	314.6	247	244 249	324 324	319 425
	370.4	106	153 235	177 225	170

Figures 9 and 10 show the examples of growth curves for different values of the pit dissolution constant, C_e, and the number of pits or cracks per unit area, n_p. It is noted that the increase in C_e corresponds to the general increase in sizes of pits or cracks at all position of the growth curve (Fig. 9). The increase in n_p corresponds to the increase in the probability of crack coalescence, and then to the acceleration of crack growth at the later stage of growth curve (Fig. 10).

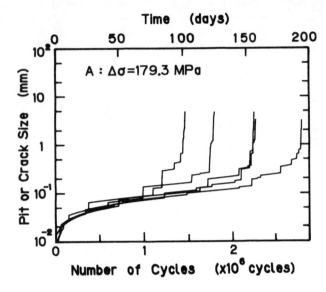

(1) $\Delta\sigma$ = 179.3MPa

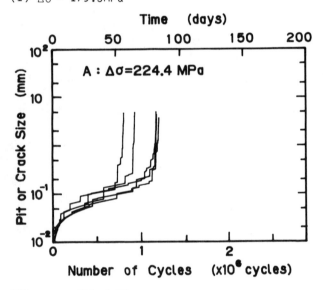

(2) $\Delta\sigma$ =224.4 MPa

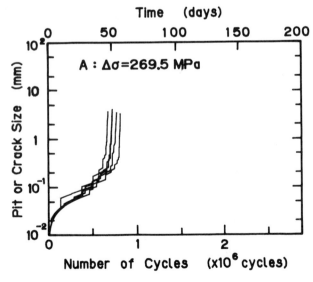

(3) $\Delta\sigma$ = 269.5 MPa

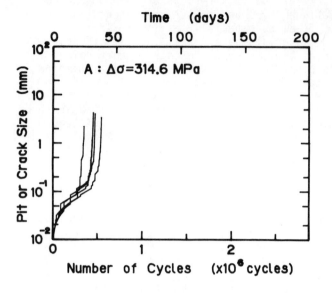

(4) $\Delta\sigma$ = 314.6 MPa

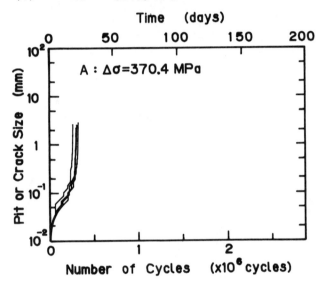

(5) $\Delta\sigma$ =370.4 MPa

Fig. 8   Simulation of the growth of pits and cracks:

$$C_e = 8\times10^{-8}$$
$$n_p = 250$$

$$n = 2$$
$$C = 1.952 \times 10^{-10}$$
$$m = 4.0$$

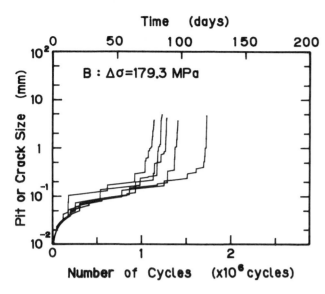

(1) $\Delta\sigma$ = 179.3 MPa

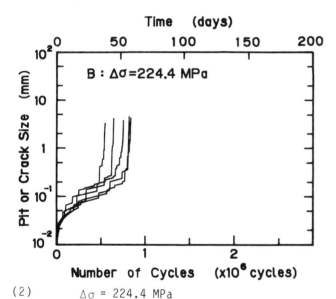

(2) $\Delta\sigma$ = 224.4 MPa

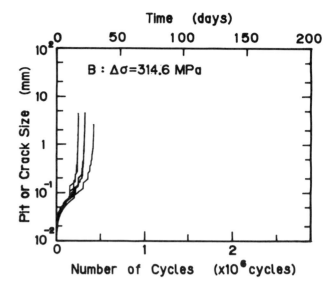

(4) $\Delta\sigma$ = 314.6 MPa

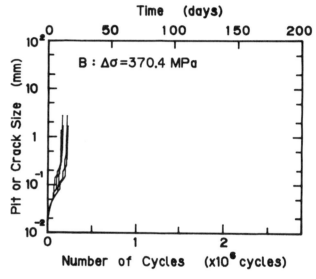

(5) $\Delta\sigma$ = 370.4 MPa

B : $\Delta\sigma$=269.5 MPa

(3) $\Delta\sigma$ = 269.5 MPa

Fig. 8   Simulation of growth of
           pits and cracks:
           (continued)

$C_p = 1.6\times10^{-7}$
$m_p = 250$

$n = 2$

$C = 1.952 \times 10^{-10}$

$m = 4.0$

135

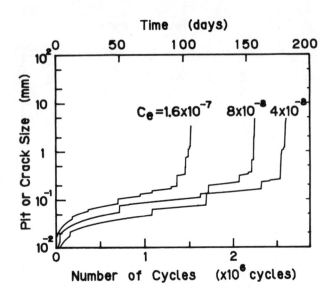

Fig. 9 The growth of pits and cracks for different values of C_e

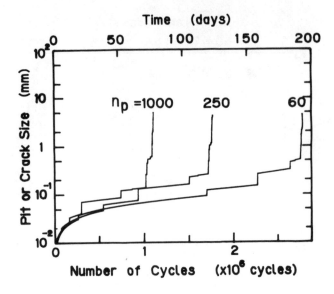

Fig. 10 Growth of pits and cracks for different values of n_p

The growth curves shape, therefore, can be regul ted by the variation of these two parameters, C_e and n_p, in addition to the propagation characteristics, C and m. It is known that the influences of metallurgi al factors upon fatigue crack growth is relatively small in material range of carbon steels. The difference in fatigue life between Steels A and B cannot be interpreted only as the difference in propagation characteristics.

The present model mentions that this difference is to be interpreted rather by the pit growth rate factors. It is to be noted that the pearlite grain sizes of Steel A are distinctly smaller than those of Steel B. It is much susceptible that the smaller the pearlite grain sizes are, the smaller the formed pits are, assuming that pits are apt to be formed around an isolated pearlite grains. The kinetics of pit growth in long time corrosion fatigue process is not yet well analysed. Further researches are required in this domaine in order to improve the understanding of the process of pit growth.

In the present analysis, the number of pits per unit area was considered as constant. Microscopic observations showed that pit population density is not homogeneous, and may largely vary even on the same specimen surface. For the purpose of fatigue life estimation, however, only higher density zones are inte ested. Examining a number of microscopical views, a constant pit density was selected.

CONCLUSION

A numerical simulations procedure were presented on the corrosion fatigue of carbon steels in seawater and obtained the following conclusion.

(1) By using a size factor common to pits and cracks, the progress of the phenomena can be characterized by four parameters; pit dissolution constant, C_e, number of pits or cracks per unit area, n_p, and crack propagation constant, C and m.

(2) This model enables to interpret the corrosion fatigue features in a series of experiments. The difference in fatigue life between two steels with different pearlite grain sizes are interpreted as the difference in the pit dissolution constant C_e.

(3) Further researches are required to improve the understandings on pit growth process which are most important points for the corrosion fatigue life estimation.

Reference

1) Kitagawa, H., T. Fujita, and K. Miyazawa, ASTM STP 642, 98-104 (1978)
2) Kitagawa, H., Y. Nakasone and S. Miyashita, ASTM STP 811, 233-263 (1983)
3) Chang, R., Eng. Fract. Mech. 16, 5, 683-693 (1982)
4) Matsuda, S., S. Okazaki, H. Hasagawa, H. Ono and K. Nihei, J. of Material Science, Japan, 33, 374, 1414-1420 (1984)
5) Kawahara, M., T. Fujita, M. Kurihara, H. Inagaki, and H. Kitagawa, Int. ASTM Symposium on Fundamental Questions and Critical Experiments on Fatigue, Paper 23, Dallas, Tex. 1984
6) Southwell, C.R. and A.L. Alexander, Material Protection, 9, 1, 14-23, (1970)
7) Kawahara, M. and M. Kurihara, Proc. 4th Int. Conf. on Fracture, Waterloo, Vol.2, 1361 (1977)

THRESHOLD PRELOAD LEVELS
FOR AVOIDING STRESS CORROSION CRACKING
IN HIGH STRENGTH BOLTS

Y. Chung
Bechtel Inc.
San Francisco, California, USA

ABSTRACT

Using a linear elastic fracture mechanics methodology and lower bound stress intensity factors for stress corrosion cracking (K_{ISCC}), threshold stress levels for avoiding stress corrosion cracking (SCC) have been calculated for high strength, low alloy bolting materials. The results have been compared with industry experience and recommendations provided for the maximum preload levels. The results of the fracture mechanics calculations for threshold preload stress levels have been found to be overly conservative when applied to bolts used in ambient environments (no more severe than humid air). Concern about the possibility of SCC exists, however, particularly in large bolts. Therefore, the maximum preload level should be evaluated for bolts larger than 38 mm (1-1/2 inches) in diameter in humid air environments and for those bolts exposed to environments more severe than humid air. For flange bolts which may experience more severe environments than humid air because of process fluid leaks, A193 Grade B7M and A320 Grade L7M bolts should be considered because of their high threshold stress values.

PRELOAD IS ONE OF THE MOST CRITICAL FACTORS in bolted joint integrity. It is the force a tightened fastener (bolt or stud) exerts on an assembly. High preloads have been used traditionally over the years, as most bolted assemblies require high preloads. Opinions vary, however, as to what a proper preload is.

To prevent leaks during hydrostatic testing, the ASME Code[1] allows actual stress levels in the bolts to be higher than the maximum allowable stress values. There are no upper limits. For civil structural applications, a preload level of at least 70 percent of the specified minimum bolt tensile strength is recommended[2] for installing high strength structural bolting materials (ASTM A325 and A490). The Society of Automotive Engineers recommends preloading bolts to 75 percent of proof load; the Air Force "... insists on limits of 80% of proof load for tensile fasteners and 30% for shear fasteners used in aircraft," and Europeans have largely abandoned the 75-percent limit in favor of 100-percent[3]. These examples illustrate the diversity of opinion about what constitutes a proper preload level. All of them point to high preloads, however.

The high preload practice was fine until the 1950s when the aerospace industry began to experience hydrogen induced cracking (HIC) with (very) high strength bolts (tensile strength greater than 180 ksi or HRC greater than 41). In the last several years, several incidents of high strength bolt failures in nuclear power plants have been reported[4]. In general, SCC has been recognized as the failure mechanism of these high strength bolts. Common factors in these high strength bolt failures in ambient atmospheric environments are the high bolt hardness (HRC greater than 40) and high preload.

Finding a proper preload level for high strength bolts is not simple. Many variables (joint and bolt geometry, bolt material mechanical properties, thread forming methods, thread lubrication, stress relaxation, etc.) affect the selection of a proper preload. Also, for high strength bolts, the maximum preload that can be sustained without developing SCC (or HIC) must be taken into account.

Stress corrosion cracking can be avoided by limiting the preload. The recent development in fracture mechanics has provided a methodology to calculate the threshold stress level under some simplifying assumptions. This report presents the results of threshold preload stress calculations for avoiding SCC in high strength bolting materials. Since the necessary data are still limited and the analytical tool provides only approximate solutions, the results should be regarded only as a guidance.

Table 1 provides a summary of ASTM specifications for carbon and low alloy steel bolting materials. The low alloy steels in this table are commonly referred to as low alloy quenched and tempered steel (LAQTS) or high strength low alloys (HSLA). Both terms are used interchangeably throughout this report.

FRACTURE MECHANICS CALCULATIONS – Linear elastic fracture mechanics (LEFM) was used to calculate the maximum preload levels without leading to SCC in high strength bolts. The LEFM equation for SCC is as follows:

$$K_{ISCC} = C \sigma \sqrt{\pi a} \qquad \text{Eq. (1)}$$

where K_{ISCC} = threshold stress intensity factor for SCC

C = shape factor

σ = nominal stress

a = crack depth

The following conservative assumptions were used:

(1) Crack Depth (a)

The thread height (a_0) was assumed to be the mouth of an initial crack at the root of a thread. The following three cases were considered.

(i) $a = a_0$ \qquad\qquad Eq. (2)

(ii) $a = a_0 + 2.5$ mm \qquad Eq. (3)
        (0.1 inch)

(iii) $a = 2a_0$ \qquad\qquad Eq. (4)

(2) Shape Factor (C)

The values for C are usually 1.1 to 1.2 for fracture mechanics evaluation of cracks in large plate. Even for a solid cylindrical body such as a bolt with a single transverse crack on a side, C is less than 1.2 for a/D ratios up to 0.26[5]. In this case, however, a higher number (1.5)

has been assigned to account for the grip end effects in tightened bolts[6].

(3) Threshold Stress Intensity Factor for SCC (K_{ISCC})

Conservative lower bound K_{ISCC} values were estimated from collective data[7]. The yield strength was estimated from a conversion curve for the relationship between hardness numbers in Rockwell C (HRC) and the yield strength of LAQTS[8]. The following three cases were evaluated.

(i)   38 HRC = 1070 MPa (155 ksi)
        yield strength

(ii)   31 HRC = 830 MPa (120 ksi)
        yield strength

(iii)   22 HRC = 655 MPa (95 ksi)
        yield strength

THRESHOLD STRESSES FOR AVOIDING SCC – Figure 1 is a plot of threshold stresses for avoiding SCC against Rockwell C hardness numbers for different bolt sizes. The four solid curves represent the threshold stress values when the initial crack depth was assumed to be two times the thread depth (i.e., $a = 2a_0$). For $a = a_0$, however, the curves will shift upwards as indicated by the broken curves for 25-mm (1-inch) and 38-mm (1-1/2-inch) diameter bolts. The threshold stress values will increase by a factor of $\sqrt{2}$ or by more than 40 percent when the postulated crack depth is assumed to be the same as the thread depth. By coincidence, the threshold stress curve for a 64- to 100-mm (2-1/2- to 4-inch) diameter bolt with $a = a_0$ is the same as that for a 25-mm (1-inch) diameter bolt with $a = 2a_0$. The top ends of the threshold curves for $a = a_0$ are limited by the yield strength of the material when the hardness is lower than 27 HRC.

The curves in Figure 1 illustrate the significant effects of crack depth assumptions on the calculated threshold stress values. Whether these crack depth assumptions ($a = a_0$, $a_0 + 2.5$ mm or $2a_0$) are reasonable or not has not been verified. The suggestion[9] that the initial crack depth be equal to one-tenth the material thickness appears overly conservative. This is particularly so when one assumes a flaw completely around the circumference. Even for a thumbnail crack, the one-tenth material thickness criterion for a 100-mm (4-inch) diameter bolt would mean a 10-mm (0.4-inch) deep by 76-mm (3-inch) long crack, which seems clearly out of a reasonable range for cracks which are assumed to exist even before installation. There seems to be no consensus on one of the important factors, the initial crack depth, in calculating a threshold stress using the fracture mechanics methodology. The $a = a_0$ assumption is severe enough when compared with the multiple groove geometry of the thread. Adding a 0.25-mm (0.01-inch) crack would not result in significant differences in calculating threshold stress levels. Adding 2.5 mm (0.1 inch) or one-tenth the bolt diameter would result in overly conservative estimates of threshold stresses, particularly when coupled with lower bound K_{ISCC}.

ASTM A325 limits bolt sizes to 38 mm (1-1/2 inches) in diameter. The hardness is limited to 35 HRC for sizes up to 25 mm (1 inch) and 31 HRC for sizes up to 38 mm (1-1/2 inches) in diameter. For $a = a_0$, Figure 1 indicates that the threshold stress values for A325 bolts are higher than the AISC minimum preload stress level (70 percent of the tensile strength = 500 MPa (73 ksi)). This would mean that A325 bolts (up to 38 mm (1-1/2 inches) in diameter) will not fail because of SCC in ambient atmospheric environments. This is supported by the industry experience that no SCC failures of A325 bolts have been reported. Therefore, there is no need for establishing a preload requirement for A325 bolts.

ASTM A490 also limits bolt sizes to 38 mm (1-1/2 inches) in diameter with a hardness requirement of 33-38 HRC (ignoring the surface hardness allowance up to 41 HRC). Even for $a = a_0$, this hardness range corresponds to a threshold stress range from 570 MPa (83 ksi) for a 25-mm (1-inch) diameter bolt to 340 MPa (50 ksi) for a 38-mm (1-1/2-inch) diameter bolt. These values are all lower than the AISC minimum preload stress level (720 MPa (105 ksi) for A490). This means that A490 bolts may fail due to SCC when the AISC installation requirement is followed. Except for a few isolated incidents of bolt failures involving extremely high hardness, however, the construction industry has had good experience with A490 bolts. This is reflected by the AISC Specification for Structural Joints Using ASTM A325 and A490 Bolts[2], which has not revised installation requirements since 1966. The few cases of A490 bolt failures since a report in 1964 have involved only extremely hard bolts.

The problem lies in the quality control of bolt heat treatment rather than in technical requirements. A recent plant-wide hardness survey of LAQTS bolting materials by one utility disclosed that only 44 percent were within the specified hardness ranges[10]. This is partly due to the current ASTM specification requirement that only one out of a lot of 800 bolts, two out of a lot of 8000 bolts, or three out of a lot of up to 35,000 bolts be checked for hardness. These requirements may change.

It appears that the calculated threshold stress values for A490 bolts within the specified hardness range using the fracture mechanics are overly conservative when viewed from the experience of the last 20 years. One of the reasons for this conservatism lies in the use of the current lower bound K_{ISCC} data, which are conservative, particularly for LAQTS with a yield strength of about 1240 MPa (180 ksi) or lower.

THRESHOLD STRESSES FOR BOLTS LARGER THAN 38MM (1-1/2 INCHES) IN DIAMETER - The larger the bolt size, the greater the thread depth. Even when using higher K_{ISCC} values for humid air environments than for the lower bound values, the threshold stress levels would be only 480 to 550 MPa (70 to 80 ksi) or lower for bolts larger than 38 mm (1-1/2 inches) in diameter with 38 HRC and $a = a_0$. Therefore, there is concern that some of these large bolts may develop SCC under high preload, particularly if the bolt hardness is high. In one plant, three out of 96 A354 Grade BD reactor pressure vessel anchor studs failed in two, five, and ten months of initial preloading (630 MPa (92 ksi))[4]. These failed studs involved extremely high hardness (45 HRC and 51 HRC).

To avoid SCC failures in high strength bolting materials, some utilities have lowered preload to 20 to 38 MPa (3 or 4 ksi)[4]. Typically, however, the low preload levels would not be applicable in most situations. It appears that a uniform preload level for a particular grade of large bolts over a wide range of applications may be neither desirable nor necessary. A 200-MPa (30-ksi) maximum preload level may be used, however, as a conservative limit for large LAQTS bolts with 38 HRC maximum hardness. If a higher preload level is required, a proper preload level for large bolts can be determined on a case-by-case basis, based' on the particular application to which a bolting material is subjected. The threshold stress curves shown in Figure 1 and the K_{ISCC} data[7] should provide some guidance in determining a proper preload level.

From the point of high threshold stress levels, the use of ASTM A193 Grade B7M or A320 Grade L7M has an advantage. These grades are being used where SCC in the presence of hydrogen sulfide is a problem, such as in gas or oil fields. Considering about 5- to 10-percent stress relaxation takes place after preloading, these grades can be preloaded to 70 to 80 percent of their minimum specified yield strength, which is 550 MPa (80 ksi) minimum, without the danger of SCC. This is even higher than the AISC minimum allowable working stress level (370 MPa (54 ksi)) for A490 bolts. These grades have sufficient strength for most flange applications. Therefore, A193 Grade B7M bolts and A320 Grade L7M should be considered for flange joints where reactor coolant leaks may create an environment more severe than humid air from the SCC point of view. These grades may be purchased with a 100-percent hardness check as a supplementary requirement.

CONCLUSIONS AND RECOMMENDATIONS

(1) The results of threshold stress calculations using the lower bound K_{ISCC} factors and the fracture mechanics equation are overly conservative for LAQTS bolts used in ambient atmospheric environments (humid air).

(2) For A325 bolts used in ambient atmospheric environment, no preload restriction is required to prevent SCC. For A490 bolts used in ambient atmospheric environment, the preload requirements in the AISC specification are safe, provided the bolt hardness is lower than the maximum in the specified range (33-38 HRC).

(3) For other LAQTS bolts such as A193 Grade B7 (4140) and A354 Grade BD (4340), the same preload practice as for A490 bolts should be used up to 38 mm (1-1/2 inches) in diameter, provided the bolt hardness is no higher than 38 HRC and the bolts are exposed only to ambient atmospheric environments.

(4) For bolt sizes 64 mm (2-1/2 inches) and larger in diameter, the maximum preload level should be 200 MPa (30 ksi), provided the bolt hardness is no higher than 38 HRC. This is a conservative limit. Significantly higher preload is possible depending on actual conditions.

(5) For LAQTS bolts exposed to environments more severe than humid air (e.g., flange joints with potential reactor coolant leaks), A193 Grade B7M or A320 Grade L7M bolts are recommended for avoiding SCC.

REFERENCES

1. ASME Boiler and Pressure Vessel Code, Section VIII, Appendix S

2. Specification for Structural Joints Using ASTM A325 or A490 Bolts Approved by Research Council on Riveted and Bolted Structural Connections, April 26, 1978, and Endorsed by American Institute of Steel Construction, Inc., and by Industrial Fasteners Institute

3. Dann, Richard T, "Machine Design," v. 47, n. 20, pp. 66-69 (August 21, 1975)

4. Electric Power Research Institute, Proceedings of "Bolting Degradation or Failure in Nuclear Plants Seminar," Knoxville, Tennessee (November 2-4, 1983)

5. Carney, M; Cartwright, D. J.; Daoud, O. E. K.: "Journal of Strain Analysis," v. 13, n. 2, p. 83 (1978)

6. Caddell, R. M., Deformation and Fracture of Solids, p. 213, (Prentice-Hall) (1980)

7. Goldberg, A; and Juhas, M. C.: "NUREG CR-2467 Lower-Bound K_{ISCC} Values for Bolting Materials - a Literature Study," pp. 57, U.S. NRC (1982)

8. Cipolla, R. C., Data presented to the AIF/MPC Bolting Task Group Meeting (1982)

9. IE Bulletin No. 82-02, "Degradation of Threaded Fasteners in the Reactor Coolant Pressure Boundary of PWR Plants," U.S. NRC, Office of Inspection and Enforcement, SSINS No. 6820, OMB No. 3150-0086, Expiration Date 05-30-1986, Washington, D.C. (June 2, 1982)

10. Slager, H. W., Data presented to the AIF/MPC Bolting Task Group Meeting (1983)

Table 1

A Summary of ASTM/SA Specifications for Carbon and Low Alloy Steel

Bolting Materials (Bars, Bolts, Screws, Studs, and Stud Bolts)

	ASTM No.	Specification Title	Grades /Types	Size (inch)	Min. Temper	Min TS	Min YS	Hardness and Remarks
BOLTS AND STUDS	A193	Alloy-Steel and Stainless Steel Bolting Materials for High-Temperature Services	B7 (4140*)	2½ and under 2½ - 4 4 - 7	1100F 1100F 1100F	125 115 100	105 95 75	TS/YS in ksi Nuts - A194
			B7M (4140*)	2½ and under	1150F	100	80	94-99 HRB (201-235HB)S3-100% Hardness Testing
			B16 (Cr-Mo-V)	2½ and under 2½ - 4 4 - 7	1200F 1200F 1200F	125 110 100	105 95 85	25-34 HRC (S4)
	A307	Carbon Steel Externally Threaded Standard Fasteners	Gr.A (General) Gr.B (Flange)	¼ - 4 ¼ - 4	- -	60 100**		69-100 HRB Galvanized when specified 69-95 HRB Threads - rolled or cut Nuts - A563
	A320	Alloy Steel Bolting Materials for Low-Temp Service	L7 (4140*)	2½ and under	LQ+T	125	105	
			L7M (4140*)	2½ and under	1150F	100	80	94-99 HRB Nuts - A194
			L43 (4340)	4 and under	LQ+T	125	105	
	A325	High-Strength Bolts for Structural Steel Joints	Type 1 (CS) Type 2 (boron Stl) Type 3 (ACR)	½ - 1 ½ - 1½ ½ - 1½	800F 800F 800F	105 105 105	81 81 81	24-35 HRC for ½ to 1 19-31 HRC for 1-1/8 to 1½ Galvanized when specified
	A354	Quenched and Tempered Alloy Steel Bolts, Studs, and Other Externally Threaded Fasteners	Gr. BC	¼ - 2½ 2½ - 4	850F	125 115	109 99	26-36 HRC For normal atmospheric temp- 22-33 HRC service
			Gr. BD	¼ - 2½ 2½ - 4	OQ+ 900F	150 140	130 115	33-39 HRC Refer to A490 for sizes up 31-39 HRC to 1½
	A449	Quenched and Tempered Steel Bolts and Studs (general application)	Medium carbon stl	½ - 1 1 - 1½ 1½ - 3	800F 800F 800F	120 105 90	92 81 58	25-34 HRC 19-30 HRC 183-235HB (89HRB - 22 HRC) Threads-rolled, cut, or ground Galv when specified (Ex Type 2 - AISC)
	A490	Heat-Treated Steel Structural Bolts, 150 ksi Minimum Tensile Strength	Type 1 (alloy stl)	½ - 1½	900F	150- 170**	130	33-38 HRC at 1/8 inch from the surface -41 HRC at 0.003 inch from the surface Refer to A354 Grade BD for sizes greater than 1½.
			Type 2 (boron stl)	½ - 1	650F			Threads-rolled or cut
			Type 3 (ACR)	½ - 1½	900F			AISC prohibits galvanizing A490 bolts.
	A540	Alloy-Steel Bolting Materials for Special Applications (for nuclear and other special applications)	B23 (E4340-H)	-6 CL. 5 6 - 8 8 - 9½	850F	120 115 115	105 100 100	241-311 HB (surface hardness) (23-33 HRC) 255-321 HB 262-321 HB
			B24 and B24V (4340 Mod and 4340V Mod) are similar to B23	- 3 CL. 4 3 - 6 6 - 9½		135 135 135	120 120 120	269-341 HB 277-352 HB 285-363 HB
				- 3 CL. 3 3 - 6 6 - 9½		145 145 145	130 130 130	293-363 HB 302-375 HB 311-388 HB (33-42HRC)
				- 3 CL. 2 3 - 6 6 - 9½		155 155 155	140 140 140	311-388 HB 311-401 HB 321-415 HB
				- 3 CL. 1 3 - 6 6 - 8		165 165 165	150 150 150	321-415 HB 331-429 HB 341-444 HB (36.6 - 47.1HRC)
	A574	Alloy Steel Socket-Head Cap Screws	Alloy steel	0.060 - 4	OQ+ 650F	170	153	39-45 HRC for ½ and under 37-45 HRC for 0.625 and larger
	A687	High-Strength Nonheaded Steel Bolts and Studs	(A4140)	5/8 - 3	800F	150	105	Galvanized when specified
THREADED ROUND STOCK	A36	Structural Steel	Carbon Steel	- 8	-	58- 80**	36	165 HB max (equiv to 80 ksi TS max) 8" dia max according to AISC Manual
	A572	High-Strength Low-Alloy Columbium-Vanadium Steels of Structural Quality (Types 1 - 4)	Grade 42 50 60 65	- 6 - 2 - 1¼ - 1¼	- - - -	60 65 75 80	42 50 60 65	Type 1 - Columbium (Cb) Type 2 - Vanadium (V) Type 3 - Cb + V Type 4 - V + N
	A588	High-Strength Low-Alloy Structural Steel with 50 ksi Minimum Yield Point to 4 in. Thick	Grade A,B,C,D,E, F,H,J,K (ACR)	- 4 4 - 5 5 - 8	- - -	70 67 63	50 46 42	

* includes 4142 and 4145   LQ = liquid quenching   ACR = atmospheric corrosion resistant   HRB = hardness in Rockwell "B"
** maximum   OQ = oil quenching   HRC = hardness in Rockwell "C"   HB = hardness in Brinell

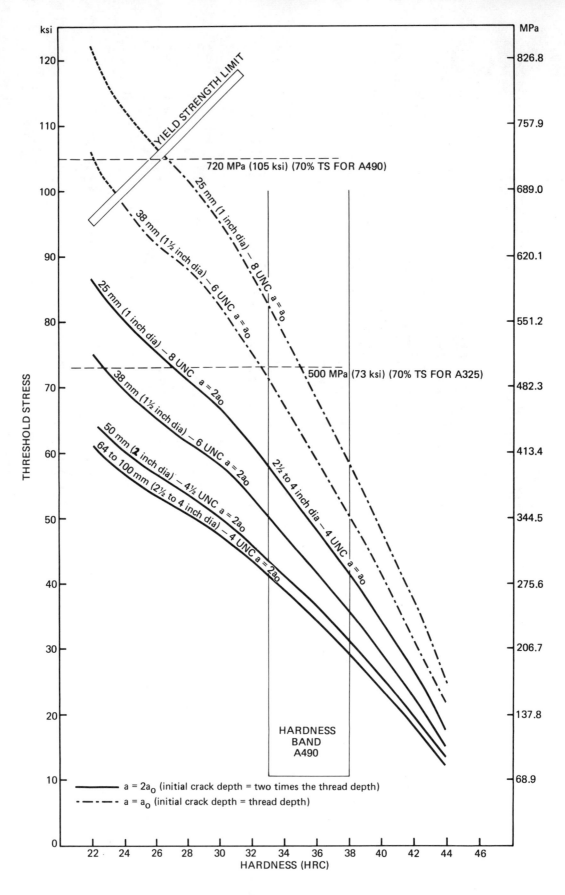

Figure 1 **A PLOT OF THRESHOLD STRESSES FOR STRESS CORROSION CRACKING IN LOW ALLOY QUENCHED AND TEMPERED BOLTING MATERIALS WITH DIFFERENT HARDNESS LEVELS**

RECOVERY OF A TYPE 304 STAINLESS STEEL PIPING SYSTEM CONTAMINATED WITH CHLORIDES

A. A. Stein, V. Zilberstein
Stone & Webster Engineering Corporation
Boston, Massachusetts, USA

W. S. Clancy, G. J. Davis
Boston Edison Company
Boston, Massachusetts, USA

ABSTRACT

Sustained chloride concentrations, even at low levels (when certain other conditions exist), can cause cracking in unsensitized austenitic stainless steel piping. Replacing chloride contaminated piping to correct this problem is expensive and can cause significant plant outages. This paper disusses the methodology and criteria by which major portions of a system contaminated with chloride were either replaced or retained.

Contamination and localized damage were discovered during pre-startup flushing and testing of a new system. Microscopic analyses of contaminated pipe samples clearly established the failure mechanism (in the most severely affected sections) as transgranular chloride stress corrosion cracking. Pertinent material and environmental parameters known to influence this failure mechanism were determined, and the potential for further damage was then evaluated. The most important parameters for explaining the observed variations in the stress corrosion of the pipe material were: (a) local residual chloride levels, (b) the existence (or absence) of local corrosion, and (c) duration of exposure to a hot (250 to 500°F) aqueous environment.

Operating guidelines to maintain system immunity to chloride-induced stress corrosion cracking (SCC) were developed. In assessing the potential for further damage (during future operation of the system), it was particularly important to recognize that the system was intended to transport only dry gases. Practical operational controls to minimize the potential for future pipe cracking were both necessary and consistent with the intended function of the system. This effort enabled about 75 percent of the system to be salvaged, thereby reducing the total cost of the recovery by as much as $1,000,000 and avoiding a potential extension of the refueling outage by as much as four weeks.

INTRODUCTION

It is common knowledge that unsensitized austenitic stainless steels in the presence of chlorides and water can fail by transgranular stress corrosion cracking (TGSCC). Furthermore, the levels of chlorides, temperatures, and stress required to produce cracking can be surprisingly low as long as certain other conditions exist (Refs. 1, 2, 3, & 4).

Given a stainless steel system, where chloride contamination exists, what can be done to continue using the system, without replacing all the contaminated material and still eliminate the concern about TGSCC? In this case, replacement cost of all the piping was estimated at over $1,000,000, not including potential cost penalties for delayed startup of the nuclear power plant. Pipe replacement costs were particularly high (approximately $2,000/ft.) due to the necessity to erect scaffolding, extensive welding in potentially contaminated (radioactive) environments, tight quarters, stringent quality controls on nuclear safety-related piping, and the need to remove and reinstall insulation and heat tracing cables. In addition, the replacement could have extended the unit refueling outage. Had this been necessary, the increased generating costs for replacement power could easily have reached $500,000/day.

In view of these high replacement costs, alternative, justifiable technical solutions were sought; and a retain or replace decision was made for each span of pipe. This paper summarizes the methodology and criteria by which Stone &Webster Engineering Corporation (SWEC) was able to recommend continued use of such a system, without total replacement of materials (including piping, socket welds and valves) (Ref. 5).

FAILURE ANALYSIS

This study began with the failure analysis of 1 in. diameter Type 304 stainless steel piping. This piping system was designed to carry containment atmosphere samples to an analyzer to monitor hydrogen and oxygen levels during operational and the design basis accident conditions that are postulated to occur in a boiling water reactor. The system was newly installed and had never been in service. Prior to start-up testing, the system had been flushed, drained, and heat traced* to temperatures greater than 200°F. Leaks were detected during subsequent hydrostatic testing. The lines were examined and evaluated for integrity and fitness for operation.

Pre-operational test data indicated that, after the system had been flushed and drained, the pipe had been heat traced to 275°F for over one month. Thermal stresses were present, although the magnitudes are not known. Chemical analyses of the inside diameter surface revealed localized chlorides in excess of 8 mg/m² in certain sections of the piping.

Figure 1 shows a thru-wall crack as seen from the outside surface of the pipe. It is circumferential and is deepest at locations near the heat tracing. The inside surface of the pipe was covered with a uniform, reddish-brown film of corrosion products.

Figure 1. Thru-wall Crack Viewed from the Outside Diameter Surface of Pipe from Line X-50. The scale from the heat tracing cable can be seen on the right.

10% Oxalic Acid   304 Stainless Steel
2.7X Magnification

Figure 2 illustrates the extensive branching of the primary crack as well as fine secondary cracking and indicates that cracking initiated on the inside diameter suface.

Figure 2. Primary and Secondary Cracking with Branching in Pipe X-50. Notice that the crack narrows as it approaches the outside diameter surface.

10% Oxalic Acid   304 Stainless Steel
50X Magnification

Figure 3 shows the crack tip, demonstrating that crack propagation was transgranular and that the material was not sensitized (as confirmed by subsequent sensitization tests performed in accordance with ASTM A262).

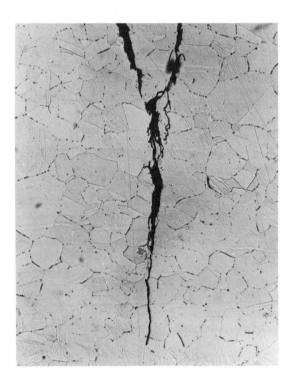

Figure 3. Crack Tip Near the Outside Diameter Surface. Notice the transgranular arrest. Pipe of Line X-50.

10% Oxalic Acid   304 Stainless Steel
200X Magnification

Close examination of the complete system showed that only one of six lines in the system had thru-wall cracks. Shallow incipient cracks were detected at the lowest elevations of one other line. The balance of the system had no signs of

SCC attack. Chlorides and corrosion deposits in varying amounts were found throughout the system, raising the possibility that more cracks could initiate and propagate if the system were wetted and heated to high temperatures.

ASSESSMENT OF CRITICAL PARAMETERS

The synergistic effects of environment, elevated temperatures, and tensile stresses on unsensitized stainless steels are complex. The critical concentration of chloride ions necessary for SCC can vary from thousands of ppm at ambient temperatures to less than 1 ppm at 500°F. Chloride SCC is not normally considered a problem below 200°F and definitely not a problem below 150°F (Ref. 1). The following four factors must be present simultaneously for TGSCC to occur: (1) chlorides, (2) elevated temperature, (3) sustained tensile stress, and (4) availability/exposure to an aggressive environment. Other factors such as pH and presence of oxidizers also influence TGSCC but are not variables in this case.

Each of the four major influencing factors are addressed below as they came into play in this problem.

(1) *Chlorides* — Local surface deposits and/or chlorides were found throughout the system in piping and valves. Amounts and distribution varied widely. Different areas of the system were examined destructively and in situ by fiberscopic examination. Optical and scanning electron microscopy, wet chemistry, and localized surface microanalyses techniques were used to identify and quantify chlorides and deposits (Ref. 5). The chloride cleanliness criterion established by the Department of Energy was used to evaluate each line. This required that noncold-worked nuclear components not exceed a surface chloride level of 8 mg/m² (Ref. 6).

(2) *Temperature* — The specified heat trace temperature was 275°F. During normal operation, when little or no moisture is present in the line, this temperature will

---

* Heat tracing was provided to prevent condensation in the pipe, thereby maintaining a constant gas composition between the sampling point and the analyzer.

insure dry conditions and will prevent SCC. Alternately, if the system is de-energized, the temperature of the piping will be below the 150°F threshold required for SCC. Peak local temperatures during pre-operational testing may have been as high as 500°F prior to adjustment and stabilization of the heat tracing circuits.

(3) *Stresses* — Residual stress was measured by X-ray diffraction on samples of installed pipe and were found to be compressive and not detrimental. Thermal stresses introduced during the operation of the heat tracing cable were the only significant stresses acting on the pipe. SWEC estimated that the thermal stresses were tensile and above the threshold for SCC. Furthermore, it appeared impractical to exercise any further control over thermal stresses during future operation of the system.

(4) *Residence Time in the Aggressive Environment* — Although it is possible that all piping within the system experienced an aqueous chloride environment and temperatures above 150°F, cracks occurred only in the lower elevation zones of 1 of the 6 sample lines. Incipient cracks were found by metallographic examination in one other line. The difference in SCC behavior between the lines was attributed to water residence time and level of chlorides present. SCC requires an incubation time for cracks to initiate. The four unaffected lines contained very little water. Upon activation of the heat tracing, the water in these lines was quickly converted to steam and expelled. Therefore, the time that these four lines experienced an aggressive environment was too short for SCC initiation. The line with shallow SCC attack contained intermediate amounts of water. In the case of the cracked line, over four gallons of water had been trapped at low elevations of the pipe for about 5 months, apparently due to incomplete draining after system flushing.

As noted previously, local metal temperature could have reached as high as 500°F during this pre-operational testing period.

The thru-wall cracked line had heavy surface deposits, the highest levels of surface chlorides and the longest water residence time of all the lines. The pipe with shallow SCC had heavy deposits and surface chlorides only at the lower elevations. These deposits can concentrate chlorides. The four remaining lines have superficial and less frequent deposits and chloride levels below those in the cracked line. Also, the water residence time in these lines was too short for SCC initiation.

ACCEPTANCE CRITERIA AND RESULTS

Long exposure times at temperatures greater than 150°F with high chlorides and tensile stresses in an oxygenated, aqueous environment can result in SCC. It was impractical to eliminate tensile stresses or to substantially reduce peak operating temperatures below 270°F, as a consequence of the necessity for heat tracing the piping. Controls could be easily established, however, to eliminate aqueous environments at elevated temperatures. Therefore, decisions to retain or replace specific sections of piping were made based on the following criteria and considerations:

1. Presence of SCC — all pipe sections which contained any indications of incipient cracking were replaced.
2. Previous exposure to significant aqueous chloride levels — all pipe sections which exhibited corrosion deposits indicative of long exposure time to high chlorides were replaced.

The gas sampling lines are now dried and will never operate with water present. During normal operation only minimal amounts of moisture are expected to enter the system. This will not be a problem because the pipe wall temperature will be either below 150°F or self-drying when heat-traced above 220°F. In either of these conditions, SCC will not occur.

As a result of the analysis, approximately 75% of the total piping believed to have been inadvertently exposed to an aqueous chloride solution was able to be retained. The additional cost of replacing the salvagable portion of the system, which might have been necessary if the analysis had not been successful, would have been about $1,000,000, excluding any cost penalties associated with delayed start-up of the plant.

CONCLUSIONS

1. The failure mechanism was transgranular, chloride, stress corrosion cracking.
2. The Type 304 piping installed is acceptable material for this application (replacement piping was also Type 304).
3. Replacement decisions were based on the presence of SCC attack or heavy corrosion deposits indicative of extended exposure time to chloride-contaminated water.
4. The existing uncracked pipe, about 75 percent of the piping in the system, was retained despite the presence of low level surface chlorides, since controls were implemented to insure that either of the following conditions is met:
   • Temperature below 150°F
   • Walls of the pipe are moisture-free or the cumulative wetted period will never exceed 30 hours.

REFERENCES

1. A. John Sedriks, Corrosion of Stainless Steel. John Wiley & Sons, 1979.
2. Cragnolino G., Lin L. F., and Szklarska — Smialowska Z. S., Stress Corrosion Cracking of Sensitized Type 304 Stainless Steel in Sulfate and Chloride Solutions at 100°C and 250°C, Corrosion, pp. 312–319, June 1981.
3. Szklarska — Smialowska Z. S. and Cragnolino G., Stress Corrosion Cracking of Sensitized Type 304 Stainless Steel in Oxygenated Pure Water at Elevated Temperatures (Review), Corrosion, pp. 653–665, December 1981.
4. Cragnolino G., and Macdonald D. D., Intergranular Stress Corrosion Cracking of Austenitic Stainless Steel at Temperatures Below 100°C — A Review, Corrosion, pp. 406–424, August 1982.
5. Stein A. A., and Zilberstein, V. A., Evaluation of Piping in the Containment Gas Sampling Systems, Report for Boston Edison Co., March 1985.
6. Cleaning and Cleanliness Requirements for Nuclear Components, RDT F 5-1T, Department of Energy, 1972.

HYDROGEN DAMAGE OF METALLIC MATERIALS

Ubbo Gramberg
Mobay Chemical Corporation
Pittsburgh, Pennsylvania, USA

ABSTRACT

Interaction between hydrogen and metallic
materials almost always results in a deterioration of mechanical properties eventually
leading to failures. This paper provides a
survey of the causes and mechanisms, and
presents representative examples of the
appearance of hydrogen-induced types of
deterioration:
- Internal mechanism: originated in the melt
supersaturated with hydrogen; such as fish
eyes, flake formation, and weld cracking.
- External mechanism: occurring in a
material exposed to a hydrogen containing
environment; such as hydrogen-induced
cracking and hydrogen embrittlement of steel,
hydride cracking of reactive metals, and
high-temperature reactions in steel and
copper.

HYDROGEN HAS BEEN A CHALLENGE for materials
engineering and equipment performance since
the early days of technical development. Due
to their small diameter hydrogen atoms are
absorbed by metals and can easily move in and
interact with the metallic lattice. Since
this interaction in most cases results in an
embrittlement of the material, hydrogen
containing environments or hydrogen generating processes can limit the application
of materials due to the eminent danger of
brittle fracture. Apart from the macroscopically low deformation of failed equipment, the appearance of hydrogen induced
damage can vary widely as influenced by·
environment and material.

The purpose of this paper is to survey
the various types of hydrogen induced deter-
iorations of metallic equipment with regard
to cause, mechanism, and appearance in order
to provide basic information about how to
recognize and prevent these failures.

THEORIES OF INTERACTION

Various theories were developed to
explain the interactions between hydrogen
and metals:
- The generation of very high internal
pressures caused by the recombination of
atomic hydrogen atoms.
- During crack initiation and propagation in
the microstructure, hydrogen is adsorbed at
the new crack surfaces thus reducing the
energy necessary for both processes.
- Due to the increased solubility for
hydrogen in strained zones of dislocations
the atoms are trapped thus impairing the
movement of dislocations.
- Hydrogen reduces the bonding forces between
metallic atoms in the lattice.
Only in rare cases can one theory alone
explain the failure satisfactorily.
Generally, a combination of the various
theories should be applied.

Hydrogen is dissolved in metals by
forming an interstitial solid solution. Due
to this interaction the toughness is reduced
considerably; the strength properties usually
are influenced to a minor degree. The
amount of hydrogen dissolved is depend- ent
upon the temperature. There is a sharp
decrease in hydrogen solubility during
solidification, which is common for

all metals. In addition to this, iron shows a reduction in hydrogen solubility during the transformation austenite/ferrite. This phenomena was the major cause for the early problems with steels in contact with hydrogen.

INTERNAL DETERIORATION

With regard to the origin of hydrogen one can differentiate between an internal and an external mechanism. While the latter one can be characterized as an absorption of hydrogen by solid metals, the internal deterioration results from supersaturated hydrogen in the solid metal immediately after solidification due to wet products used during processing of the ore.

FLAKE FORMATION - In the cast structure, despite supersaturation, no cracking is to be expected. Problems can arise only after rolling or forging operations in a specific range of temperatures. During deformation hydrogen is trapped by lattice defects. Microsegregation of alloying elements shifting the transformation of austenite/ferrite (with resulting reduction in hydrogen solubility) to lower temperatures is of special importance. Reduced ductility can cause brittle cracking of zones containing trapped hydrogen, appearing as bright spots on the fracture surface after fracture of the cross section, fig. 1. These so-called flakes often are responsible for in-service fatigue fracture initiation. Prerequisites for flake formation thus are; supersaturated hydrogen, segregations of specific alloying elements, and rapid cooling after deformation.

FISH EYES - A similar process can cause an embrittlement of welds. Hydrogen is dissolved during welding due to moisture contained in powder or the welding electrode coating, again resulting in weld metal supersaturated with hydrogen. Local brittle cracking occurs during deformation of the joint initiated at defects in the cast weld structure. The appearance of the primary failing sites is responsible for the name of this local damage, fig. 2. This embrittlement is reversible. If the weld metal is annealed prior to deformation, hydrogen will leave the metal structure and the weld metal will regain its toughness.

WELD CRACKING - Brittle cracking of welds caused by an absorption of hydrogen during welding is also observed without any subsequent deformation. Fig. 3 shows an example. During welding of 1" thick cross sections of an austenitic-ferritic stainless steel, cracking occurred in columnar ferrite grains due to humid electrode coating. Other steels are also sensitive to this type of failure. Sometimes delayed fracture can be observed; the influence of time will be discussed later.

EXTERNAL DETERIORATION

Embrittlement caused by hydrogen absorption during service is much more prevalent that the aforementioned internal deterioration. External deterioration includes a variety of failure types ranging from hydrogen induced corrosion (HIC) of steel to high temperature cracking of copper. Due to its technical impact HIC will be discussed first.

HIC covers the following hydrogen-related cracking and embrittlement processes at ambient or slightly elevated temperatures as influenced by environment, steel, and stress-time conditions: Blistering, internal cracking, hydrogen-induced stress corrosion cracking (HSCC), and hydrogen embrittlement.

BLISTERS - Hydrogen absorption from the environment causes disbonding of rolled steel along segregations close to the surface. The recombination theory is applicable since the disbonded material is raised locally thus forming a surface blister, fig. 4. Preconditions are a high supply of atomic hydrogen in an aqueous solution and low-strength rolled steel with segregations. Blistering is observed after electrolytic pickling and galvanizing processes.

INTERNAL CRACKING - There is a certain similarity of crack appearance and orientation to an initial state of blistering, see fig. 5. In contrast, internal cracks are initiated at sulfide or oxide inclusion lines and occur all over the cross section of the damaged part. After separation along the inclusions, the material bridges between different cracks fail due to transgranular cracking. This type of damage is observed in the absence of any external load of the equipment. However, if the part is subjected to external stresses, all the cracking becomes transgranular and propagates orthogonally to the main tensile stresses. Failures of this kind are observed in medium strength steels in acidic environments containing compounds called promotors, such as H_2S. These compounds delay or prevent the recombination of generated atomic hydrogen thus increasing its supply to the metallic surfaces.

HSCC - If high-strength steels are subjected to very high cyclic stress levels in acidic, promotor-containing environments, hydrogen-induced stress corrosion cracking with characteristic branching of cracks can be observed, fig. 6. Opposite to the anodic stress corrosion cracking of austenitic SS, which is initiated at the surface due to failing passive layers, HSCC cracks can start below the surface depending on the location of critical stress states. Since this type of failure is often observed in H_2S containing environments, it is called sulfide stress cracking as well. The identification of this failure type is not always easy, since the presence of hydrogen in the structure is difficult to prove. Additionally, different fracture modes of HSCC complicate recognition in the scanning electron microscope. The topography of fracture can show step-wise cracking as well as cleavage or intergranular cracking, fig. 7. The latter type of propagation is observed especially in high-strength steels.

INFLUENCE OF TIME - The diffusion of absorbed hydrogen to attractive sites in the microstructure is dependent on time. Therefore, delayed fractures can be observed at constant stress levels in the presence of hydrogen. Due to the interaction with hydrogen, the critical stress level for crack initiation is reduced, eventually causing fracture. Fig. 8 shows a galvanized high-strength steel bolt which cracked approximately 1 hour after installation in the body structure of a car. Unless fracture occurs, this type of embrittlement is reversible by sufficient annealing.

HYDROGEN EMBRITTLEMENT - The degree of steel embrittlement caused by the interaction of hydrogen with the microstructure is influenced by strength properties. Increasing strength values usually increase the tendency of a reduction of ductility. This can be demonstrated by means of slow strain rate tensile testing in specific environments. Fig. 9 shows a tensile test specimen made from fine-grained, medium-strength constructional

steel after fracture in 100 bar (1420 psi) hydrogen at ambient temperature. At a sufficiently high hydrogen pressure the ductility attains its lowest level. Increasing strain rates limits the diffusion range of absorbed hydrogen thus improving the ductility.

Additions of small amounts of oxygen to the environment also increase the ductility due to the formation of an oxide film at the surface of the steel. However, an addition of promoting compounds such as H_2S can even reduce the low ductility level measured in high-pressure hydrogen.

HYDRIDE EMBRITTLEMENT

The interaction between absorbed hydrogen and metals of Groups IVa and VIa of the Periodic System is characterized by the following steps. At low hydrogen contents, the mechanical strength is improved due to the formation of an interstitial solid solution. Increased contents cause the precipitation of very small hydride particles leading to an additional hardening effect, however, impairing ductility as well as chemical resistance. At even higher contents, needle-like hydrides are formed in the structure close to the attacked surface, causing embrittlement, high internal stresses, and eventually fracture.

Opposite to iron and nickel alloys, metals such as titanium, zirconium, and tantalum are susceptible to hydride precipitation at low hydrogen partial pressures. Fig. 10 demonstrates the destruction of a titanium plate in a plate heat exchanger due to extensive hydride formation. The appearance of the cracking is similar to that of HSCC failures.

The tendency of hydrogen absorption and the resulting interaction increases with elevated temperatures. Therefore, the metals have to be protected by sufficient inert gas shielding during welding or heat treatment. The amount of hydrogen dissolved in tantalum can be estimated through measurement of the electrical conductivity; increasing hydrogen content results in impaired electrical conductivity. In addition, the high electrochemical potential of tantalum causes a sensitivity to hydrogen absorption due to galvanic coupling with a less noble metal in acidic environments with no access of oxygen.

HIGH-TEMPERATURE EMBRITTLEMENT

With increasing temperature the amount of atomic hydrogen increases due to the dissociation of the molecules. In addition, diffusivity and tendency to interact with metallic structures are increasing as well. The reactions with two metals at high temperatures are of special technical interest.

STEEL - High-pressure, hydrogen containing environments are responsible for a reaction between absorbed hydrogen and the iron-carbide phase (cementite) in carbon steels at high temperatures. As a result of this reaction, methane is formed in the microstructure. Therefore, the steel is decarburized resulting in a decrease of mechanical strength. Further stages of the process are characterized by the effect of high-pressure methane trapped in the microstructure. First, there is some blistering of areas close to the surface, similar to blisters formed at ambient temperatures as discussed before. In a second step, disbonding of grain boundaries occurs causing fracture due to intergranular cracking in the final step, see fig. 11. This reaction is initiated in carbon steels in environments exceeding hydrogen partial pressures of 50 bar (700 psig) and temperatures of 220°C (430°F). To shift the limits of application to higher pressures and temperatures the steel must contain elements such as chromium or molybdenum forming carbides with higher bonding energy than the iron carbide.

COPPER - In oxygen containing copper such as tough-pitch copper grades with low residual deoxidizer contents a failure type similar to that in carbon steel is observed in high temperature atmospheres containing hydrogen. Examples of deteriorating environments include reducing atmospheres in furnaces as well as reducing conditions during oxy fuel gas welding.

If insufficiently deoxidized, copper can contain up to 1% cuprous oxide. At temperatures above 400°C (750°F) absorbed hydrogen reacts with Cu_2O forming steam in the microstructure. Due to this reaction, local high pressure causes grain boundary fissures eventually leading to intergranular failure of the unit, see fig. 12. Therefore, only oxygen-free copper grades should be considered if heat treatment or oxy-fuel gas welding is necessary.

In case of any doubts regarding a possible sensitivity, ASTM B 379 recommends 20 min. annealing at 800-875°C (1470-1600°F) in atmospheres containing min. 15% hydrogen. The test specimen has to pass subsequent bending without cracks.

SUMMARY

Hydrogen interacts with metallic materials in the atomic state. The gas can be introduced into the metal through the melt (internal deterioration), or through solid state absorption (external deterioration). In most cases the absorption of hydrogen results in an embrittlement of the metal eventually leading to failure. Types of interaction include: recombination of atomic hydrogen, trapping of hydrogen by lattice defects, hydride formation, reactions with carbides or oxides. Examples of internal deterioration are flakes, fish eyes, and weld cracking of steel. Examples of external deterioration are blistering, transgranular cracking, hydrogen-induced stress corrosion cracking of steel, hydride-formation in reactive metals, and high-temperature embrittlement of steel and copper.

├─┤ 10 mm

Fig. 1 - Flake Formation in Plain Carbon Steel
        Courtesy: D. Horstmann

1 mm ├─┤

Fig. 2 - Fisheyes in a Tensile Test Specimen Machined
        from Plain Carbon Steel Weld
        Courtesy: D. Horstmann

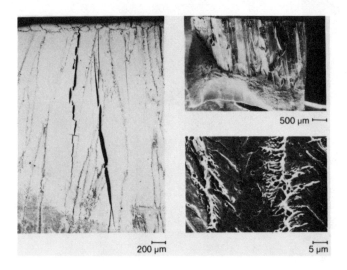

Fig. 3 – Cleavage Cracks in Ferritic–Austenitic
         Stainless Steel Weld

Fig. 4 – Blisters in Plain Carbon Steel
         Courtesy: G. Herbsleb

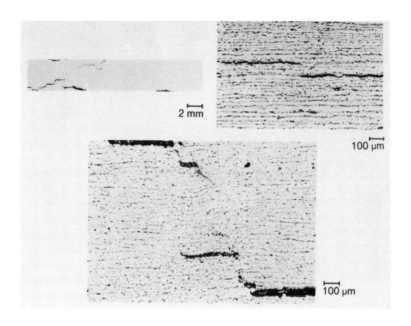

Fig. 5 - Internal Cracks in Fine-Grained, Medium-Strength Steel
Courtesy: G. Herbsleb

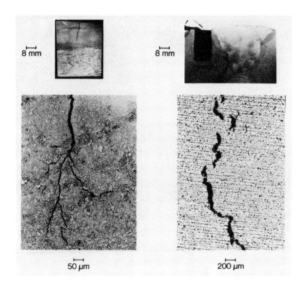

Fig. 6 - Hydrogen-Induced Stress Corrosion Cracking in
the Weld and Base Material of a
Fine-Grained, Medium-Strength Steel

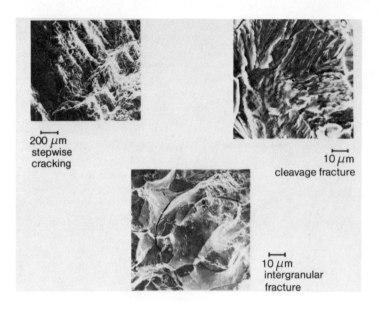

Fig. 7 - Fracture Modes of HSCC

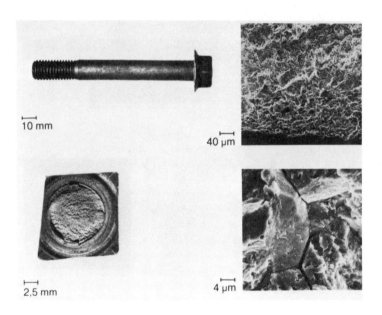

Fig. 8 - Delayed Fracture of a Galvanized Bolt
Courtesy: H. Streng

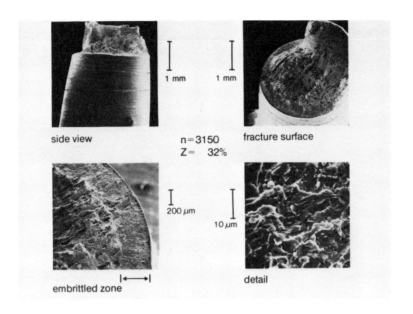

side view   n=3150   fracture surface
            Z= 32%

1 mm   1 mm

200 μm   10 μm

embrittled zone   detail

Fig. 9 – Fracture of a Tensile Test Specimen in
        100 bar Hydrogen

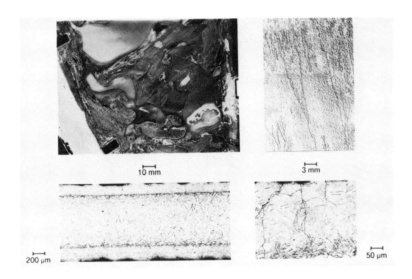

10 mm   3 mm

200 μm   50 μm

Fig. 10 – Hydride Embrittlement of a Titanium Plate

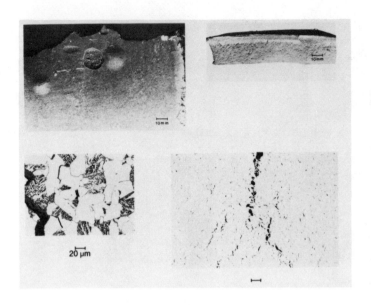

Fig. 11 – High Temperature Embrittlement of Plain
        Carbon Steel

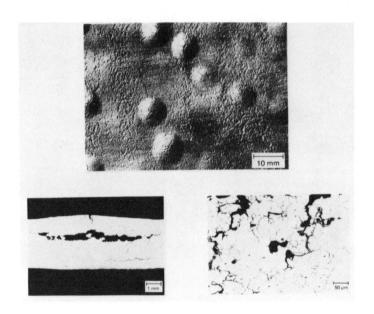

Fig. 12 – High Temperature Embrittlement of Copper

EFFECTS OF SALT WATER ENVIRONMENT AND LOADING FREQUENCY ON CRACK INITIATION IN 7075-T7651 ALUMINUM ALLOY AND Ti-6Al-4V

D. E. Gordon, S. D. Manning
General Dynamics
Ft. Worth, Texas, USA

R. P. Wei
Lehigh University
Bethlehem, Pennsylvania, USA

ABSTRACT

Strain - controlled experiments were conducted on 7075-T7651 aluminum alloy and beta annealed Ti-6Al-4V in both dry air and salt solution. A new method for acquiring crack initiation data was used. For the 7075-T7651 aluminum alloy, accelerated fatigue crack initiation was observed in the 3.5% NaCl environment at both high and low strain amplitudes. In beta-annealed Ti-6Al-4V, only slightly accelerated crack initiation was observed in the 3.5% NaCl environment. Nonlinearity was observed in strain-initiation life plots for both alloys. This non-linearity was observed whether plastic strain amplitude $(\Delta\varepsilon_p/2)$ or plastic energy per unit cycle $(\Delta\omega_p)$ was plotted versus initiation life (reversals required to obtain a 0.250 mm deep crack). No appreciable effect of frequency on crack initiation was observed in 7075-T7651 aluminum alloy in either environment. However, results for beta-annealed Ti-6Al-4V indicated a strain rate dependence in both dry air and 3.5% NaCl environment.

THE STRAIN LIFE APPROACH has been used extensively for analytically predicting the time-to-failure (TTF) of fatigue-critical components. Typical procedures for making these predictions have been reviewed by Mitchell [1]. To implement this approach, strain life allowable curves (i.e., strain amplitude versus reversal-to-failure) are normally generated for a specific test condition and environment. This approach has also recently been applied for predictions of TTCI (time-to-crack-initiation) [2,3]. To make analytical predictions for TTCI,

strain life data must be acquired in a cycles-to-initiation format. For example, Dowling [2] proposed using the strain life concept for TTCI predictions and strain cycle fatigue data for the number of cycles to initiate a specified crack size. He found that steels with fatigue lives below the transition fatigue life, cracks 0.25 mm long, were formed at approximately one-half of the failure life of the specimens.

Experimentally, acquiring crack initiation data is much more difficult than obtaining TTF data. Crack initiation strain life data has been obtained using cellulose acetate replicas to measure surface crack lengths [2]. This inspection procedure is very accurate, but it is time consuming because the test has to be stopped in order to take replicas. Also, replicas can be difficult to obtain in an environment such as salt water.

Advantages of the strain-life approach for characterizing fatigue behavior have been documented [4]. Advantages over the traditional S-N approach to fatigue characterization of alloys are: (1) more clearly able to distinguish initiation from propagation stages of fracture, (2) less scatter in data, (3) more successful in predicting failure of fracture critical parts, and (4) normally, fewer tests are required to characterize the fatigue behavior for a given material.

Two materials, 7075-T7651 aluminum alloy and beta annealed Ti-6Al-4V were selected for this investigation. There is a heavy reliance on exfoliation resistant 7075-T76 for sheet and light gage plate on the S-3A Viking, F-14 and F-18 aircraft. This material was also used in the coordinated AGARD program on corrosion fatigue [5]. There is a continuing interest in beta-processing of alpha-beta titanium alloys. Fracture toughness and fatigue crack propagation resistance improvements are often associated with the coarse, colony microstructure produced by such beta processing [6,7].

The objectives of this paper are to: (1) describe and discuss a load-shedding technique recently developed for acquiring strain life crack initiation data for strain controlled specimens (hour-glass type) in both dry air and 3.5% NaCl environments, (2) present and discuss the strain life crack initiation data acquired for 7075-T7651 aluminum and beta-annealed Ti-6Al-4V and (3) evaluate the effects of loading frequency and environment on crack initiation for 7075-T7651 aluminum and beta-annealed Ti-6Al-4V.

MATERIAL AND EXPERIMENTAL PROCEDURES

Both alloys used in this investigation were obtained from plate material. The 7075-T7651 aluminum alloy complies with federal specification QQ-A-00250/24A [8] and was obtained from a plate of 12.7 mm thickness. The beta-annealed Ti-6Al-4V alloy complies with General Dynamics specification FMS1109 [9] and was from plate of 22.2 mm thickness.

All strain-controlled fatigue testing was conducted using an hour-glass type specimen (Fig. 1). Diameters for the specimen were 6.35 mm in the reduced section and 12.44 mm in the grip area. Total length of the specimen was 177.8 mm. Specimen designs for low cycle fatigue testing have been described [10]. The hour-glass specimen is commonly used in strain-controlled experiments, where large strain amplitudes are required ($\Delta\varepsilon_\tau/2 \geq \pm 1.5\%$). This specimen geometry is less susceptible to buckling than the commonly used longitudinal specimen. In our experiments, we also felt that small cracks could be more easily detected using the hour-glass design specimen at low strain amplitudes.

Total strain-controlled fatigue tests were performed on a closed loop hydraulic MTS machine (MTS model 810.13) controlled by fully reversed ($R' = \varepsilon_{min}/\varepsilon_{max} = -1$) sinusoidal strain amplitude waves. Hydraulic operated grips were used with a self aligning feature to relieve possible bending stresses in the test coupons.

Experiments were conducted in both dry air and 3.5% NaCl environment. The environmental chamber for both dry air and 3.5% NaCl consisted of clear tygon tubing of 19.0 mm o.d. and 12.7 mm i.d. (Fig. 1). The tubing was sealed at both top and bottom with polysulfide sealant. For the dry air tests, desiccant was poured into the container before the top was sealed. In the 3.5% NaCl experi-

Figure 1 - Test Coupon Immersed in 3.5% NaCl Shown Mounted in Test Frame.

ments, salt water was added at an opening near the top of the chamber after the container had been sealed.

Normally, diametrical strain measurements are made on the hour-glass specimen. However, since many of the tests were conducted in a 3.5% NaCl environment, a 50.8 mm gage length extensometer was mounted outside of the environmental chamber (Fig. 1). Extensometer voltage output (axial deformation) was measured as a function of axial strain in the reduced section of the specimen. These measurements were made by correlating extensometer voltage readings to strain gage readings for strain gages mounted axially in the minimum area section of the test coupon. A calibration curve based on these measurements was established. This calibration curve was used to select extensometer voltages to obtain specified strain amplitudes during the experiments.

Preliminary investigations established that load shedding techniques were more sensitive to early crack detection than other NDE techniques such as eddy current surface inspection. Continuous monitoring of the decrease in the tensile stress amplitude, due to load shedding, was particularly sensitive in the cyclically stable 7075-T7651 aluminum alloy where changes in the tensile stress amplitude with cycling due to cyclic softening or hardening did not occur.

After fatigue testing of specimens to obtain different percentages of tensile stress amplitude decreases with cycling, tension overloads were then applied to fail the specimens. The size of the fatigue cracks present were then measured and plotted as a function of decrease in tensile stress amplitude. If multi-cracks were present, the depth of the largest crack was used for comparison. As shown in Fig. 2, a calibration curve was established between the decrease in tensile stress amplitude and crack depth.

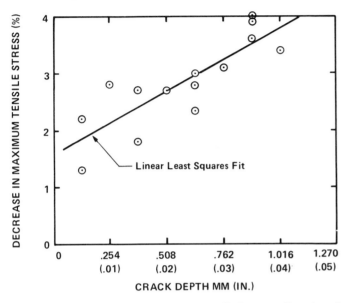

Figure 2 — Decrease in Maximum Tensile Stress as a Function of Crack Size in Strain-Controlled Specimens (7075-T7651 Aluminum Alloy)

A macroscopic crack depth initiation size of 0.250 mm was used in these studies. This crack size was consistent with minimum crack depths commonly detectable using fractographic techniques in spectrum fatigue testing. Also, this crack size is a reasonable initial flaw size for applying linear elastic fracture mechanics methods for crack growth predictions in mechanically fastened joints.

Cyclic softening occurred in the beta-annealed Ti-6Al-4V alloy at higher strain amplitudes. Both the tensile stress and compressive stress amplitudes decreased as a function of cycling. The percentage decrease in compressive stress amplitude was used to measure cyclic softening occurring and thus allowing the effects of "load-shedding" and cyclic softening to be separated in the measurements of maximum tensile stress. The onset of a 0.250 mm deep fatigue crack was defined as the number of cycles when the tensile stress amplitude showed a 2% greater decrease than the compressive stress amplitude.

RESULTS AND DISCUSSION

MONOTONIC PROPERTIES - Tensile properties and fracture toughness properties for both alloys are given in Table 1. Tensile properties for both alloys conformed to General Dynamics specifications for these alloys [8,9]. Since plane-strain conditions were not met, K_Q values are presented.

Table 1. Monotonic Properties for 7075-T7651 Aluminum Alloy and Beta-Annealed (BA) Ti-6Al-4V

Monotonic Properties	7075-T7651	7075-T2651 (MIN. REQ.)*	(BA)Ti-6Al-4V	(BA)Ti-6Al-4V (MIN REQ.)**
Ultimate Tensile Strength, MPa (ksi)	528.8(76.7)	489.5(71.0)	937.7(136.0)	875.7(127.0)
Yield Strength, MPa (ksi)	457.1(66.3)	413.7(60.0)	875.7(127.0)	772.2(1120)
Elongation, Pct	12	6	10	10
Reduction of Area, Pct	31.6	–	17.7	15
Fracture Toughness K_Q, MPa $\sqrt{mm}$ (ksi$\sqrt{in}$)	36.5(33.2)	–	151.5(137.7)	–

* Min. requirements for QQ-A-250/24A specification [8]

** Min. requirements for FMS 1109D specification [9]

CYCLIC STRESS-STRAIN BEHAVIOR

7075-T7651 Aluminum Alloy - Monitoring the stress amplitude as a function of cycles indicated that the 7075-T7651 aluminum alloy was in general, cyclically stable. Normally, after the first few cycles were completed, both the tensile and compressive stresses remained approximately constant until crack initiation. Typical plots for four different total strain amplitudes are shown in Fig. 3. The cyclic stability of this material was also verified by the "incremental step" test method [11], i.e., the total strain

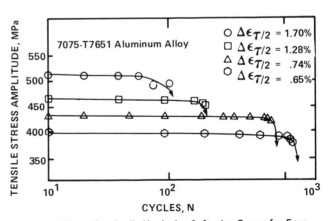

Figure 3 — Cyclic Hardening-Softening Curves for Four Different Total Strain Amplitude Values (7075-T7651 Aluminum Alloy).

amplitudes are increased and decreased in steps, until the hysteresis loops have stabilized. No effect of the 3.5% NaCl environment on the cyclic hardening or softening in this alloy was observed.

Previous tests of alloys 7075 and 7050 in peak-strength and overaged tempers have demonstrated that these alloys are cyclically stable (stress amplitude remains essentially constant in strain- controlled tests [1,12]. However, cyclic hardening and/or cyclic softening has been observed in modified 7075 compositions where purity level and dispersoid type were varied [13]. Cyclic hardening and softening behavior in the modified 7075 compositions were found to be dependent on heat treatment. Residual stresses produced during quenching from the solutionizing temperature in material not stretched prior to ageing can produce cyclic softening. Such behavior was observed in high purity ITMT 7475 aluminum alloy [14].

Beta-annealed Ti-6Al-4V - at higher strain amplitudes, cyclic softening was observed in the beta-annealed (BA) alloy. For total strain amplitudes greater than 0.7%, both the tensile and compressive stress amplitudes decreased with cycling. Typical plots are shown in Fig. 4. Decreases in the stress amplitude with cycling were accompanied by increases in the widths of the hysteresis stress-strain plots (increases in $\Delta\epsilon_\rho$). Nearly identical softening occurred in both dry air and 3.5% NaCl environments indicating that the effect was mechanical rather than environmental. For strain amplitudes below 0.7%, the material was cyclically stable.

Initial cyclic strain hardening has been found to occur in Ti-6Al-4V material forged and heat-treated below the beta-transus [15]. This initial cyclic strain hardening, caused by an increase in dislocation density, was followd by sub-

sequent strain softening. In the beta-annealed material, we did not observe the initial strain hardening. Softening is often associated with localized strains. This effect can be enhanced for material with large slip length. In the beta-annealed (BA) titanium microstructure condition, the controlling microstructural parameter appears to be the Widmanstatten-alpha colony size [16]. The alpha platelets in a single colony tend to act as a single slip unit especially in the absence of a significant amount of beta phase between the alpha platelets as shown by Hack and Leverant [17]. In our beta-annealed material, the Widmanstatten alpha colony size was quite large ($\approx$100 μm), thus the significant softening that occurred could be related to the large Widmanstatten alpha colony size.

An anomalous endurance enhancement has been observed in alpha-beta titanium alloys in which a large amount of inverted strain hardening occurs [18]. This behavior, termed "inverted strain hardening", is characterized by cyclic softening in the first half of each excursion and it is reversed by extra strain hardening in the latter half with the peak restored to its normal value. This inverted strain hardening can be observed by comparing shapes of stress-strain hysteresis loops after different intervals of cycling. Initially, the cyclic hysteresis loops exhibit a normal Bauschinger effect, a fully convex outward shape, but as cycling proceeds in some titanium alloys, the loops become more concave near the maximum tensile stress and the difference in slopes is related to inverted strain hardening.

Some amount of inverted strain hardening was observed in our material. In the beta-annealed alloys, oxygen content has been found to have a strong influence [18]. Improved LCF performance has been observed in low oxygen, (BA) material where a large amount of inverted strain hardening occurred [18]. For the material we tested, the O_2 percentage was slightly higher than that for (BA) alloys where a significant inverted strain hardening has been observed.

STRAIN-INITIATION LIFE CURVES - Cyclic strain versus initiation life curves are shown for both alloys in Fig. 5. Total strain amplitude, $\Delta\epsilon_T/2$, is plotted as a function of reversals, $2N_i$ which occur before a fatigue crack of 0.25 mm depth is formed. Curves are shown for both dry air and 3.5% NaCl environment.

In the 7075-T7651 aluminum alloy, accelerated crack initiation is observed in the 3.5% NaCl environment at both high and low strain amplitudes. The general shape of the curve is basically the same

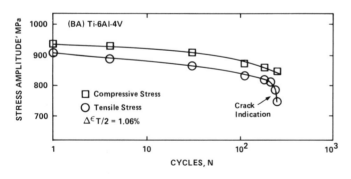

Figure 4 — Typical Cyclic Stress Response Curves in Beta Annealed (BA) Ti-6Al-4V.

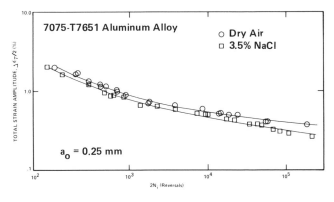

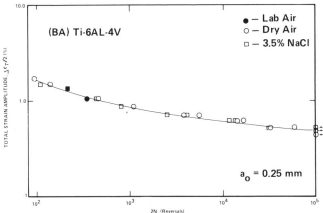

Figure 5 — Total Strain Amplitude Versus Reversals to Crack Initiation in Both Dry Air and 3.5% NaCl Environments; (a) 7075-7651 Aluminum Alloy and (b) Beta Annealed (BA) Ti-6Al-4V.

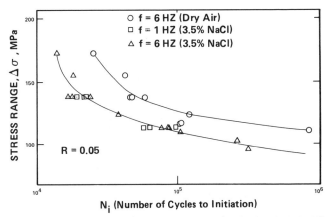

Figure 6 — Comparison of $\Delta\sigma$ Versus N_i Curves (Dry Air & 3.5% NaCl) for 7075-T7651 Aluminum Alloy (a_o = 0.25 mm).

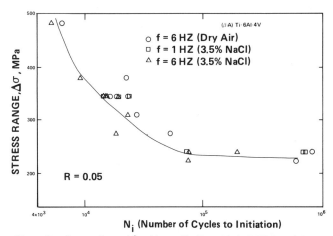

Figure 7 — Comparison of $\Delta\sigma$ Versus N_i (Dry Air & 3.5% NaCl) for β-Annealed Ti-6Al-4V Alloy (a_o = 0.25 mm).

have a larger effect on strain-life results than the salt water environment.

Again, similar results were obtained for stress-controlled tests of beta-annealed Ti-6Al-4V dog-bone coupon specimens tested under constant amplitude, R = 0.05, tension-tension tests (Fig. 7). Stress-controlled test data showed little difference in susceptibility to crack initiation for specimens exposed to the 3.5% NaCl environment as compared to dry air. Scatter in test data, again was more prevalent in the stress-controlled test data.

for both dry air and 3.5% NaCl. The results for 7075-T7651 in dry air agree well with those for 7075-T7 materials tested in lab air [19].

Similar results were obtained under stress control for 7075-T7651 dog-bone coupon specimens tested under constant amplitude, R = 0.05, tension-tension tests [11]. Open-hole, dog-bone specimens with 6.35 mm diameter holes (K_T = 2.4) were tested under both dry air and 3.5% NaCl environments. The number of cycles to produce a 0.250 mm deep fatigue crack was determined using eddy current techniques.

As observed in Fig. 6, crack initiation resistance in the 3.5% NaCl environment was substantially less than in dry air, over the range of stresses investigated. Considerably more scatter was observed in the stress-controlled, open-hole test data, however.

In the beta-annealed Ti-6Al-4V alloy, only a slight degradation in fatigue crack initiation life was observed in 3.5% NaCl solution as compared to dry air (Fig. 5b). The effect of the 3.5% NaCl environment on crack initiation is considerably less in this alloy than in the 7075-T7651 aluminum alloy. The effect of strain rate (test frequency) was found to

PLASTIC STRAIN BEHAVIOR - Coffin-Manson type plots (plastic strain amplitude versus reversals to crack initiation) were plotted for both alloys in dry air and 3.5% NaCl environment. In the 7075-T7651 aluminum alloy, similar plots were obtained in dry air and salt water environments (Fig. 8). In the salt water environment the number of reversals to crack initiation was lower than that in the dry air environment. These re-

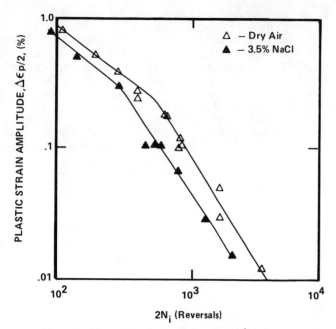

Figure 8 — Plots of Plastic Strain Amplitude, $\Delta \epsilon_p/2$, Versus Reversals to Initiation in 7075-T7651 Aluminum Alloy

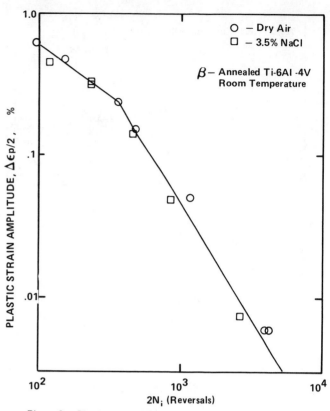

Figure 9 — Plastic Strain Amplitude Versus Reversals to Initiation in β- Annealed (BA) Ti-6Al -4V.

sults indicate that the fatigue initiation mechanism is the same in the two environments but that it is accelerated in 3.5% NaCl.

Non-linear slope behavior was obtained in the 7075-T7651 aluminum alloy. Failure of the Coffin-Manson relation is commonly observed in aluminum based alloys [20-22]. Failure of the relation to hold has also been found when reversals to initiation are plotted instead of reversals to failure [22].

It has been suggested that the hysteretic plastic energy per cycle, $\Delta \omega_p$, may be a better parameter for a Coffin-Manson type plot during a total-strain controlled experiment than $\Delta \epsilon_p$ [23]. For materials which cyclically hardened or soften during testing, $\Delta \omega_p$, may be more nearly constant during the experiment. However, results of the plastic energy per unit cycle, $\Delta \omega_p$, versus reversals to initiation, $(2N_i)$, plots revealed similar non-linearity as observed in the $\Delta \epsilon_p/2$ versus $2N_i$ curves.

Since steady state or "saturated" hysteresis loops were not achieved at higher total strain amplitudes in beta-annealed Ti-6Al-4V, then the question of which $\Delta \epsilon_p$ to plot in a Coffin-Manson type relationship must be considered. By using the final or largest value of plastic strain amplitude obtained, a Coffin-Manson type curve as shown in Fig. 9 is obtained. Non-linear behavior is observed in this alloy, also. Again, little difference is observed in initiation life in the 3.5% environment as compared

to dry air.

Occasionally, different test results are obtained from S-N curves (Wohler diagrams) obtained under stress-controlled conditions and strain controlled test data using Coffin-Manson type plots [24]. This can be especially true at low plastic strain amplitudes where comparisons of materials having different cyclic hardening behavior can yield different results in strain-controlled experiments compared to stress-controlled tests. However, since no significant difference in cyclic hardening behavior was observed between the dry air and 3.5% NaCl environment, no discrepancy between stress-controlled and strain-controled tests would be expected.

EFFECT OF TEST FREQUENCY ON CRACK INITIATION -

7075-T7651 Aluminum Alloy - The possible effect of test frequency on crack initiation in a 3.5% NaCl environment was examined. Most of these studies were conducted at lower strain amplitudes where the test coupons were exposed to a salt water environment longer. Frequencies ranging from 0.1 Hz to 5.0 Hz were used. Results shown in Fig. 10 indicate no appreciable frequency effect. All of the data could be fitted to one general curve. No frequency effect on fatigue crack initiation was found [11] for

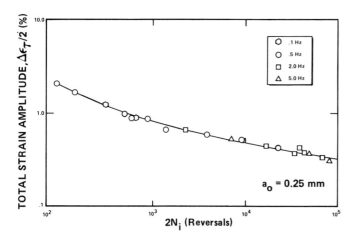

Figure 10 — Total Strain Amplitude Versus Reversals to Crack Initiation for 7075-T7651 Aluminum Alloy in a 3.5% NaCl Environment for Different Loading Frequencies.

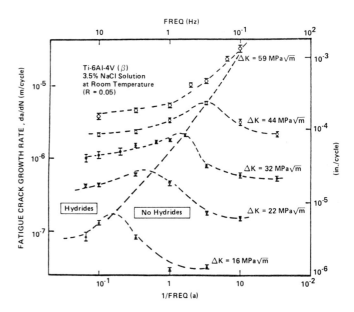

Figure 12 – The Influence of Frequency on Fatigue Crack Growth Rate of β-Annealed Ti-6Al-4V Alloy in 3.5% NaCl Solution at Room Temperature (R = 0.05).

stress-controlled constant amplitude tests conducted on the same material (Fig. 6). Frequency range tested in the stress control tests were similar to that used in the strain-controlled tests (1 Hz and 6 Hz).

Beta-Annealed Ti-6Al-4V - Studies conducted at frequencies ranging from 0.3 Hz to 5.0 Hz in beta-annealed Ti-6Al-4V indicate that both the dry air and 3.5% NaCl environment data are strain rate dependent (Fig. 11). Results generally showed enhanced initiation life at higher frequencies. Similar results have been obtained in the crack growth behavior of this alloy where an effect of frequency on fatigue crack growth rate has been found (Fig. 12 [25]). The results show that the crack growth rate at a given ΔK increased with the decreasing frequency, reached a maximum, and then decreased with decreasing frequency. Diffusion control is considered to be the rate controlling process. Enhancement of crack growth is attributed to hydride formation and rupture, which is dependent on strain rate. The observed frequency effect is explained in terms of diffusion control in conjunction with a strain rate dependent hydride embrittlement mechanism. This explanation is supported by analyses of fracture morphology and crack path in relation to the alloy microstructure [25].

Methods for predicting fatigue crack propagation behavior from low cycle fatigue properties and microstructural parameters are currently receiving considerable interest [26]. Both involve the concept of cyclic accumulated damage. A model, developed by Chakrabortty [27] and modified by Heikkenen [28] for near-threshold growth, can be used to predict fataigue crack growth behavior from low-cycle fatigue (LCF) properties and a characteristic miocrostructural parameter depending on slip length. This model was used successfully for predicting crack growth behavior in aluminum alloys by Starke, et. al. [26] and for titanium alloys by Chesnutt and Wert [16] and Chakrabortty and Starke [29].

A frequency effect on fatigue crack growth data can be predicted from the Chakrabortty model if a similar dependence is observed in low cycle fatigue parameters as was observed in the present experiments. To determine if a quantitative fit exists between LCF data and crack growth measurements, low cycle fatigue parameters, c, ϵ_f', n', k', and $\Delta \epsilon_{th}$ are required for each frequency tested.

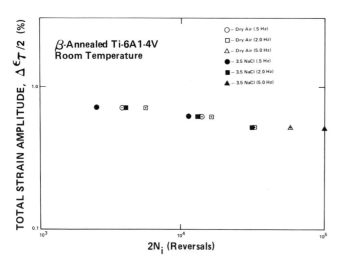

Figure 11 — Total Strain Amplitude Versus Reversals to Initiation in β-Annealed Ti-6A1-4V Showing Effect of Frequency.

SUMMARY

1. The load-shedding technique is very promising for acquiring crack initiation strain life data using hour-glass type specimens for both dry air and 3.5% NaCl environments.

2. Accelerated time-to-crack-initiation was observed for the 7075-T7651 aluminum alloy in a 3.5% NaCl environment at both high and low strain amplitudes.

3. In beta-annealed (BA) Ti-6Al-4V, only slightly accelerated crack initiation was observed in the 3.5% NaCl environment.

4. Low cycle fatigue data for beta-annealed (BA) Ti-6Al-4V indicated a strain rate dependence for both dry air and 3.5% NaCl environment. These results are consistent with fatigue crack propagation data for this same alloy. No strain rate dependence was observed in the 7075-T7651 aluminum alloy.

5. Qualitative agreement was obtained between strain-controlled low-cycle-fatigue initiation tests and stress-controlled constant amplitude (R = 0.05) dog-bone coupon tests.

ACKNOWLEDGMENT

This research was sponsored by the Naval Air Development Center (Warminster, PA) under NADC Contract N62269-81-C-0268. Mr. P. Kozel was the project engineer. The authors appreciate the support of J. W. Hagemeyer of the General Dynamics/Fort Worth Division during this program.

REFERENCES

1. Mitchell, M. R., "Fundamentals of Modern Fatigue Analysis for Design," in Fatigue and Microstructure, Collections of Papers presented at the 1978 ASM Materials Science Seminar, St. Louis, MO, 385-437 (1978).

2. Dowling, N. E., "Cyclic Stress-Strain and Plastic Deformation Aspects of Fatigue Crack Growth," ASTM-STP 637, American Society for Testing and Materials, 97-121 (1977).

3. Socie, D. F., J. D. Morrow, and W. C. Chem, Journal of Eng. Fracture Mech., 11, 851-860 (1979).

4. Bucci, R. J., Alcoa Report No. 57-77-24, Sept. 1, 1977.

5. Wanhill, R. J. H., and J. J. DeLuccia, "An AGARD-Coordinated Corrosion Fatigue Cooperative Testing Programme," AGARD Report No. 695, February 1982.

6. Rosenberg, H. W., J. C. Chesnutt, and H. Margolin, "Fracture Properties of Titanium Alloys," Application of Fracture Mechanics for Selection of Metallic Structural Materials, Amer. Soc. of Metals, 213-252 (1982).

7. Eylon, D., J. A. Hall, C. M. Pierce, and D. L. Ruckle, Met. Trans. A, 7A(12), 1817-1826 (1976).

8. Federal Specification, "Aluminum Alloy, 7075, Plate and Sheet (Improved Exfoliation Resistant)" QQ-A-00250/24A (Navy AS) Nov. 1972."

9. General Dynamics, "Specification FMS-1109D, Titanium Alloy, 6Al-4V, Beta Annealed Bar, Forged Billet, and Plate," May 1981.

10. Raske, D. T., and J. D. Morrow, "Manual on Low Cycle Fatigue Testing", ASTM STP 465, Amer. Soc. for Testing and Materials, 1-25 (1969).

11. Kim, Y. H., D. E. Gordon, S. M. Speaker, and S. D. Manning, "Development of Fatigue and Crack Propagation Design and Analysis Methodology in a Corrosive Environment for Typical Mechanically-Fastened Joints," Volume I-Phase I Documentation, Naval Air Development Center (Warminster, PA) Report No. NADC-83126-60-Vol. I, March 1983.

12. Sanders T. H., R. R. Sawtell, J. T. Staley, R. J. Bucci, and A. B. Thakker, "Effect of Micro-structure on Fatigue Growth of 7XXX Aluminum Alloys Under Constant Amplitude and Spectrum Loading," Naval Air Development Center Contract No. N00019-76-C-0482, Final Report (April 1978).

13. Santner, J. S., Met. Trans. A, 12A, 1823-1826 (1981).

14. Sanders, R. E. and E. A. Starke, Met. Trans. A, 9A, 1087 (1958).

15. Wells, C. H. and C. P. Sullivan, Trans. Am. Soc. Metals, 263-270 (1972).

16. Chesnutt, J. C., and J. A. Wert, "Effects of Micro-structure and Load Ratio on K_{th} in Titanium Alloys" in Fatigue Crack Growth Threshold Concepts, 83 (1983).

17. Hack, J. E., and G. R. Leverant, Met. Trans. A, 13A(10), 1729-1738 (1982).

18. Krafft, J. M., Fat. of Engineering Materials and Structures 4, 111-129 (1981).

19. Sanders, T. H., and J. T. Staley, "Review of Fatigue and Fracture Research on High-Strength Aluminum Alloys," in Fatigue and Micro-structure, Collections of Papers presented at the 1978 ASM Materials Science Seminar, St. Louis, Mo., 467-522 (1978).

20. Quesnel, D. J., and M. Meshii, "Investigation of Cyclic Deformation and Fatigue Failure in High Strength Low-Alloy Steel" Proc., 2nd Int. Conf. on Mechanical Behavior of Materials (ICM-II), Boston, Mass., August 16-20 (1976).

21. Sanders, T. H., and E. A. Starke, Met. Trans. A., 7A, 1407-1418 (1976).

22. Heikkenen, H. C., F. S. Lin, and E. A. Starke, Materials Science and Eng., 51, 17-22 (1981).

23. Santner, J. S. and M. E. Fine, Unpublished Research, Northwestern University, Evanston, Ill. (1976).

24. Starke, E. A., and G. Lutjering, "Cyclic Plastic Deformation and Microstructure," in Fatigue and Microstructure, Collections of Papers presented at the 1978 ASM Materials Science Seminar, St. Louis, Mo., 205-243 (1978).

25. Chiou, S., and R. P. Wei, "Corrosion Fatigue Cracking Response of Beta Annealed Ti-6AL-4V Alloy in 3.5% NaCl Solution," Naval Air Development Center (Warminster, PA), Report No. NADC-83126-60-Vol. V, June 1984.

26. Starke, E. A., F. S. Lin, R. T. Chen, and H. C. Heikkenen, "The Use of the Cyclic Stress Strain Curve and a Damage Mode for Predicting Fatigue Crack Growth Thresholds," in Fatigue Crack Growth Threshold Concepts, 43 (1983).

27. Chakrabortty, S. B., Fat. Engr. Mat. and Struct., 2, 331 (1979).

28. Heikkenen, H. C., E. A. Starke, and S. B. Chakrabortty, Scripta Met. 16, 571 (1982).

29. Chakrabortty, S. B., and E. A. Starke, Met. Trans. A, 10A, 1901 (1979).

INFLUENCE OF C, N AND Nb CONTENTS ON INTERGRANULAR CORROSION, STRESS CORROSION CRACKING AND MECHANICAL STRENGTH OF ALLOY 600

Yoshito Fujiwara, Rikio Nemoto
Koichiro Osozawa, Kazuo Ebato
Nippon Yakin Kogyo Co., Ltd.
Kawasaki, Japan

Kiyoshi Yamauchi
Tomoaki Okazaki
Babcock-Hitachi K.K.
Hiroshima, Japan

Abstract
Dependence of C, N and Nb contents on intergranular corrosion(IGC), stress corrosion cracking (SCC) and mechanical strength of alloy 600 were examined. To evaluate the corrosion resistance of heat affected zone (HAZ), post weld heat treated specimens were used. To evaluate IGC and SCC susceptibility, Ferric Sulfate-Sulfuric Acid Test and High Temperature Water Test were used. The tendency of SCC resistance in high temperature water is correspond to that of IGC resistance. Carbon is harmful to the corrosion resistance. When C content is extraordinarily lowered the corrosion resistance becomes exellent in case of having low content of other harmful element such as P and S, but yield strength at 0.2 % elongation becomes lower than 25 Kg/mm². Nitrogen is not harmful to the corrosion resistance and beneficial for improving mechanical strength. Paticularly Nb is effective as stabilizer of C. Consequently to prevent corrosion attack on HAZ in high temperature water Nb should be added more than 100(% C − 0.005)% and to maintain mechanical strength, Nb should be added more than 3 + 75(%C + %N)%. Modified alloy 600 satisfing above condition was manufactured commercially and its various properties includ ing weldability were confirmed to be excellent.

ALLOY 600 is one of the oldest of the commercially significant Nickel- Chromium- Iron general purpose alloy and it has been used widely in nuclear steam generator because the chloride stress corrosion cracking resistance is superior. But it is of known that inter-granular stress corrosion cracking (IGSC C) occures in high temperature pressurized water.[1] The behavior of stress corrosion cracking of alloy 600 in high temperature water largely depends on its heat treatment and its dependence is very complex and therefore there were many urgument[2] about the cracking mechanism. But recently [3][4][5] it has been cleared that cracking in high temperature pure water is mainly due to the existence of chromium depleted zone around grain boundary and its complex behavior is due to the low Carbon solubility in Nickel base alloy. Therefore, SCC resistance in high temperature water can be evaluated by IGC test. It is also of known that SCC resistance of alloy 600 in high temperature water can be improved by paticular heat treatment[6]. During welding, although, the effect of particular heat treatment will dispel in heat affeced zone, particularly in high temperature heat affected zone. Therefore improvement of SCC resistance including weld should be necessary.

In this study, the effect of elements such as Carbon, Nitrogen and Niobium content on IGC resistance and SCC resistance of alloy 600 base composition was revealed. IGC sensitivity was estimated by using Modified Ferric Sulfate-Sulfuric acid test (Modified F S Test) which has good correspondence to high temterature water SCC resistance[7] and finally SCC resistance was evaluated in high temperature water SCC test.

Element such as C, N and Nb also affects on mechanical strength and these dependence were also evaluated.

Based on these experimental results, Modified alloy 600s were produced in commercial process and they were tested on full size.

1 Experimental Procedures
1.1 Test Materials
The chemical composition range of test materials is shwon in table 1. Contents of C, N and Nb were varied from the viewpoint of improvement of resistance to grain boundary sensitization.

These alloys were melted into 6 to 10 Kg alloy ingots by using an induction furnace in

Table 1 Chemical Composition of Test Materials (wt.%)

C	Si	Mn	P	S	Ni	Cr	Fe	Nb	N
0.003	≤0.5	≤0.5	≤0.02	≤0.005	Bal	16.0	6.0	0.0	0.004
0.06						17.0	8.0	4.0	0.05

atmosphere and these ingots were forged into pieces each 10mm thick and 70 to 100mm wide. These pieces were heated at 1100°C for an hour and then cooled into water. They were further heated at 870°C for 2 hours, and then cooled into water.

Test pieces for mechanical tests were prepared from the obtained alloy pieces. As shown in Fig.1 a groove was prepared in each of the alloy pieces and padded in layers with a filler metal having a composition as shown in table 2 by TIG arc welding. These alloy pieces were heated at 600°C for 20 hours, then cooled in air, further heated at 500°C for 40 hours and cooled in air. These heat treatments correspond to stress relief (SR) and low temperature sensitization respectively. All of the above test pieces were cut to form crosssections for welds to which the final finishing was applied by wet polishing with #800 emery paper. Test pieces for a corrosion test were prepared from these treated alloy pieces.

Intergranular corrosion test and high temperature water SCC test were performed. In order to evaluate the intergranular corrosion resistance, modified FS tests were conducted in boiling 50% Sulfuric Acid + 83g/1 Ferric Sulfate for 24 hours.

To estimate the intergranular corrosion resistance specimens were cut along the doted line as indicated in Fig.1b and measured the maximum penetration depth which was better correspondence to high temperature SCC susceptivility than the weigt loss.

High temperature water SCC tests were conducted on CBB[8] test in 288°C pure water containing 40 ppm disolved oxigen for 500 hours.

Tension strength were conducted at room temperature.

2 Results and Discussion
2.1 IGC and SCC Resistance
Fig.2 shows the dependence of C content on

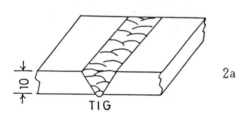

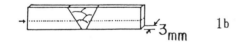

Fig.1 Corrosion Test Sample

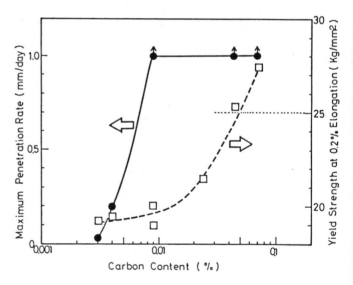

Fig.2 Dependence of Carbon content of alloy 600 on penetration depth of itergranular corrosion and yield strength at 0.2% elongation.

Table 2 Chemical Composition of Filler Metal (wt.%)

C	Si	Mn	P	S	Ni	Cr	Fe	Nb	Ti
0.01	0.11	2.96	0.005	0.002	Ba	20	0.90	2.6	0.36

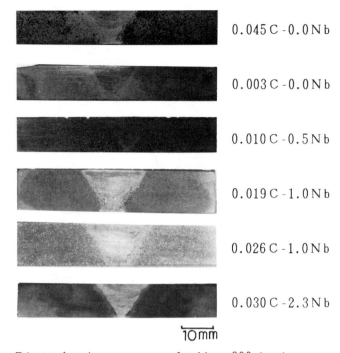

0.045 C - 0.0 Nb

0.003 C - 0.0 Nb

0.010 C - 0.5 Nb

0.019 C - 1.0 Nb

0.026 C - 1.0 Nb

0.030 C - 2.3 Nb

|← 10mm →|

Photo.1 Appearance of alloy 600 having differant Carbon and Niobium content tested by modified   FS   Test.

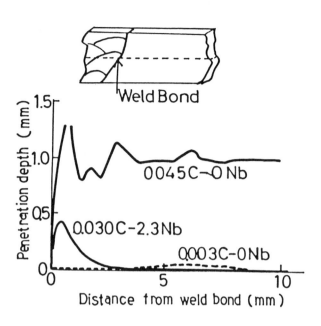

Fig.3 Penetration depth profiels of 600 base alloys tested by modified   FS   Test.

corrosion resistance and mechanical strength of alloy 600.   Penetration depth of inter-granular attack decreases with decrease in C content and when C content is below 0.005%, corrosion resistance becomes fairly good. But the mechanical strength conversely decrease, particularly the yield strength at 0.2% elon-gation is lowered to below 25Kg/mm², which is the specification for alloy 600, therefore the alloy is lacking in one of the most im-portant properties for engineering materials. It is necessary to investigate the depenedence of another elements such as N or Nb for improving the corrosion resistance.

Photo.1 shows the appearance of after inter-granular corrosion tested samples. Even though there are small differences in type of attack between these alloys, the maximum penetration depth of all samples are the same in heat affected zone occured during welding.

Fig.3 shows the penetration depth profiles of some of Photo.1. Addition of Nb improve corrosion resistance in matrix but heat affect ed zone is attacked when Nb content is not sufficient.   Fig.4 shows the dependence of C content on penetration depth of the alloys hav ing various Nb content.   In case of alloys without Nb, when C content is over 0.005%, the penetration depth increase abruptly. This critical C content increase with increase in Nb content.

Fig.5 shows the relationship between crack depth of high temperature water SCC

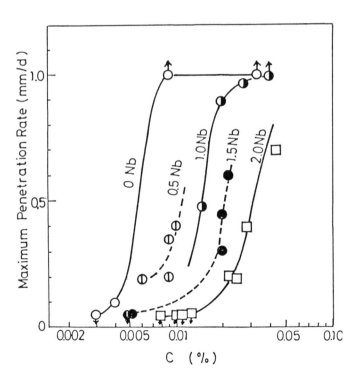

Fig.4 Dependence of C content on pene-tration rate of intergranular corrosion by Modified FS test for various Nb containt alloys.

and penetration depth of IGC. Tenedncy of SCC susceptibility is correspond to that of IGC and when penetration rate of IGC was less than 0.5 mm/day, SCC did not occured.

Fig.6 is a diagram showing a relationship between the intergranular corrosion and the contents of C and Nb wherein corrosion resistance of alloys were distinguished in three grades. The figure shows that in order to obtain a good corrosion resistant alloy showing a maxium penetration rate of below 0.5mm/day, it is necessary to add at least the following related amount of Nb

$$\%Nb \geqq 100(\%C - 0.005) \qquad (1)$$

The reason why this large amount of Nb is necessary to obtain a good corrosion resistance is that C solubility in alloy 600 at 600°C or lower temperature is less than 0.005 %.[3)9)]

During welding, precipitated carbide redistribute in heat affected zone and when Nb content is not sufficient, the amount of super satulation of C increases. Photo.2 shows the microstructure of weld of Nb containing alloy Region 2 is high temperature heataffected zone where grain growth occured and NbC dispeled. By post weld sensitization this heat affected zone is sensitized. In intergranular corrosion test, its region is selectively attacked as already shown in Fig2.

To prevent this selective attack, the soluble C in heat affected zone should be reduced to below 0.005%, during on cooling. Cooling rate of weld is comparatively fast

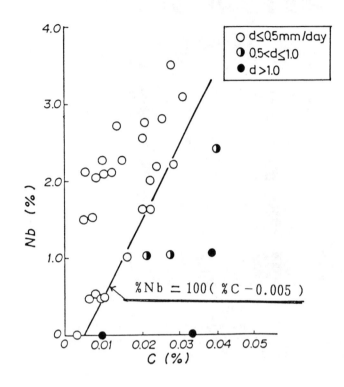

Fig.6 Diagram of relationship between IGC resistance and the content of C and Nb.

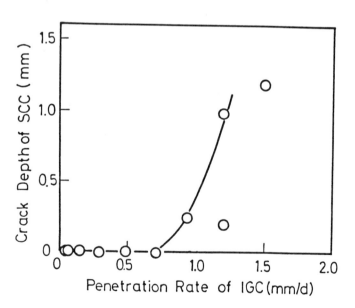

Fig.5 The relationship between high temperature water SCC depth and IGC penetration rate. SCC tests were conducted by using CBB test pieces in 288 °C pure water containing 40ppm disolved oxizen.

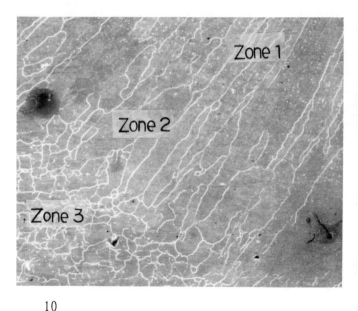

10

Photo.2 Scanning electron micrograph of weld of 0.025% C−1.0%Nb containing alloy. Zone 1 is deposited metal, zone 2 is high temperature heat affected region where the grain growth occured and zone 3 is relatively low temperature heat affected region.

170

so that it is necessary to add a larg amount of Nb indicated in eq.(1) for pretending sensitization.

Nitrogen is also important element on sensitization of Nickel base alloy. Fig.7 shows dependence of N content on intergranular corrosion. Nitrogen addition improves the IGC resistance of alloys having more than 0.005% C.According to the study of Nagano et al.[10] N is detrimental to intergranular corrosion resistance in alloy 600. On the contraly, in this study, N rather surrpress the formation of Chromium depletion due to carbid precipitation at lower temperature.

Photo.3 shows the electron probe micro analysis (E P M A) of precipitation on Nb and N added alloy. This indicates that precipitation consists of Nb(C, N). NbN precipitates at higher temperature than NbC and promotes the carbide precipitation at higher temperature. Consequently,for the same C and Nb content, the higher N content is more effective than the lower N content for the reduction of soluble C in heat affected zone on cooling during welding.

Fig.8 shows the effect of grain size number on intergranular penetration rate. The increase of N content resulted in grain refin-ing and the alloy of smaller grain size has a lower penetration depth at the same C and Nb content. As G. S. Was[11] discussed, a smaller grain material will have a higher grain boundary concentration because in case of shorter diffusion path the depleted element more rapidly approach to equiliblium.

Since Nitride hardly precipitates at 600° C or lower temperature, the harmful effect of N due to the low temperature sensitization is small. Therefor the effect of N in reducing of soluble C and refining of grain becomes significant.

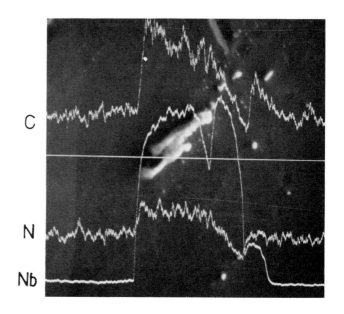

Photo.3 E P M A result of precipitation in N added 0.02C-1.6Nb-0.028N allloys.

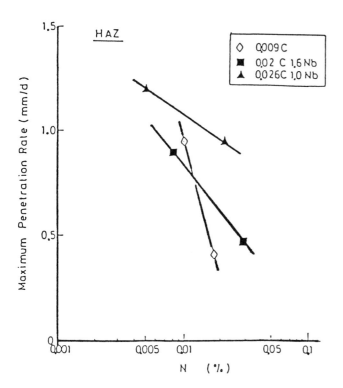

Fig.7 Dependence of N content on penetration rate of intergranular corrosion.

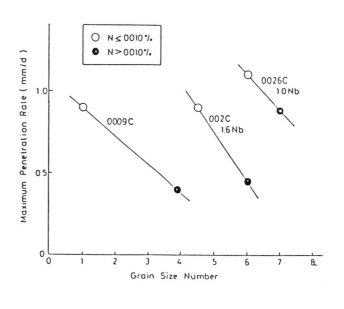

Fig.8 Dependence of grain size on penetration rate of intergranular corrosion.

2.2 Mechacical Properties

The most important one of mechanical properties is yield strength at 0.2% elongation as indicated in Fig.2.

Fig.9 is a diagram showing a relationship between the yield strength and the content of Nb and $(N+C)$. This figure shows that in order to obtain the alloy showing 0.2% yield strength exceeding $25 Kg/mm^2$, it is necessary to add at least the following related amount of Nb.

$$\%Nb \geqq 3.0 - 75 (\%C + \%N) \qquad (2)$$

3 Test Results of Modified Alloy 600 in Full Size

Modified alloy 600 whose composition satisfy both eq.(1) and (2) were produced in commercial process and shaped into sleeve in full size as shown in Fig.10. Their compositions were listed in table 3. Production procedures are also indicated in table 3.

The test results are indicated in table 4. Modified alloy 600 has a excellent IGC resistance, high temperature pure water SCC resistance and mechanical properties.

Thus, it is comfirmed that modified alloy 600 is superior to conventional alloy 600 in many properties.

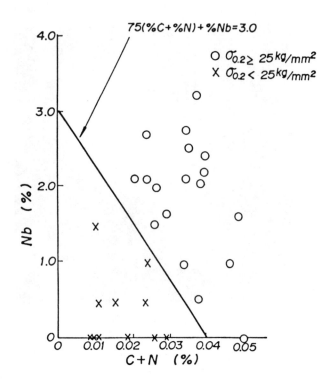

Fig.9 Diagram of relationship between IGC resistance and yield strength at 0.2% elongation.

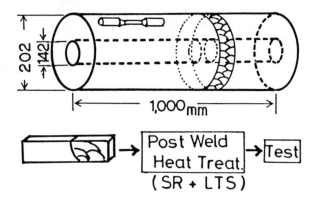

Fig.10 Sample of Modified Alloy 600 in Full Size.

Table 3 Chemical composition and Production Process of Modified Alloy 600

Heat	C	Mn	Ni	Cr	Fe	Nb	N (%)	Production Process
NO.1	0.005	0.2	74.3	16.2	6.8	2.3	0.006	18^{TON} VOD $->6^{TON}$ Ingot $->$H.F.$->$H.T.
NO.2	0.015	0.2	73.2	16.5	7.1	2.7	0.010	1^{TON} VA $->1^{TON}$ Ingot $->$H.F.$->$H.T.

H.F.: Hot Forged   H.T.: Heat Treatment

Table 4 Test Results of Modified Alloy 600

Heat	Mechanical Properties at Room Temperature			Corrosion Resistance	
	0.2% Yield Stress (Kg/mm²)	Tention Strength (Kg/mm²)	Elongation (%)	IGC (Modified FS) Max.Penetration Rate (mm/day)	IGSCC (in Pure Water at 288 C, DO=40ppm)
NO.1	27.6	68.5	57.6	< 0.1	no SCC
NO.2	29.0	72.6	53.6	< 0.1	no SCC

3. Summary

The following conclusions can be drown from the results obtained in the present investigation.

1 In Alloy 600 base composition, Carbon is the most harmful element to IGC resistance and high temperature water SCC reistance against post weld low temperature sensitization.

2 In order to obtain a good corrosion resistance including weld , it is necesary to add larg amounts of Nb as indicated in eq. (1)

3 Nitrogen does not always promote the grain boundary sensitization. Grain refinement by adding N surppress the grain boundary sensitization due to the carbide forming.

4 Modified alloy 600s whose composition were determined from viewpoint of improving I G S C C resistance and mechanical strength were produced in commercial process and tested on full sized forging and they were confirmed that modified alloy 600 is superior to conventional alloy 600 in many properties, particularly post weld low temperature sensitization.

References

1) H.R.Compson et al. Corrosion 24 No.3 (1968)55
2) D.VAN Rooyen Corrosion 31(1975)327
3) M.Kowaka et al. The 26th annual synposium on corrosion and protection (1979)196
4) I.Hamada et al. ibid. (1979)200
5) Y.Sakai et al. ibid. (1979)204
6) M.Kowaka et al.Sumitomo Kinzoku Technical report 34(1982)101
7) K.Yamauchi et al. The JSCE annual meeting in '82 No.A-313 (1982)156
8) M.Akashi et al. The JSCE annual meeting in '78 No.B-26 (1978)177
9) J.Blanchet et al. Firminy Conference prepeint No.G-13(1973)
10) H.Nagano et al. The Spring Japan Metal Institute No.478 (1981)March p268
11) G.S.Was et al. Acta Metall.33 (1985)841

FIELD PROCEDURE DEVELOPMENT
OF HEAT-SINK WELDING
AND CORROSION-RESISTANT CLADDING

Shane J. Findlan
J. A. Jones Applied Research
Charlotte, North Carolina, USA

INTRODUCTION

Remedies of intergranular stress corrosion cracking (IGSCC) prevention in boiling
water reactor (BWR) Type 304 stainless steel
recirculation-loop piping systems are intended to eliminate one or more of the factors
leading to IGSCC. These factors are a sensitized microstructure, tensile stresses, and
an agressive environment. Two remedies,
heat-sink welding (HSW) and corrosion-resistant cladding (CRC) are discussed herein.
Both of these remedies were developed under
Electric Power Research Institute (EPRI)-
sponsored research projects. The work performed at J. A. Jones Applied Research Company was aimed at providing proven equipment
and procedures for the field application of
these remedies. HSW results in a compressive stress state in the weld region if a
sufficient through-wall thermal gradient is
created between the cooling water on the
inside of the pipe and the heat from the
welding torch on the outside of the pipe.
CRC places a deposit of weld metal on the
pipe i.d. adjacent to the girth weld location.
This protects the piping from IGSCC by providing a corrosion-resistant duplex stainless steel boundary layer over the girth weld
heat-affected zone (HAZ) where IGSCC is typically located.

HEAT-SINK WELDING

HSW of stainless steel piping is performed with techniques similar to conventional welding except for the addition of
cooling water on the pipe i.d. The intent
of HSW is to create a compressive residual
stress state by establishing a temperature
gradient across the pipe wall sufficient to
allow yielding of the outside-diameter material heated by the welding torch. On cooling,
the pipe i.d. at the weld zone will be in a
compressive stress state if process conditions
were favorable. An additional benefit is that
HSW increases the cooling rate of the weld
nugget, thus reducing shrinkage and subsequent
distortion. Weld distortion creates bending
movements at the root of the weld and adjacent base metal. These tensile stresses may
offset the compressive stresses produced by
the thermal gradient effects. By reducing
or eliminating the tensile residual stresses,
HSW influences one of the essential conditions for IGSCC. A series of trial welds were
performed to develop HSW parameters and techniques for 324-mm (12-in.) and 711-mm (28-in.)
pipe with automatic gas tungsten arc welding.
After the welds were performed, several test
methods were used to determine the results
including residual stress measurements, degree
of sensitization studies, and metallographic
evaluation.

The test welds have provided data that
show a relationship between pipe wall thickness and the residual stress state for a pipe
butt weld. In thicker wall 711-mm (28-in.)
diam, 30.5-mm (1.2-in.) wall piping, it is
difficult to achieve a compressive stress
state on the i.d. of the weld zone. This problem is largely due to the difficulty in
achieving a large thermal gradient through
the wall of the pipe. Recent tests have shown
that with special welding techniques large-
diameter heavy wall piping may be addressed.
For 324-mm (12-in.)-diam Schedule 80 17.5-mm
wall-thickness stainless steel piping, HSW has
been very effective in producing significant
compressive residual stress on the pipe i.d.
provided heat inputs of 80 kJ/cm (35kJ/in.) or
more are used for HSW fill passes.

A process specification for heat-sink
welding with automatic gas tungsten arc welding systems was prepared based on the trial
welds and information from EPRI development

work. This specification details limitations and requirements for HSW including water flow and coverage, materials, heat input limits, and other process variables.

FIELD CORROSION-RESISTANT CLADDING

Field corrosion-resistant cladding (CRC) is a BWR pipe remedy that provides a protective weld-deposited cladding which covers the interior pipe surface in the area of the pipe weldment HAZ. A new HAZ is created by the cladding, but it is located away from the high tensile stress field of the subsequent pipe weldment. Procedure development for CRC has been concentrated primarily on 324-mm (12-in.) pipe. A Dimetrics Gold Track II gas tungsten arc welding system was selected for use in performing the CRC development and application. To perform field CRC on 324-mm (12-in.)-diam stainless steel pipe and fittings with the Dimetrics Gold Track II, specialized fixturing and mounting equipment was required. By using a restraint ring and a pipe extension, the shrinkage normally associated with this operation was reduced significantly. Process specifications and a welding procedure were completed. This information detailed the requirements for the mounting fixture, restraint ring design, welding parameters, and shrinkage control techniques.

The objective of this program was to provide a field-usable, easy-to-fabricate, rugged mounting fixture with suitable welding parameters that would permit CRC of 324-mm (12-in.)-diam pipe, reducers, and sweepolets utilizing conventional Dimetrics equipment. A mounting fixture was fabricated using a clamping ring which attaches to the forward edge of the pipe to be clad. With minor modifications, a Dimetrics Gold Track II welding head can be installed on the i.d. track and perform the cladding operation. Although CRC would be beneficial to all pipe sizes, physical limitations due to i.d. access restrictions tend to limit its application to piping of 219-mm (8-in.) diam and larger with automatic GTAW equipment.

CONCLUSION

Equipment techniques and procedures for the field application of HSW and CRC to BWR recirculation-loop piping were successfully developed during this project. Both HSW and CRC are remedies that may be practically applied in the field during repair or replacement activities. These remedies each remove one of the factors influencing the initiation and growth of IGSCC. HSW was found to be beneficial in reducing or eliminating i.d. tensile residual stresses in the weld zone for Type 324 (12-in.) Schedule 80 pipe. Although

not effective in eliminating tensile stresses in 711-mm (28-in.) piping without the use of special welding techniques, other benefits including reduced shrinkage were realized. HSW is most effective when measures (joint geometry, number of welding passes, etc.) were taken to reduce the levels of weld shrinkage and distortion that affect the pipe i.d. stress state. CRC may be applied to large-diameter stainless steel piping by using automatic gas tungsten arc welding equipment. It effectively eliminates one of the factors contributing to IGSCC, sensitization, by placing a boundary layer of weld metal between the coolant water and the weld-sensitized heat-affected zone. As a result of this work, these remedies may be effectively applied in the field for protection against IGSCC in BWR recirculation-loop piping.

ENVIRONMENTAL CRACKING OF ALLOY 2205
IN BOILING LiCl

R. B. Griffin, M. N. Srinivasan
L. R. Cornwell
Texas A&M University
Mechanical Engineering Department
College Station, Texas, USA

G. S. Gordon
Cortest Laboratories
Cypress, Texas, USA

ABSTRACT

Sheet tensile specimens of a duplex stainless steel, alloy 2205, were tensile tested in LiCl at 120 C, in a constant load test apparatus under controlled potential conditions. The polarization curve and corrosion rate were obtained at 120 C. During anodic polarization the specimens failed in direct proportion to the anodic current. Under cathodic polarization no failure occurred for the duration of the tests. The failure mode is discussed, and metallographic evidence is presented to support the discussion.

INTRODUCTION

Duplex stainless steels appear to be an attractive alternative to either austenitic or ferritic grades for certain applications. In particular, the duplex grades appear to have better resistance to chloride stress corrosion cracking than do the austenitic grades (1). It is thought that the ferrite helps to act as a crack arrestor in the chloride environment. Additionally, the ferrite is anodic with respect to the austenite, and this provides a certain amount of electrochemical protection (2,3).

The duplex stainless steel used in this study was alloy 2205 (The material was donated by Eastern Stainless Steel, Baltimore, MD.) Its composition is shown in Table 1. The Schaeffler diagram was used to estimate the amount of ferrite and austenite present. The nickel equivalent was 26.6 and the chromium equivalent was 10.9 which corresponds to approximately 50% austenite and 50% ferrite.

This study will report on the environmental cracking behavior of sheet tensile specimens of 2205 in a LiCl solution maintained at 120 C. The polarization curves were obtained, the corrosion current was determined by linear polarization resistance, and the time to failure was measured for specimens tested under anodic polarization.

EXPERIMENTAL PROCEDURE
SPECIMENS - All of the specimens were sheared from a 0.043 inch thick sheet of alloy 2205. Figure 1 shows the dimensions of the test specimen which had their long direction in the longitudinal direction of the original sheet. The sheared specimens were stacked together in a fixture and machined to size.
ENVIRONMENTAL CRACKING TEST - The constant load frame was modelled after one used by Kwon(4), Troiano, and Hehemann(5). A schematic diagram of the apparatus is shown in Figure 2.

Samples were tested in two solution annealed conditions: as-received and further annealed. The tensile properties for the two conditions are listed in Table 2. All specimens were lightly abraded and then coated with Stoner's Mudge (a vinyl acetate based stop-off lacquer) except for the gage length.

Each of the specimens was tested in a 29.4 wt% LiCl solution maintained at 120 C. As Figure 2 shows, the cell used two graphite counter electrodes, and a SCE reference electrode connected to the LiCl solution by a salt bridge with a Luggin capillary that was placed close to the exposed metal surface in the gage length of the specimen. Upon reaching 120 C, the solution was held at temperature for 1/2 hour before the tests were begun. The

Table 1. Composition of alloy 2205 used in this study.

C	N	Cr	Ni	Mo
0.016	0.152	22.57	5.57	3.35

Mn	Si	Cu	Co	P	S
0.56	0.46	0.18	0.10	0.017	0.011

Table 2. Tensile properties for as-received and annealed (1066C(1950°F) for 45 minutes and water quenched) alloy 2205.

	As-received	Annealed
Yield Stress (0.2%)	592 MPa (85,850 psi)	523 MPa (75,850 psi)
Tensile Strength	802 MPa (116,341 psi)	690 MPa (100,000 psi)
Percent Elongation	39.6	48.7
Percent of Yield Stress Engineering Stress Used During Test	100 592 MPa (85,900 psi)	87 439 MPa (66,600 psi)

specimens required about 1 hour to reach temperature, and as Figure 3 shows the potential became more active as it decreased from −170 mV to E_{corr} after 90 minutes. Except for the tests conducted at the freely corroding potential, all other tests were conducted at constant potential. A Model 363, Princeton Applied Research potentiostat maintained the potential. The extension of the specimen was measured by an LVDT with the output recorded on a strip chart recorder. For the controlled potential tests, the potentiostat was turned on and then the load was applied as quickly as possible.

Potentiostatic polarization tests were also performed in the same solution, at the same temperature, and with the same test setup. For this case, the entire specimen was coated with a stop-off lacquer except for a 1 cm square area just below the gage length. The polarization curves were determined without the specimen being loaded. For the as-received specimen, the corrosion rate was determined by linear polarization resistance.

RESULTS

The polarization curve for the alloy is shown in Figure 4. The freely corroding potential is approximately −330 mV (SCE). The linear polarization resistance plot for the as-received material is shown in Figure 5. The polarization resistance, R_p, is 0.57 mV/μA. There was no correction made for any IR drop in the cell.

The elongation versus time curves for the two materials are shown in Figure 6A and B. The potentials are all anodic with respect to E_{corr}, except for the specimen tested at E_{corr} (Figure 6A).

There is a significant change in the current as the sample is taken to failure. Figure 7 shows how the current and elongation vary for a specimen held at −325 mV. Initially the current was about 22 mA; it then decreased to about 5 mA; and finally increased to greater than 30 mA as the specimen began to fail.

DISCUSSION

The freely corroding potential for alloy 2205 in boiling LiCl is −330 mV. When the sample is loaded, the potential becomes much lower as Figure 8 illustrates. The potential dropped instantaneously to less than −410 mV. This drop in potential as the load is applied is associated with the cracking of the protective layer and then the rise is associated with the repassivation process. The potential recovers to about −325 mV within 5 minutes and recovers to E_{corr} in approximately 25 minutes.

The results of controlled potential tests for the specimens under load are shown in Figure 6A for the annealed specimens and in 6B for the as-received material. Both Figures clearly show that the more anodic the potential, the shorter the time to failure. Contrast the above with a loaded specimen at the freely corroding potential, where failure did not occur in a loaded specimen in over 2000 minutes at E_{corr}. However, on examining the specimen, there were several corrosion pits on the sides as well as the flat surface. It is our opinion that if the sample was held for a long enough time it would have failed eventually.

The failure mode for all of these anodically polarized samples appears to be the following. First, a pit or trench forms(4). This occurs uniformly over the gage length of the specimen. Eventually the trenches grow to a point where they could be designated as cracks. Second, aggressive local attack occurs at these cracks. This is shown in Figure 9 where the localized attack is clearly shown. Third, final failure occurs when the cross sectional area is reduced to the point that ductile overload occurs. Figure 10 shows the regions of ductile failure as evidenced by the dimples. The shorter time to failure for the more anodic potentials corresponds to the increased anodic current density which increases the rate of local attack.

Examination of the micrograph, Figure 11, shows the local attack. It appears to be integranular and tending to follow the austenite/ferrite boundaries. There is evidence for secondary cracking as seen in Figure 11. Qualitatively there appear to be more cracks, over the gage length, at higher potentials than at potentials closer to E_{corr}. The specimen annealed at −1066 C was tested at a cathodic potential of −600 mV without failure for 24 hours, while the as-received material was tested at −1200 mV for 36 hours without failure. The intent was to induce hydrogen embrittlement, although no recombination poison was added.

The results of these tests suggest that E (critical cracking) is more noble than E_{corr}

A comparison of the potential as a function of the time to failure is shown in Figure 12. The squares represent the annealed specimens, or lower strength, and the circles represent the as-received material, or higher strength material. In examining this figure, it is tempting to say there was a beneficial effect of the 1066 C anneal. However, the stress level was different: 100% of 0.2% yield stress for the as-received compared to 87% for the annealed specimens. This difference may have accounted for the shortening of the time to failure. At E_{corr} the time to failure increases to much longer times. In fact, for these tests no failure occurred for specimens tested at E_{corr}. (We have chosen to stay with E_{corr} instead of using E open circuit for the freely corroding potential under load.) It remains to be seen if such behavior could be observed in a stronger LiCl solution. If this is the case, then 2205 alloy would be quite close to E-Brite in performance in chloride environments(5).

CONCLUSIONS

Constant load stress corrossion cracking studies conducted on alloy 2205 in boiling 27.5% lithium chloride solution indicates the following:

1. The susceptibility to failure, as measured by the time to failure, is greater the higher the anodic potential.

2. No failures occurred at cathodic potentials, indicating that E (critical cracking) is equal to or more noble to E_{corr} in this environment. The alloy therefore, appears to be safe for applications at E_{corr} potential, the latter having a value of about −330 mV.

3. Anodic dissolution seems to control failure at highly anodic potentials, the resulting fracture being due to ductile overload.

4. As the potentials approach E_{corr}, the crack front formed by multiple intergranular paths seems to control the fracture, which involves cleavage in the final stages.

5. The localized attack at these potentials seems to follow the austenite/ferrite boundaries.

The polarization curves were obtained for alloy 2205 in boiling LiCl. Both i_{corr} and E_{corr} were determined. During the potentiostatic polarization sweep, upon going from the cathodic direction to the anodic direction, the alloy repassivated at potentials much less than E_{corr}.

REFERENCES

1. J. E. Truman, _Int. Metals Reviews,_ No. 6, p. 301 (1981).

2. T. Suzuki, H. Hasegawa, and M. Watonabe, _Nippor Kinzoku Gakaishi,_ Vol. 32, p. 1171, (1968).

3. S. Shimodaira, M. Takano, T. Takizawa, and H. Kamide, Stress Corrosion Cracking and Hydrogen Embrittlement of Iron Based Alloys, NACE-5, p. 1003 (1977).

4. H. S. Kwon, Ph.D. Thesis, Case Western Reserve University, (1982).

5. A. R. Troiano and R. F. Hehemann, "Stress Corrosion Cracking of Ferrite and Austenitic Stainless Steels," _Hydrogen Embrittlement and Stress Corrosion Cracking,_ ed. R. Gibala and R. F. Hehemann, ASM, 1984.

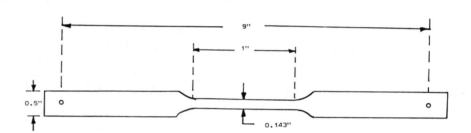

Figure 1 — Sheet tensile specimen used in the constant load test.

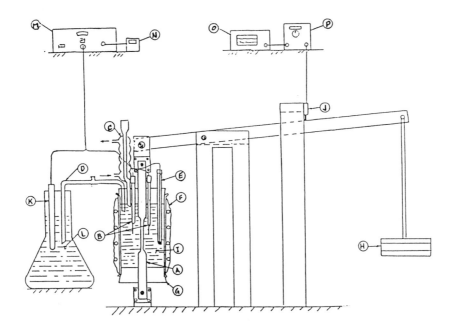

LEGEND

A - SPECIMEN (AISI 2205)
B - GRAPHITE COUNTER ELECTRODE
C - REFLUX CONDENSOR
D - AGAR-AGAR SALT BRIDGE
E - THERMOCOUPLE
F - HEATING COIL (WITH THERMOSTAT)
G - STOPPER
H - WEIGHTS

I - 29.5% LITHIUM CHLORIDE SOLUTION
J - LVDT (G. L. COLLINS #LMT-25104)
K - SATURATED CALOMEL ELECTRODE
L - POTASSIUM CHLORIDE SOLUTION
M - POTENTIOSTAT (EG&G PRINCETON #363)
N - VOLTMETER (HICKOK #LX-303)
O - CHART RECORDER (GOULD #220)
P - DEMODULATOR (VALIDYNE #CD12)

Figure 2 - Constant load environmental cracking apparatus.

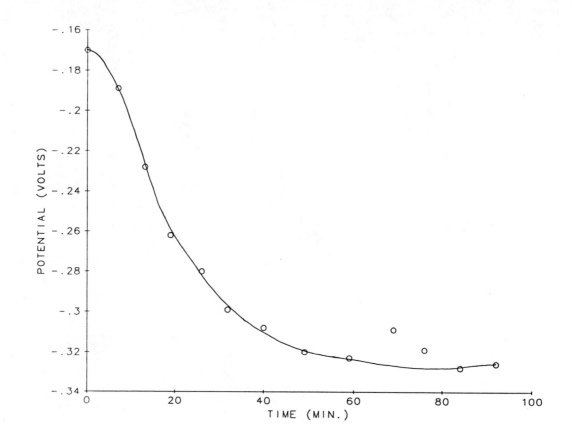

POTENTIAL VS. TIME

Figure 3 - Variation in potential as the
specimen is heated up to 120 C

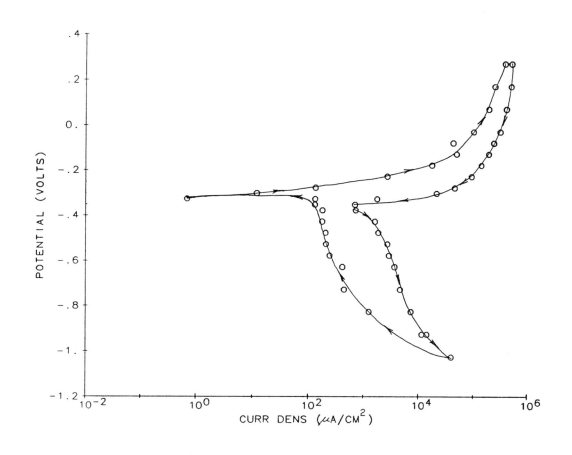

POTENTIAL VS. CURRENT DENSITY

Figure 4 - The polarization curve for alloy
2205 (as-received).

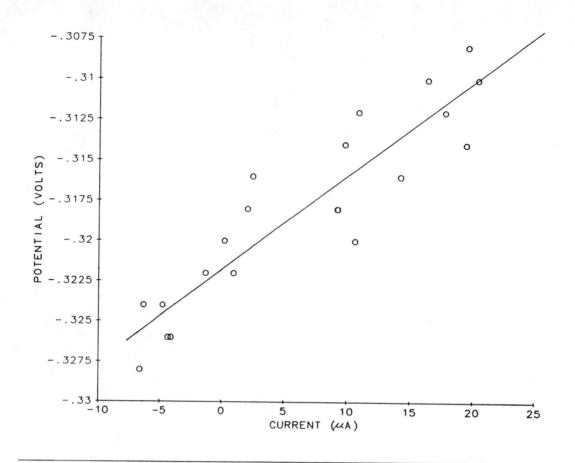

POTENTIAL VS. CURRENT

Figure 5 - Linear polarization resistance
curve for alloy 2205 in the as-received
condition.

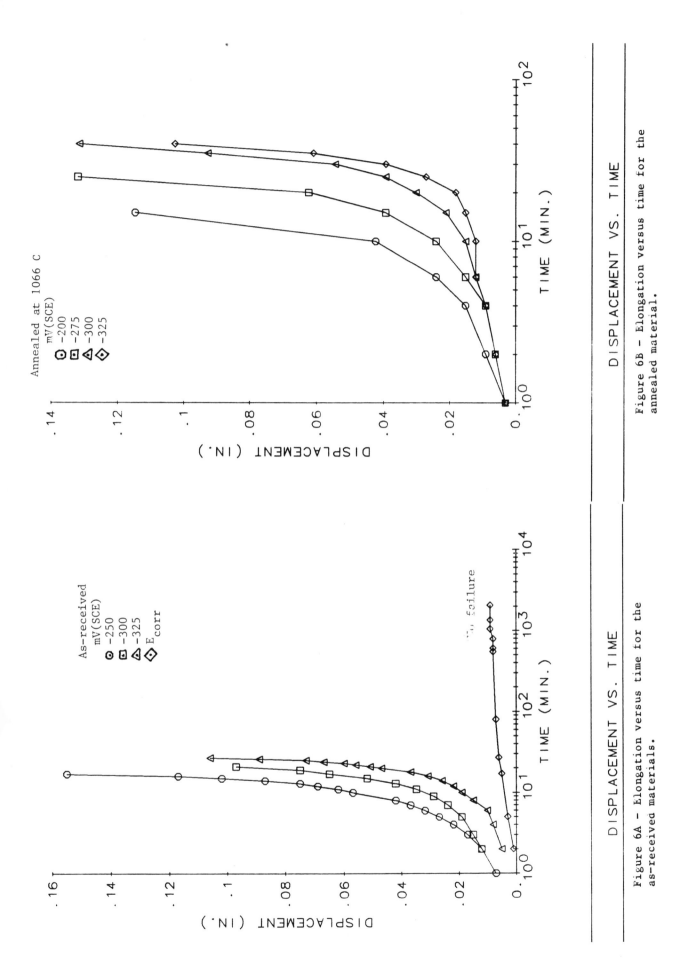

DISPLACEMENT VS. TIME

Figure 6A - Elongation versus time for the as-received materials.

DISPLACEMENT VS. TIME

Figure 6B - Elongation versus time for the annealed material.

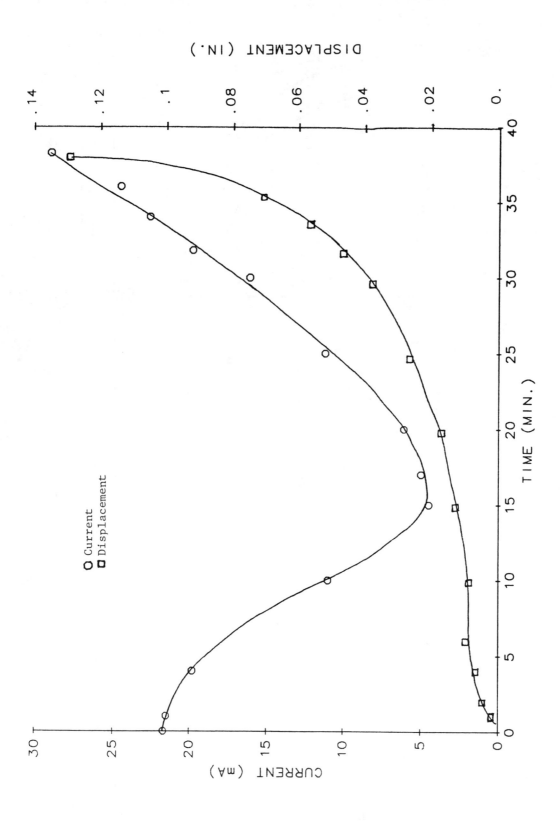

CURRENT VS. TIME

Figure 7 - The current and elongation for a
specimen taken to failure at -325 mV.

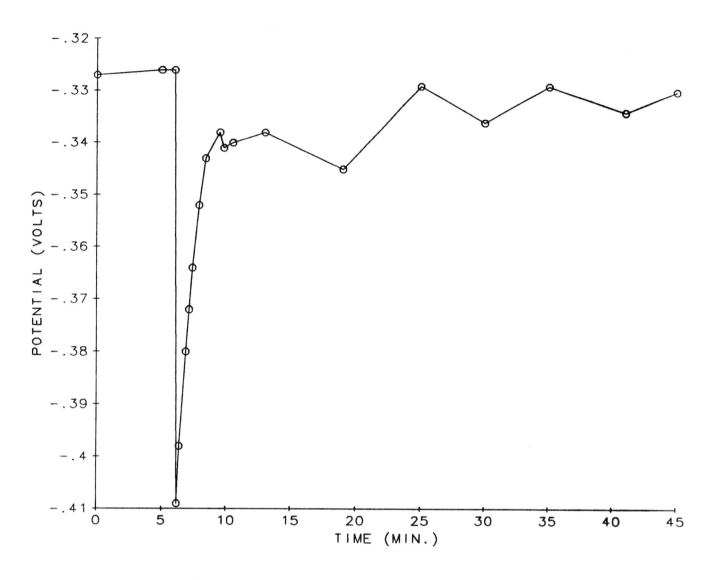

POTENTIAL VS. TIME

Figure 8 – The variation of the potential
as a function of time after the load has
been applied.

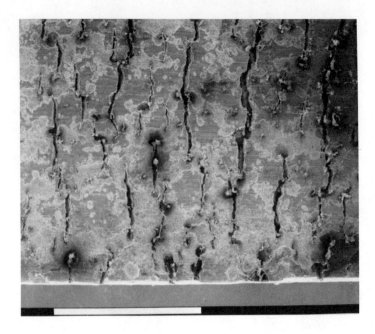

1mm

Figure 9 - Scanning electron micrograph of
the cracks produced on the surface of alloy
2205 tested in boiling LiCl at -325 mV.

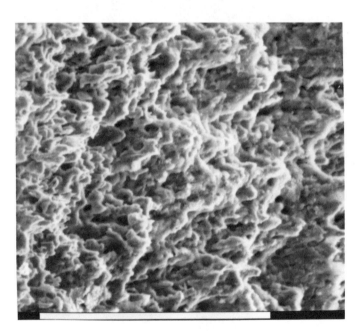

0.1mm

Figure 10 - Regions of ductile rupture for
the same sample as shown in Figure 9.

1μm

Figure 11 - Annealed specimen showing
regions of local attack. Note secondary
cracks. Specimen tested at -325 mV.

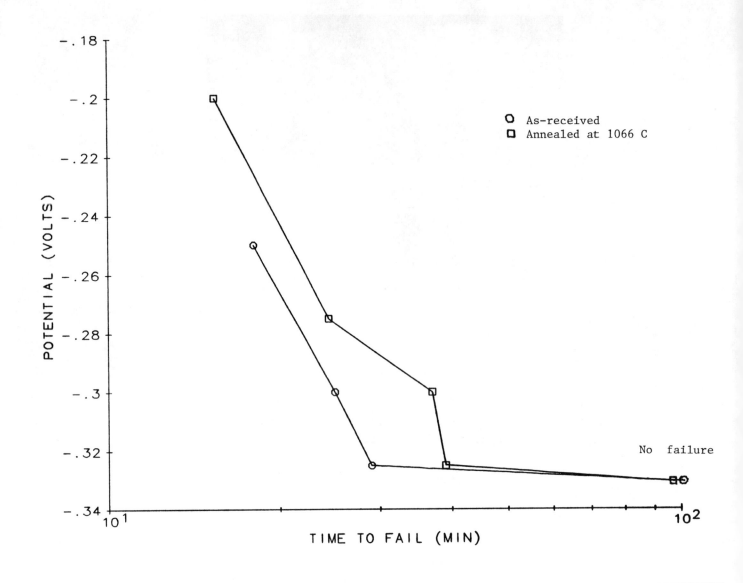

POTENTIAL VS. TIME TO FAILURE

Figure 12 - A comparison of the times to
failure for the annealed and the
as-received material.

CRITICAL ASSESSMENT OF PRECRACKED SPECIMEN CONFIGURATION AND EXPERIMENTAL TEST VARIABLES FOR STRESS CORROSION TESTING OF 7075-T6 ALUMINUM ALLOY PLATE

Marcia S. Domack
NASA Langley Research Center
Hampton, Virginia, USA

Abstract

A research program was conducted to critically assess the effects of precracked specimen configuration, stress intensity solutions, compliance relationships and other experimental test variables for stress corrosion testing of 7075-T6 aluminum alloy plate. Modified compact and double beam wedge-loaded specimens were tested and analyzed to determine the threshold stress intensity factor and stress corrosion crack growth rate. Stress intensity solutions and experimentally determined compliance relationships were developed and compared with other solutions available in the literature. Crack growth data suggests that more effective crack length measurement techniques are necessary to better characterize stress corrosion crack growth. Final load determined by specimen reloading and by compliance did not correlate well, and was considered a major source of interlaboratory variability. Test duration must be determined systematically, accounting for crack length measurement resolution, time for crack arrest, and experimental interferences. This work was conducted as part of a round robin program sponsored by ASTM committees G1.06 and E24.04 to develop a standard test method for stress corrosion testing using precracked specimens.

NOMENCLATURE

a	Specimen crack length measured from load line
ASTM	American Society for Testing and Materials
B	Specimen thickness measured on smooth faces
B_n	Specimen thickness measured at base of machined notches
COD	Clip-on crack mouth opening displacement gage
da/dt	Crack growth rate
$DB(W_b)$	Abbreviation for double beam bolt-loaded specimen
E	Young's modulus
F	Denotes fatigue precracked test condition
G	Denotes face-grooved specimen
H	One-half the $DB(W_b)$ specimen height
K_I	Stress intensity factor for Mode I failure
K_{Ic}	Plane strain fracture toughness
K_H, K_L	High and low target starting stress intensity levels for environmental exposure, respectively
K_{Iscc}	Threshold stress intensity level for stress corrosion cracking
M	Denotes mechanical precracked test condition
$MC(W_b)$	Abbreviation for modified compact bolt-loaded specimen
P	Specimen bolt load
S	Denotes smooth-face specimen
$2V_{11}$	Total crack opening displacement at the load line
$2V_0$	Total crack mouth opening displacement
W	$MC(W_b)$ specimen dimension from the load line to the back edge of the specimen

STRESS CORROSION CRACKING (SCC) of aluminum alloys has long been recognized as a significant contributor to environmentally assisted failures in structural applications. The ranking of SCC susceptibility in structural aluminum alloys as estimated from currently standardized test procedures is at best only qualitative.[1] These standards have been developed for smooth specimen techniques and provide only time to failure data. More quantitative evaluation of material susceptibility can be made by using fracture mechanics principles and precracked specimens. There has been much characterization done with various precracked specimen configurations but there is little agreement among researchers concerning test controls.[2] As such, a widespread, accepted standard for stress corrosion evaluation using precracked specimens has not been developed.

The development of a standard test procedure requires thorough evaluation of potential specimen configurations and analyses, as well as an evaluation of the possible effects of each test variable. A research program was conducted to evaluate the effect of several test variables on measured values of threshold stress intensity factor for stress corrosion cracking, K_{Iscc}, and crack velocity, da/dt, as a function of stress intensity level, K_I. This program was conducted as part of a round robin study sponsored by ASTM committees G1.06 on Corrosion of Metals and E24.04 on Fracture Testing. Precracked specimen configurations were selected for the advantage of analysis with linear elastic fracture mechanics. A constant deflection, K-decreasing test was conducted and the effects of specimen configuration, starting stress intensity level, method of precracking, and presence of face grooves on measured values of K_{Iscc} and da/dt were evaluated. The results of a single laboratory are discussed in detail and a summary of data from eight of the thirteen round robin laboratories is presented. Miniature load cells were employed on selected specimens to monitor load decay behavior during stress corrosion crack growth. Also included is an evaluation of available stress intensity solutions for the specimen configurations, and compliance relationships for determination of deflections for each configuration.

SPECIMENS AND PROCEDURES

EXPERIMENTAL PROGRAM - The research program was based on a factorial design experiment with four variables; specimen configuration, method of precracking, surface grooving, and starting stress intensity; each having two levels. The factorial test is statistically designed to avoid replication of test conditions and allows evaluation of test variables with a minimum number of specimens. Test specimens were machined from a 6.35 cm thick plate of aluminum alloy 7075-T651 such that loading would be in the short transverse direction with crack growth in the longitudinal direction. The plate had a yield strength of 433 MPa and plane strain fracture toughness, K_{Ic}, of 22.53 MPa-m$^{1/2}$. The modified compact (MC(W_b)) and double beam wedge-loaded (DB(W_b)) specimens were machined in two configurations, one with smooth faces (fig 1) and the other having five percent thickness v-shaped face grooves.

SPECIMEN PRECRACKING PROCEDURES - Specimens were precracked by either fatigue cycling at low stress intensity levels or by mechanical overload (pop-in) to initiate a crack 0.25 to 0.38 cm beyond the chevron notch and then bolt-loaded to one of two target starting stress intensity levels, K_H=19.78 or K_L=13.19 MPa-m$^{1/2}$. Stress intensity levels during fatigue loading

were limited to two-thirds of the starting stress intensity for environmental exposure. The stress intensity solution for the DB(W_b) specimen is based on an exact solution by Fichter[3] and is given by:

$$K_I = \frac{Pa}{B(H)^{3/2}} (12)^{1/2} \left(1 + 0.673 \frac{H}{a}\right) \quad (1)$$

and is a function of crack length, a, measured from the bolt load line, the specimen bolt load, P, which can be determined by compliance from the load line displacement, $2V_{11}$, and specimen geometry. The solution for the MC(W_b) specimen was developed by Newman[4] from boundary collocation and is given by:

$$K_I = \frac{P}{B(W)^{1/2}} \frac{\left(2 + \frac{a}{W}\right)}{\left(1 - \frac{a}{W}\right)^{3/2}} C_3\left(\frac{a}{W}\right) \quad (2)$$

$$C_3\left(\frac{a}{W}\right) = 1.308 + 5.278 \left(\frac{a}{W}\right) - 19.67 \left(\frac{a}{W}\right)^2 +$$

$$24.57 \left(\frac{a}{W}\right)^3 - 10.27 \left(\frac{a}{W}\right)^4 \quad (3)$$

and is again a function of crack length, bolt load, and specimen geometry. Newman's MC(W_b) solution accounts for crack line loading and load concentration effects, and both solutions allow for proportional scale up of specimen dimensions.

Computer control was used to monitor the fatigue precrack length during cycling by using experimentally determined compliance relationships of the form a/W=fn(EB2V_0/P) for the MC(W_b) specimen and a/H=fn(EB2V_0/P) for the DB(W_b) specimen. The quantity B is replaced by $(BB_n)^{1/2}$ for the face grooved specimen configuration. During fatigue precracking a clip-on crack mouth opening displacement (COD) gage was used to measure the displacement, $2V_0$. Visual measurements made to verify the final precrack length agreed with values estimated by the current experimentally determined compliance solution to within two percent.

Mechanical precracking resulted in crack tip stress intensity levels greater than the starting target values for environmental exposure. Mechanically precracked specimens targeted for exposure at K_H were left at these higher K values. Mechanically precracked specimens targeted for testing at the low starting stress intensity were partially unloaded to the K_L value. This unloading resulted in a compressed region at the crack tip. As SCC requires the application of a sustained tensile stress, this condition was included to evaluate whether any crack incubation phenomena could generate forces sufficient to initiate a stress corrosion crack regardless of the compressed zone.

192

ENVIRONMENTAL EXPOSURE PROCEDURE - Specimens were continuously immersed in a 3.5 percent NaCl solution in controlled temperature and humidity. The solution was acetate buffered to pH=4 and dichromate inhibited to reduce the amount of general corrosion product accumulation during specimen exposure. Corrosion product wedging can generate self-loading forces sufficient to exceed the intended bolt load.[5] Specimens were removed from solution at periodic intervals during the 2000 hour test and crack lengths measured visually on each surface with a travelling stage microscope. This crack length versus time data was used to determine crack growth rate, da/dt. A stress intensity value was calculated for each measured crack length, and the combined data plotted to determine the SCC crack growth rate as a function of calculated stress intensity.

At test termination crack mouth displacement, $2V_0$, was measured with a COD gage as the loading bolts were removed. Final specimen load, P_f, was indirectly measured by reloading to the measured displacement in a tensile test machine. Specimens were fatigue cycled to mark the end of the stress corrosion crack, and then loaded to failure in tension to expose the crack faces. An accurate measurement of final crack length, a_f, was made by averaging five measurements across the specimen width. A final stress intensity value, K_{If}, was calculated based on P_f and a_f.

RESULTS AND DISCUSSION

CORRELATION OF STRESS INTENSITY AND COMPLIANCE SOLUTIONS - The stress intensity solutions developed by Newman and Fichter for the $MC(W_b)$ and $DB(W_b)$ specimen configurations, respectively, were generally found to be in good agreement with other published solutions. Fichter presents a correlation of his solution with others available in the literature.[3] These solutions were determined for other double beam specimen configurations subjected to splitting forces, and while each of these solutions required prior assumptions about the functional relationship between K and a/H, Fichter's solution makes no prior assumptions. The solutions generally agreed to within two percent, but Fichter presents his solution to be the most accurate expression of the functional relationship between K and a/H.[3] Newman's solution for the $MC(W_b)$ specimen agreed within one percent with solutions developed by LeFort and Mowbray[6] by finite element analysis and by Novak and Rolfe[7] by a fracture mechanics based three dimensional analysis.

Compliance values, $2V_0/P$, were experimentally determined for both smooth and face-grooved specimens with crack length ranges 2.0<a/H<8.0 and 0.3<a/W<0.8 for the $DB(W_b)$ and $MC(W_b)$ specimens respectively. The resulting compliance solutions, of the form $EB2V_0/P=fn(a/W)$ and $EB2V_0/P=fn(a/H)$ were compared with solutions developed by Newman through boundary collocation for the $MC(W_b)$ specimen previously[4] and for the $DB(W_b)$ specimen for this study. The $MC(W_b)$ solution was also compared with other solutions found in the literature. Compliance data for the $DB(W_b)$ specimen configuration is summarized in table 1. The solutions developed from experimental data differed from Newman's approximation by less than four percent for the smooth face specimen and less than seven percent for the face grooved specimen for 2.0<a/H<8.0. The experimentally determined solutions for the smooth and face grooved specimens agreed to within three percent for all a/H values. No other compliance solutions were available for the $DB(W_b)$ specimen.

The compliance calibration for crack mouth opening displacement for the $MC(W_b)$ specimen are summarized in table 2 for the range of crack lengths given by 0.3<a/W<0.8. The solutions generally showed better agreement at higher a/W values. The current solution agrees with Newman's boundary collocation solution to within only ten percent at a/W=0.4 but improves to within three percent at a/W=0.65 for both the smooth and face grooved specimens. Experimental compliance solutions have also been developed by Novak and Rolfe[7] for smooth and face grooved specimens and by LeFort and Mowbray[6] for the face grooved specimen. The current solutions agree with the Novak and Rolfe results to within only ten percent at a/W=0.4 with agreement improving to within five percent at a/W=0.8. The current solution for the face grooved specimen agrees with the result of LeFort and Mowbray to within four percent for all values of a/W. These differences in experimental results may be due to the method used to simulate an advancing crack. In the current analysis a 1.52 mm slot was incrementally advanced by electric discharge machining, Novak and Rolfe incrementally advanced a 60-micron sawcut, while LeFort and Mowbray advanced a crack by fatigue cycling. LeFort and Mowbray[6] also report a solution developed by finite element analysis which agreed with the current experimental solution to within three percent at a/W>0.4.

Newman's boundary collocation analysis includes an expression which correlates specimen displacements at the crack mouth, V_0, and the load line, V_{11}, for both the $MC(W_b)$ and $DB(W_b)$ specimen configurations. Load line displacements calculated with this relationship at the start of environmental exposure are the basis for subsequent stress intensity calculations using rigid bolt analysis. Values of V_0 and V_{11} were measured during the current experimental compliance evaluation for both specimen configurations. The experimental measurements agreed with Newman's analysis to within two percent for the $DB(W_b)$ specimen configuration but only within ten percent for the $MC(W_b)$ specimen configuration. A previous correlation suggested by Novak and Rolfe[7] for the $MC(W_b)$ specimen, based on proportional deflection of stiff beams about the crack tip as a pivot point, did not agree well with the experimentally generated data.

EVALUATION OF SINGLE LABORATORY RESULTS - Crack length versus time data is shown in figures 2 and 3 for each factorial test condition for both the MC(W_b) and DB(W_b) specimens, respectively. The data does not exhibit the standard smooth, decreasing-slope curves usually shown schematically but rather contains several inflection points. Regression analysis was used to determine the slope of the curve, da/dt, at each measured value of crack length, a. The perturbations in the curve may be due in part to difficulties associated with visual crack length measurements on the specimen surfaces, especially on the face grooved specimens. Ultrasonic measuring devices have been used successfully to measure an effective crack length based on the average of internal measurements.[8] The curves in figures 2 and 3 exhibit positive slopes near the end of the test, indicating crack arrest was not achieved, and consequently the final stress intensity values calculated were not threshold values. Examination of these curves reveals both the crack growth rate and the amount of crack growth at a given time is greater for the high stress intensity specimens for both specimen configurations. In addition, the final da/dt was greater for the specimens exposed at high starting stress intensity levels for both specimen configurations, indicating the high stress intensity specimens are even farther from crack arrest at 2000 hours than the low starting stress intensity specimens. Crack arrest was defined in the ASTM procedure as a crack growth rate lower than 10^{-6} cm/hr, but final crack growth rates were typically on the order of 10^{-4} cm/hr.

Final load determined by specimen reloading was found to be significantly different from that predicted by rigid bolt analysis. Miniature load cells were placed between the faces of the loading bolts in several additional test specimens to investigate the source of this difference. These specimens were loaded to very high initial stress intensity values, and the load decay was monitored during stress corrosion crack growth. A plot of crack length and load cell output versus time is shown in figure 4a. The changes in load cell output parallel crack growth events quite well, indicating the perturbations in the typical crack length versus time curve may be real and should not be attributed only to errors in crack length measurement. A slight increase in load cell output was observed for all specimens at the beginning of environmental exposure, as shown in figure 4b. This region of the curve corresponds to the incubation period prior to stress corrosion cracking. The load begins to decrease as soon as crack growth begins. In some cases the load calculated by rigid bolt analysis agreed within five percent of the load cell output but the

difference in other cases was quite large, nearly a factor of two. This may be due to large regions of plasticity introduced at the crack tip by inadvertent mechanical overloads which occurred during bolt loading in these tests. The large amount of plastic deformation associated with the mechanical overload introduces considerable uncertainty in the approximation of loading conditions by linear elastic fracture mechanics based stress intensity solutions. The final bolt load determined by specimen reloading, P_f, was always substantially less than the final load indicated by the load cell, which would be expected if the crack is not fully closing on specimen unloading. For a particular specimen in which the load measured by the load cell and the load calculated by rigid bolt analysis agreed within two percent, the final load determined by specimen reloading differed by nearly 40 percent, suggesting that the reloading technique using crack opening displacement gages for determination of final load needs additional analysis and refinement.

Records of specimen reloading to failure are shown for several MC(W_b) and DB(W_b) specimens in figures 5a and 5b, respectively. The amount of crack growth, Δa, which occurred for each test case during stress corrosion cracking is shown beside each curve, and increases from left to right. Each curve exhibits a nonlinear region at small values of load and becomes linear well before specimen failure. This region of nonlinearity may be attributed to crack closure caused by corrosion product wedging or by uneven fracture surfaces due to regions of delamination. Both specimen configurations exhibit more nonlinear reloading behavior for cases of greater stress corrosion crack growth. This may be due to the larger area over which corrosion product wedging and delaminations may be effective.

SUMMARY OF ROUND ROBIN RESULTS - Final stress intensity, K_{If}, values shown in figure 6 are averaged for each test condition for data received from eight of the thirteen laboratories participating in the round robin program. K_{If} values were calculated based on final load determined by specimen reloading, P_f, and by rigid bolt analysis, RBA, based on load line displacement. Measured values of P_f are sensitive to artifacts such as corrosion product wedging and crack surface morphology. The assumption that the loading bolts are rigid in RBA is not exact, but the bolts selected were sufficiently stiffer than the specimen and the error associated with RBA should be much smaller than that associated with the reloading technique.

The effect of each test variable on final stress intensity values can be evaluated by examining figure 6. The effect of specimen configuration is observed by making an overall comparison for each group of bars. K_{If} values calculated by rigid bolt analysis for the

DB(W_b) specimen were generally less than for the MC(W_b) specimen for each test condition. The difference in K values at test termination may be attributed to the fact that total crack arrest was not achieved and that the rate of stress intensity decay varies with crack growth for the two specimen configurations. Selective comparisons in the figure allow evaluation of the remaining test variables. Comparison of appropriate test conditions indicates that K_{If} values were consistently greater for specimens exposed at K_H than for specimens exposed at K_L. The effect observed is consistent for both specimen configurations and is due to the fact that crack arrest was not achieved in these tests. The method of precracking had no significant effect on K_{If} in the MC(W_b) specimen configuration, but values determined from mechanically precracked DB(W_b) specimens were significantly less than for the fatigue precracked condition. Further comparisons show that, as expected there is no significant, reproducible difference in K_{If} values due to surface grooving.

Statistical variation in the final stress intensity values calculated by rigid bolt analysis are averaged for each test condition for data from eight laboratories participating in the round robin program and shown in table 3. Comparing specimen configurations, the mean values were slightly lower for the fatigue precracked DB(W_b) specimens, but significantly lower for the mechanically precracked specimens. Standard deviation was greater in all cases but one for the DB(W_b) than MC(W_b) specimens and were generally greater for the mechanical precrack and high starting stress intensity test conditions. This is a reflection of the error introduced by large plastic deformation caused by mechanical precracking, visual measurement difficulties, and the fact that crack arrest was not achieved. Statistical variations in final K values for each laboratory are shown in table 4. Since starting stress intensity level and method of analysis were identified as significant sources of variability, the data presented is averaged for all other test variables. Final K values determined by both rigid bolt analysis and based on final load determined by specimen reloading are averaged for all low K and high K specimens. Data from some of the participating laboratories have noticeably greater standard deviations than others, and reflect that some laboratories had more difficulty following the complex test procedure. The mean values were generally greater for specimens exposed at high K than those started at low K, but again this is due to crack arrest not occurring. The mean values were generally, and in some cases significantly, greater for data based on P_f than for that calculated by rigid bolt analysis. The range of standard deviations was much greater for the data based on P_f than on rigid bolt analysis, and the magnitude of these standard deviations

is an indication of the error inherent in the reloading technique.

CONCLUDING REMARKS

A research program was conducted to evaluate stress corrosion test methods using precracked specimen configurations. Stress intensity solutions and compliance relationships were evaluated for both the MC(W_b) and DB(W_b) specimen configurations. The effect of several test variables on stress corrosion data are evaluated and interlaboratory statistical variation presented.

Stress intensity solutions developed by Newman and Fichter for the MC(W_b) and DB(W_b) specimen configurations were found to agree well with previously defined solutions, but these solutions also allow for proportional scale up of specimen dimensions. Additionally, Newman's solution accurately accounts for the loading geometry. Compliance relationships determined experimentally for both specimens were compared with other solutions determined from experimental data and with solutions determined by both boundary collocation and finite element analysis. The solutions were found to be in agreement for the practical ranges of a/W and a/H. The relationships developed by Newman to correlate load line displacement with crack mouth opening displacement were found to be very accurate when compared to experimentally generated values.

Crack growth data from the current study indicate stress corrosion cracks may grow in a non-continuous manner, exhibiting several perturbations in crack length versus time curves. Data from miniature load cells indicated that these changes in crack velocity during stress corrosion cracking may be real and should not be totally dismissed as error in the crack length measurements. The load cell data identified significant differences between final load determined by specimen reloading and final load indicated by the load cells. The final load technique is experimentally difficult due to crack closure effects caused by corrosion product wedging or by mismatch of delaminated crack faces. Closure effects are more significant for larger amounts of stress corrosion crack growth.

Statistical evaluation of data from the round robin results indicated similar trends in test variable effects as predicted from the single laboratory results. Both specimen configurations appear suitable for stress corrosion evaluation for high strength materials, and the difference in final stress intensity values should be attributed to test procedures. The presence of face grooves had no effect on final stress intensity values. Mechanical precracking resulted in greater variability in final K values than did fatigue precracking. This is associated with the large plastic deformations and uneven crack faces which occur during mechanical overload. Starting stress intensity level for environmental exposure had a significant effect on final K values and cannot be selected arbitrarily. The stress intensity effect may be due to crack arrest not occurring in 2000 hours with these specimen configurations. Test duration appears to be a function of the actual crack growth behavior, the desired target arrest rate, and the resolution of the crack length measurement technique and will not be a constant value regardless of test conditions.

The data generated in the interlaboratory program provides information for use in establishing a standard test method for evaluation of stress corrosion susceptibility of structural alloys. Precracked specimen configurations evaluated with fracture mechanics principles provide a method of quantifying SCC behavior.

REFERENCES

1. 1984 Annual Book of ASTM Standards, Vol. 03.02 Metal Corrosion, Erosion, and Wear (1984)

2. B. F. Brown, ed., Stress-Corrosion Cracking in High Strength Steels and in Titanium and Aluminum Alloys, Naval Research Laboratory, Washington, D.C. (1972)

3. Fichter, W. B., "The Stress Intensity Factor for the Double Cantilever Beam", Int. Journal of Fracture, 22, 133-143 (1983)

4. Lisagor, W. B., "Influence of Precracked Specimen Configuration and Starting Stress Intensity on the Stress Corrosion Cracking of 4340 Steel", Environment-Sensitive Fracture: Evaluation and Comparison of Test Methods, ASTM STP 821, 80-97 (1984)

5. Dorward, R. C., Hasse, K. R., and Helfrich, W. J., "Marine Atmosphere Stress Corrosion Tests on Precracked Specimens from High-Strength Aluminum Alloys: Effect of Corrosion Product Wedging", Journal of Testing and Evaluation, JTEVA, Vol. 6, No. 4, 268-275 (1978)

6. LeFort, Paul, and Mowbray, D. F., "Calibration of the Side-Grooved Modified Wedge-Opening-Loaded Specimen", Journal of Testing and Evaluation, JTEVA, Vol. 6, No. 2, 114-119 (1978)

7. Novak, S. R., and Rolfe, S. T., "Modified WOL Specimen for K_{Iscc} Environmental Testing", Journal of Materials, JMLSA, Vol. 4, No. 3, 701-728 (1969)

8. Hasse, K. R., "Precracked Specimen Stress Corrosion Tests of 7075-T651 Plate - ASTM G01.04.06 Subcommittee Round Robin", Kaiser Aluminum and Chemical Corporation, Report CFT RR 84-5 (1984)

Fig. 1 - Precracked specimen configurations.

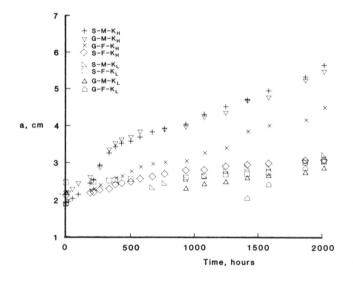

Fig. 2 - Crack growth record for MC(W_b) specimens.

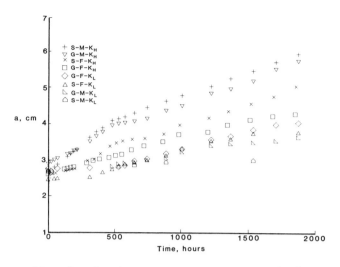

A/H	CURRENT STUDY		BOUNDARY COLLOCATION
---	S_M	G_R	
2.00	224.735A	222.475B	214.021C
3.00	507.719	496.720	516.581
4.00	977.802	952.129	1012.946
5.00	1698.912	1651.735	1751.430
6.00	2745.470	2669.466	2780.131
7.00	4203.077	4090.920	4147.087
8.00	6169.235	6014.197	5900.317

(A) EXPERIMENTAL, S   (B) EXPERIMENTAL, G

(C) BOUNDARY COLLOCATION

Table 1 - Compliance ($EB2V_O/P$) for DB(W_b) specimens.

Fig. 3 - Crack growth record for DB(W_b) specimens.

A/W	CURRENT STUDY		NEWMAN	LEFORT & MOWBRAY		NOVAK & ROLFE	
0.30	37.33A	33.22B	26.98C	30.95D	34.22B	29.20A	29.50B
0.40	47.70	46.84	42.23	46.55	45.37	42.59	44.18
0.50	69.92	69.71	64.76	68.80	67.00	65.03	66.77
0.60	109.34	108.64	102.47	106.54	105.45	100.98	102.79
0.70	187.66	186.28	182.57	186.98	183.54	173.30	177.90
0.80	402.64	390.58	413.87	408.14	397.51	399.71	471.15

(A) EXPERIMENTAL, S (B) EXPERIMENTAL, G (C) BOUNDARY COLLOCATION, (D) FINITE ELEMENT

Table 2 - Compliance ($EB2V_O/P$) for MC(W_b) specimens.

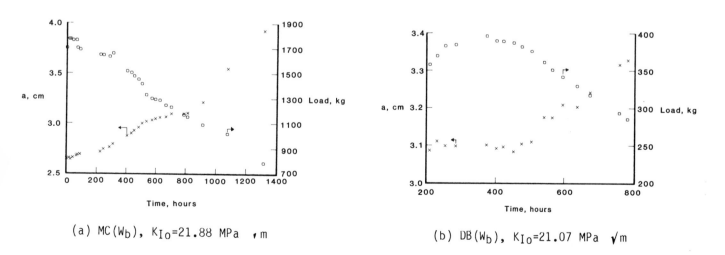

(a) MC(W_b), K_{IO}=21.88 MPa $\sqrt{}$m

(b) DB(W_b), K_{IO}=21.07 MPa $\sqrt{}$m

Fig. 4 - Load decay associated with crack growth for MC(W_b) and DB(W_b) specimens.

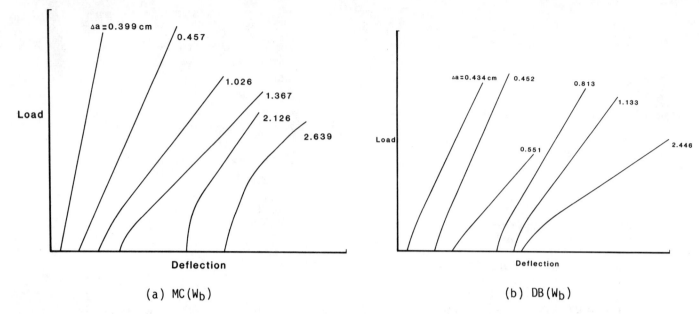

(a) MC(W_b)

(b) DB(W_b)

Fig. 5 - Reloading to failure curves for specimens with various amounts of crack growth.

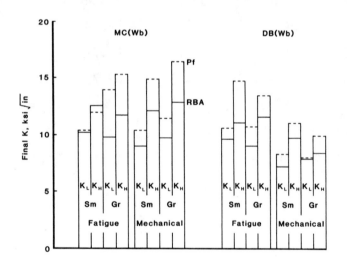

Fig. 6 - Final stress intensity values averaged for each test condition.

TEST	MC(W_b)		DB (W_b)	
CONDITION	MEAN	STD·DEV·	MEAN	STD·DEV·
F-S-K_L	10·170	±1·666	9·628	±1·884
F-S-K_H	12·531	1·510	10·999	2·761
F-G-K_L	9·777	2·082	9·027	1·566
F-G-K_H	11·714	2·600	11·603	3·068
M-S-K_L	9·002	0·599	7·140	2·014
M-S-K_H	12·092	1·321	9·658	2·606
M-G-K_L	9·719	1·818	7·983	4·502
M-G-K_H	12·834	3·690	8·338	3·444

Table 3 - Statistical variation in final K values for each test condition.

LAB. IDENT.	FINAL LOAD		RIGID BOLT ANALYSIS	
	K_L	K_H	K_L	K_H
3	10·364 ±1·205	14·423 ±1·842	9·575 ±1·721	12·908 ±3·280
4	8·751 ±1·329	11·176 ±4·186	8·187 ±1·352	9·689 ±2·627
8	5·224 ±1·049	6·695 ±4·666	7·283 ±3·155	9·243 ±6·934
9	13·285 ±5·543	12·597 ±2·366	9·418 ±1·773	13·112 ±1·103
10	11·125 ±0·686	14·391 ±0·668	10·741 ±0·656	13·108 ±1·253
15	13·438 ±1·286	15·186 ±1·756	10·521 ±2·214	10·886 1·393
16	10·985 ±2·045	14·057 ±3·454	7·752 ±1·177	9·693 ±1·547
17	11·580 ±1·052	15·070 ±3·582	11·777 ±0·395	13·108 ±1·124

Table 4 - Statistical variation in final K values for each laboratory.

STRESS CORROSION CRACKING OF 4340 STEEL IN AIRCRAFT IGNITION STARTER RESIDUES

Kevin J. Kennelley, Raymond D. Daniels
University of Oklahoma
Norman, Oklahoma, USA

Abstract

Military aircraft use a cartridge ignition system for emergency engine starts. Premature failures of steel (AISI 4340) breech chambers in which the solid propellant cartridges are burned is a serious concern. Analysis of a number of breech chamber failures identified corrosion as one problem with an indication that stress corrosion cracking may also be occurring. A study was made for stress corrosion cracking susceptibility of 4340 steel in a paste made of the residues collected from used breech chambers. The constant extension rate test (CERT) technique was employed and SCC susceptibility was demonstrated. The residues, which contain both combustion products from the cartridges and corrosion products from the chamber, were analyzed using elemental analysis and x-ray diffraction techniques. Electrochemical polarization techniques were also utilized to estimate corrosion rates.

MILITARY AIRCRAFT use a cartridge ignition system for emergency engine starts. A solid propellant cartridge is placed in a cartridge chamber/breech chamber assembly and electrically ignited. The expanding gases generated from the burning cartridge provide the necessary energy to start the engines. Premature failures of the breech chambers have become a concern to the U.S.A. F. A view of the domed portion of a failed breech chamber is shown in Figure 1. The interior surface has been heavily oxidized and pitted. The breech chambers are constructed from AISI 4340 steel heat treated to a specified hardness value of Rockwell C 40-45. Metallurgical failure analyses on several failed breech chambers conducted at the University of Oklahoma, have identified

stress corrosion cracking as a probable failure mechanism (1,2).

This paper describes the results of constant extension rate tests (CERT) carried out at slow strain rates to determine the susceptibility of 4340 steel to stress corrosion cracking in an environment consisting of the residue collected from used breech chambers. The residue contains both combustion products from the cartridges and corrosion products from the breech chamber. The electrochemical behavior of the residue was investigated using polarization techniques. X-ray diffraction studies were undertaken to determine the compounds present in the residue. The results of these studies are also presented.

EXPERIMENTAL PROCEDURES

TEST APPARATUS AND TESTING PROCEDURES – For the constant extension rate testing an Instron tensile testing machine, model TT-C-L, was modified using a 248:1 gear reducer to produce crosshead extension rates on the order of 10^{-5} to 10^{-7} cm/s (3). Tests were performed on round tensile specimens which conformed to NACE TM-01-77 tensile test specifications with the exception of the gage length diameter. Due to the maximum load limitations of the Instron, the gage diameter was reduced from 6.4mm to 5.1mm. The gage length was 25.4mm. Test specimens were machined to final dimensions from 13mm bar stock (AISI 4340), and then heat treated to Rockwell C 40-45. The heat treatment consisted of austenizing at 843°C, quenching in oil, and tempering at 482°C. The hardness obtained was Rockwell C 39-40. Prior to each test, the test sample was polished in an engine lathe using 180, 240, 320, 400, 600, E4, and E10 abrasive paper, respectively. The specimen was then polished with steel wool in the longitudinal direction, and rinsed with ethanol. The test cell consisted of a 19mm O.D. quartz glass cylinder which surrounded the tensile test specimen. An o-ring was used for a seal at the bottom of the cylinder to contain the test environment. The residue was packed into the cylinder and kept moist by adding distilled water. A schematic of the test setup is shown in Figure 2.

Tests were conducted at four different extension rates. These extension rates can be converted to nominal strain rates in units of s^{-1} if the extension rates are divided by the gage length (25.4mm) of the specimen. Nominal strain rates for each extension rate are listed below:

EXTENSION RATE(mm/s)	NOMINAL STRAIN RATE(s^{-1})
3.4×10^{-4}	1.4×10^{-5}
3.4×10^{-5}	1.4×10^{-6}
1.7×10^{-5}	6.7×10^{-7}
8.6×10^{-6}	3.4×10^{-7}

Tests were performed in an inert environment (vegetable oil) and in the residue environment at each extension rate. Mechanical properties (UTS, FS, %RA) obtained in the residue environment were compared to those obtained in the inert environment, conducted at the same extension rate, to indicate the presence and severity of SCC. Cross-sectional metallography and scanning electron microscopy were used to identify the mode of failure.

ELECTROCHEMICAL TESTING - Investigation into the electrochemical behavior of AISI 4340 steel in the residue paste was conducted using potentiodynamic polarization and linear polarization techniques. A standard three electrode cell was used with the steel as the working electrode, a saturated calomel electrode as the reference electrode, and a platinum electrode as the auxiliary electrode. A PAR 175 universal programmer, PAR 173 potentiostat with model 376 logarithmic current converter interface, and a H.P. 7044 x-y recorder comprised the test equipment. ASTM standard G-59-78 entitled "Standard Practice for Conducting Potentiodynamic Polarization Resistance Measurements" was followed as the test procedure, with the exception that the three electrodes were placed in the same beaker and all three tests were performed in the same paste. This was necessary because of the limited supply of residue product available. The order of testing in the residue paste was (1) linear polarization, (2) cathodic polarization, and (3) anodic polarization.

X-RAY DIFFRACTION AND ELEMENTAL ANALYSIS - To identify the compounds and aggressive agent or agents present in the residue product both powder x-ray diffraction and microprobe analysis were conducted. Powder x-ray diffraction was performed using a Rigaku/Geigerflex XRD system. For the microprobe analysis a Princeton Gammatech EDX system attached to an ETEC Autoscan electron microscope was used.

RESULTS

CERT - Constant extension rate tests performed in the inert environment (vegetable oil) showed a tensile strength (UTS) of 1240MPa, a fracture strength (FS) of 1900MPa, and a 51-52% reduction in area (RA). The mechanical properties data (UTS, FS, %RA) obtained with the four different extension rates were in good agreement with a standard deviation of less than 3 percent. All of the tests conducted in oil exhibited a classic cup-cone fracture, characteristic of ductile failure. Figure 3 is a scanning electron micrograph of the fracture surface of a specimen tested in oil at a nominal strain rate of $1.4 \times 10^{-6} s^{-1}$. Failure initiated at the center of the cross-section by microvoid coalescence and propagated radially outward. Final separation occurred in shear, producing a shear lip around the circumference. A high magnification view of the center of the specimen (Figure 4) shows dimple rupture and tearing which is indicative of ductile failure in high strength steel.

Results of the constant extension rate tests performed in the residue paste are presented in tabular form in Table 1. The %RA/ %RA (oil) column represents the %RA of each test normalized to the %RA of a test conducted in oil at the same nominal strain rate. From the table it is clear that a severe reduction in ductility had occurred. A graph of %RA/%RA (oil) vs. nominal strain rate for the residue paste tests, shown in Figure 5, indicates that the most severe loss in ductility occurred at a nominal strain rate of $1.4 \times 10^{-6} s^{-1}$. At both slower and faster strain rates, the ductility is increased.

Fracture stress (FS) as shown in Table 1 is calculated as load at fracture divided by the minimum cross-sectional area measured after fracture. In normal tensile testing this fracture stress is greater than the ultimate tensile stress (UTS). However, at the slow strain rates employed in the constant extension rate tests, the calculated FS is often less than the UTS. This is explained by the fact that necking ceases when cracks growing from the surface concentrate the stress at the crack tip. The load bearing capacity is then reduced without further reduction in the apparent cross-sectional area.

In all tests conducted in the residue paste, fracture initiated at the metal/solution interface. Figures 6, 7, and 8 are electron micrographs taken of the fracture surface of a specimen tested in the residue at $6.7 \times 10^{-7} s^{-1}$. There were multiple fracture origins. These multiple origins can be seen at the 1, 2, 3, and 6 o'clock positions in Figure 6. Progressively higher magnification views of the fracture origin at the 6 o'clock position (Figures 7 and 8) show an intergranular fracture mode. At the far left portion of Figure 7, the metal/solution interface at the fracture origin can be seen. At all of the origins, the mode of fracture was intergranular.

The crack initiation process for the samples tested in the residue paste is associated with the development of corrosion pits on the sample surface. Figure 9 shows an area near the fracture surface of a sample tested at $6.7 \times 10^{-7} s^{-1}$. Several localized pits had developed along the gage length. Cracks can be seen emanating from the pits. Coalescence of these cracks, with crack growth in a direction perpendicular to the tensile axis, was observed.

Secondary cracks which had developed from surface pits were examined in cross-section. Figure 10 is an unetched optical photograph of one of the secondary cracks which had initiated at a pit. The crack propagated into the specimen perpendicular to the tensile axis. These cracks are indicative of SCC.

ELECTROCHEMICAL TESTING - Table 2 summarizes the results of Tafel extrapolation and linear polarization tests performed in distilled water and the residue. The residue was diluted with distilled water only enough to accommodate the electrodes in the test cell. Addition of the distilled water to the residue did not change the pH value indicating that the solution was saturated. Corrosion rates were determined using anodic Tafel extrapolation, cathodic Tafel extrapolation, and the linear polarization technique. For the corrosion rates calculated using linear polarization, Tafel slopes obtained from experimental data were used. The polarization curves did not have linear regions which extended over at least one full decade of current. Thus, the Tafel slopes quoted in Table 2 are only approximate "best fit" values. This introduced some uncertainty when trying to obtain absolute corrosion rates.

The corrosion rates listed in Table 2 show that the residue paste is considerably more corrosive to the steel than is distilled water. This indicates that an aggressive agent or agents were present in the residue which provided a corrosive environment necessary for SCC. In addition, the anodic polarization curves did not indicate formation of a "passive" film. The pH of the residue paste was 2.8.

ELEMENTAL ANALYSIS AND X-RAY DIFFRACTION - Elemental analysis using an EDX microprobe indicated that Fe, Cl, K, Ni, and Cr were present in the residue. The Ni and Cr are alloying elements in the steel and the heat shields in the chamber are Inconel, a Ni base alloy containing Cr. Elements with an atomic number of less than nine such as hydrogen, nitrogen, and oxygen cannot be detected with this system.

From the x-ray powder diffraction pattern obtained on the residue, the compounds identified include Fe_3O_4, FeOOH, and $FeCl_2$. A sizeable background radiation was present in the spectrum. This was produced by Fe x-ray fluorescence produced by the incident CuK_α radiation. In an attempt to reduce the background radiation in the pattern, the rust colored particles were selectively removed from the residue. The background was dramatically lowered, and the diffraction pattern provided good matches for the compounds KCl, $FeCl_2$, NH_4NO_3, and Fe_3O_4.

DISCUSSION AND CONCLUSIONS

The constant extension rate tests showed that the breech chamber steel (4340) is susceptible to SCC in a paste made from the residues removed from used chambers. The most severe reduction in ductility occurred at a nominal strain rate of $1.4 \times 10^{-6} s^{-1}$. Since the residue had a pH of 2.8, cracking by a hydrogen embrittlement mechanism is a possibility. However, the increase in ductility at the slowest strain rate $3.4 \times 10^{-7} s^{-1}$ is indicative of a SCC mechanism (4). The crack initiation process involves corrosion pits which develop on the metal surface. Cracks appear to initiate at isolated pits and propagate through the sample in a direction perpendicular to the tensile axis. Linkage of small surface cracks into larger cracks was also observed.

The electrochemical testing showed that the residue was corrosive to the steel. The residue which consists of combustion products from the solid propellant cartridges and corrosion products from the chamber are quite complex. Due to the complex nature of the residue, not all of the compounds present have been identified. From the XRD study ammonium nitrate and potassium chloride were found to be present in the residue. It is well documented that ammonium nitrate is an aggressive agent responsible for SCC in steel (5, 6, 7). The presence of chloride would explain the occurrence of pitting on the metal surface.

ACKNOWLEDGEMENT

This work was supported under U.S. Air Force Contract No. F34601-83-C-3448.

REFERENCES

1. Coleman, W.R., Block, R.J. and Daniels, R.D., Proceedings 1980 Tri-Service Corrosion Conference, AFWAL-TR-81-4019, Vol. II, 241-270, (1981).
2. Perkins, P.C., Daniels, R.D. and Gillies, A.B., "Failure Analysis of Steel Breech Chambers Used With Aircraft Cartridge - Ignition Starters", Int. Conf. on Fatigue, Corrosion Cracking, Fracture Mechanics, and Failure Analysis, Salt Lake City, UT, (1985).
3. Nutter, W.T., Agrawal, A.K., Staehle, R.W., "Design and Construction of an Inexpensive Multispecimen Slow Strain Rate Machine", ASTM STP 665, G.M. Ugiansky, J.H. Payer, Editors, pp. 375-387, (1979).
4. Kim, C.D., Wilde, B.E., "A Review of the Constant Strain Rate SCC Test", ASTM STP 665, G.M. Ugiansky and J.H. Payer, Editors, pp. 97-112 (1979).
5. Al-Brady, F.S., 6th International Congress on Metallic Corrosion B754016 Sydney, Australia, pp. 629-633, (1975).
6. Biefer, G.J. Canada Dept. of Energy, Mines and Resources, Mines Branch, Tech Bul. 111, (1969).
7. Cabicciotti, D. and Boyer, W., Weld. J. Res. Suppl, 15, pp. 140-5.

Table 1 - Results of Testing Performed in Corrosion Product Paste

Nominal Strain Rate (s^{-1})	Failure (hours)	Time to Maximum Load (hours)	Ultimate Tensile Strength (MPa)	Fracture Stress (MPa)	%RA	%RA/%RA(oil)
1.4×10^{-5}	2.3	1.8	1270	1430	25.1	0.49
1.4×10^{-6}	17.1	16.9	1230	1230	2.7	0.05
*1.4×10^{-6}	18.3	18.2	1170	--	--	--
6.7×10^{-7}	22.8	22.7	1160	1130	1.9	0.04
6.7×10^{-7}	27.4	26.4	1240	1200	6.0	0.17
3.4×10^{-7}	74.4	67.0	1270	1210	9.2	0.18
3.4×10^{-7}	76.0	66.9	1240	1270	11.5	0.23

*Specimen broke at the bottom of the gage length.

Table 2 - Corrosion Rate Data From Electrochemical Testing

Environment	Anodic Tafel Slope (mv/Decade)	Cathodic Tafel Slope (mv/Decade)	Corrosion Rate		
			Anodic Extrapolation (MPY)	Cathodic Extrapolation (MPY)	Linear Polarization (MPY)
Distilled H_2O	63.3	57.0	0.37	0.37	0.97
Residue From Used Breech Chambers	46.0	23.3	29.60	11.80	12.73

Fig. 1. Overall View of a Failed Breech Chamber.

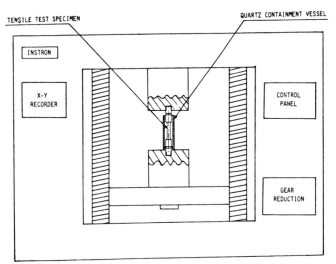

Fig. 2. Schematic of Experimental CERT Setup.

Fig. 3. SEM Photograph Showing a Cup-Cone Fracture of a Specimen Tested in Oil at Nominal Strain Rate of $1.4 \times 10^{-6} s^{-1}$.

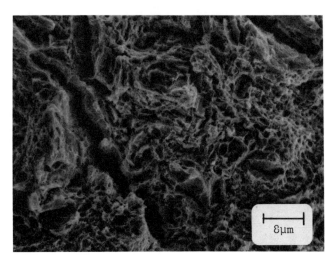

Fig. 4. High Magnification View of the Center Portion of Figure 3 Showing Dimple Rupture and Tearing.

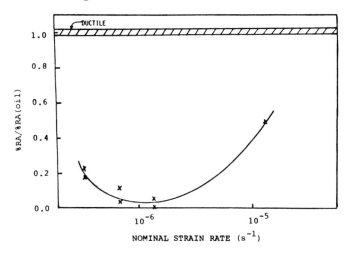

Fig. 5. Graph of %RA/%RA(oil) vs. Nominal Strain Rate for the Tests Conducted in the Residue Paste.

Fig. 6. SEM Photograph Showing Multiple Fracture Origins at the 1, 2, 3, and 6 o'clock Positions. Test Conducted in the Residue Paste at $6.7 \times 10^{-7} s^{-1}$.

Fig. 7. A Higher Magnification View of the Fracture Origin at the 6 o'clock Position in Figure 6.

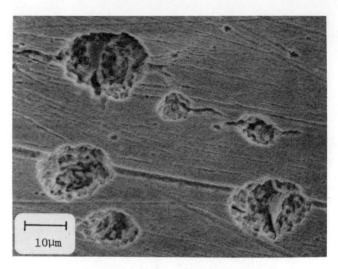

Fig. 9. SEM Photograph of the Gage Length Near the Fracture Surface of a Specimen Tested in the Residue Paste at $6.7 \times 10^{-7} s^{-1}$. Cracks Emanating from Corrosion Pits are Shown.

Fig. 8. High Magnification View of the Fracture Origin at the 6 o'clock Position in Figure 6. Fracture Mode is Intergranular.

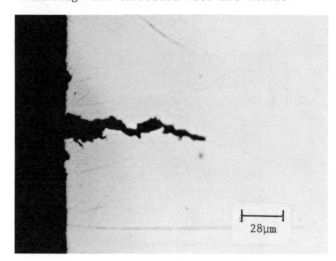

Fig. 10. Unetched Optical Metallograph Showing one of the Secondary Cracks which had Initiated at a Corrosion Pit.

INVESTIGATION OF INTERGRANULAR STRESS CORROSION CRACKING IN THE FUEL POOL AT THREE MILE ISLAND UNIT 1

Carl J. Czajkowski

Brookhaven National Laboratory
Upton, New York, USA

ABSTRACT

An intergranular stress corrosion cracking failure of 304 stainless steel pipe in 2000 ppm B as H_3BO_3 + H_2O at 100°C has been investigated. Constant extension rate testing has produced an intergranular type failure in material in air. Chemical analysis was performed on both the base metal and weld material, in addition to fractography, EPR testing and optical microscopy in discerning the mode of failure.

Various effects of Cl^-, O_2 and MnS are discussed. The results have indicated that the cause of failure was the severe sensitization coupled with probable contamination by S and possibly by Cl ions.

INSPECTIONS OF PIPING in the Three Mile Island Unit #1 spent fuel system in April 1979 revealed five through wall cracks near welded joints. Reports indicated that the cracks were circumferential in nature and located in the weld heat-affected zone adjacent to girth welds. Several months later, two more through wall cracks were detected in the spent fuel system which were reported as having the same cracking characteristics as the first five leaking welds.

In order to analyze the cracking phenomena in more detail, a section of radioactively contaminated 8" diameter pipe was sent to Brookhaven National Laboratory (BNL). The pipe (H.T. #334165) measured approximately 30" in length and contained two circumferential welds, SF 265 (through wall cracked-field weld), and SF 261 (uncracked-shop weld), as well as three uncracked longitudinal seam shop welds (Figure 1). (Note: Six of the seven through wall cracks were from material of this heat number 334165.)

The following program was initiated to analyze the failure:

1) Visual inspections and photographs of both the inner and outer surfaces
2) Optical microscopy
3) Scanning Electron Microscopy (SEM)
4) Energy Dispersive Spectroscopy (EDS)
5) Bulk chemistry of weld metal/base materials
6) Constant Extension Rate Testing (CERT)
7) Electrochemical Potentiokinetic Reactivation Analyses (EPR)

EXAMINATIONS

VISUAL/PHOTOGRAPHY - Preliminary visual examination of the pipe disclosed that the crack (near SF 265) was adjacent to what appeared to be a weld repair area. In addition, GPU also communicated that the weld was a single "V" type with a shallow ($\simeq$ 20°) edge preparation. The filler metal used was an E308 Grinnell Consumable Insert and welded with ER308 electrode (Table 1). The welding process used was GTAW (Gas Tungsten Arc Welding). The repair area was in its R-2 stage (two repairs completed). The pipe had also been counterbored prior to installation for ease in pipe fit-up prior to welding. The crack on the outside surface of the pipe was approximately five inches long, ran parallel to the girth weld and was reasoned to be in the weld heat-affected zone. The pipe was then longitudinally sectioned and the inside surface visually inspected. The inside of the pipe contained a crack 5.5" long running parallel to the girth weld for half its length, with its direction varying into the counterbore at its terminal ends (Figure 2). The inside surface was covered with an oxide film and showed evidence of significant heating during welding. Eight sections were then cut bracketing the crack to facilitate further examinations.

OPTICAL MICROSCOPY - One specimen was cut and mounted in epoxy in such a manner that the cross section of the pipe's circumference was perpendicular to the plane of examination. The surface was polished and then electrolytically etched with a 10% oxalic acid etch for macrostructural depiction. The weld insert, weld deposit, crack and counterbore area are well defined showing the through wall crack and a smaller non-through wall crack on the weld side of the main crack (Figure 3). The specimen was then repolished and examined after ASTM A-262 Practice A was performed. As can be seen from the photomicrograph (Figure 4), the crack was intergranular in nature with secondary cracks branching out from the body of the main crack. The microstructure across the face of the specimen showed a fine grained structure in the pipe's central cross section, with increasing grain size outward towards the pipe's outside and inside surfaces. This grain size variance was visible in the specimen outside the weld HAZ, which may be the result of cold working operations prior to the solution annealing heat treatment in the pipe's manufacture. The area surrounding the crack and in the heat-affected zone exhibited a ditched structure (ASTM A262), which is evidence of considerable material sensitization. In view of the above observations, the failure is considered to be intergranular stress corrosion cracking in the sensitized region of the weld repair heat-affected zone.

Examination of the base material outside of the heat-affected zone after the ASTM A262 Practice A test revealed virtually no evidence of material sensitization, which would indicate that proper heat treatment (solution annealing) was applied by the supplier.

There were also a few indications of incipient cracking on the inside surface of the weldment in the counterbore area. These indications were visible on the counterbore of both the cracked side of the weld and the uncracked side.

Microscopy also discerned the appearance of stringer-like inclusions throughout the specimen's cross section. Later examination (EDS) disclosed these stringers to be MnS.

SEM/EDX - In an effort to identify any corrosive species present and to further evaluate the mode of cracking, various cracks were opened and the fracture faces evaluated by Scanning Electron Microscopy. The fracture faces examined had a coarsened grain structure at both the inner and outer edges of the specimens' cross sections with a finer grain structure at their mid-sections, similar to that found by optical microscopy. The fracture faces also had a typically intergranular stress corrosion cracking topography.

Several Energy Dispersive X-ray Spectroscopy (EDS) scans were made across the fracture face of the specimen in order to determine qualitatively the species present. The normal components of Type 304 stainless steel, Fe, Cr and Ni, were present with trace amounts of Al, Si, S and Ca present in many of the scans. In addition, chlorine (Cl) was present in trace amounts on the fracture surface near the inner surface of the piping.

Various particulates in the fracture face were also examined by EDS. Typically, in addition to Fe, Cr and Ni they also contained traces of Si and S, while some had traces of Cl, Ca and K present.

An examination was also made of the inner surface of the pipe where there appeared to be etched grain boundaries (Figure 5). The base metal adjacent to the grain boundary etching showed Al, Si, S, Ca and a trace of Cl in the scan (Figure 6).

EDS scans were then performed on the stringer-like inclusions which were originally discerned by optical microscopy. The scans showed both high S and Mn peaks (probably MnS) and at least one scan exhibited a Ti peak.

CHEMISTRY - Table 1 shows the results of chemical analyses on the base material and weld metal. Since the carbon content was higher than specifications allowed, two more specimens from the base metal were sent out to an independent laboratory for analysis. The additional two samples were both within specification for alloying elements of 304SS. The weld metal analysis varied somewhat from those reported by GPU, specifically in the carbon and chromium contents of the weldment. The disparities could have been caused by the difference in the testing methods used or possible constituent segregation in the material.

CONSTANT EXTENSION RATE TESTING - In order to investigate further the environmental conditions under which the weldment/pipe would fail, various tensile-type specimens were cut from the remaining material in both the field and shop welds. These specimens were subjected to constant extension rate tests (CERT) in various environments. The CERT method is basically the application of slow dynamic straining to a stress corrosion (tensile type) specimen while it is exposed to the environment. This particular mode of testing has the advantage of always resulting in a fracture at its completion and normally does so in a relatively short period of time (a few days). Another advantage of this type of testing is the ability to adjust the strain rate in the most probable range of stress corrosion crack velocities, namely, 10^{-6} to 10^{-9} sec^{-1}. This strain rate is quite important, as too high a strain rate would result in ductile failure by void coalescense prior to the development of the necessary corrosion reactions; while too slow a strain rate could again fall out of the range for SCC failures to occur. For these reasons, it is important to realize that the absence of cracking in a given medium does not necessarily preclude the possibility of stress corrosion cracking until additional strain rate tests (both higher and lower values) have been completed.

Twelve CERT tests were performed in different environments, as defined in Table 2. Ten of the specimens were machined in such a manner that the counterbore and weldment were left intact, while two of the specimens were machined flat. The reason for leaving the counterbore and weld root intact was to maintain the inside surface condition of the pipe at the pipe/liquid interface. This method would also give an indication of the effect of any inside surface imperfections on the susceptibility of the material to IGSCC.

Nine of the twelve specimens were tested in a 2000 ppm B (as H_3BO_3) solution with additions of Cl^- (varying from 1 to 15 ppm) and MnS additions, and strain rates were in the range of 1.3 to 1.7 x 10^{-6} sec^{-1}. The solution temperature was maintained at 30°C in an open beaker and the solutions kept stagnant. After each test, a new solution was prepared and the entire apparatus thoroughly cleaned. The tenth and eleventh specimens were tested in air at a strain rate of approximately 8.7 x 10^{-6} sec^{-1}, while the twelfth specimen was tensile-tested at a strain rate of 1" min^{-1}. After the completion of the CERT testing (to fracture) all specimens were cut in order to view the fracture face by SEM, and the opposite side of the specimen mounted in hard epoxy for viewing by optical microscopy. Prior to the optical microscopic examination, all of the specimens were etched by ASTM A262-Practice A (oxalic acid test) for determination of degree of weld sensitization.

Table 2 outlines the results of the CERT testing. From the table it can be seen that five of the twelve failures were identified as either fully or partially intergranular in nature. Four of these occurred in the field weld, while only one intergranular-type failure occurred in the shop weld.

The CERT test, with the MnS addition, resulted in a ductile-type failure with no areas of intergranular fracture. It was noted that during the mixing of the solution with MnS + H_3BO_3, H_2S was liberated. The faces of the intergranular-type fractures were characterized by grains which have a dimpled-ruptured appearance on their facets. EDS examination of the dimples showed them to be composed of either MnS or a Ti/Mn (probacarbide) inclusion. These facts suggest that the intergranular-type failure is not a brittle-type failure but more ductile in nature than that observed in the original failed crack.

The microstructures exhibited in the field weld were predominantly of the ditched type (ASTM A262), which is indicative of a highly sensitized material; while the shop weld displayed a predominantly dual type structure, which would indicate a lesser degree of sensitization. This correlates reasonably well to the results of the CERT testing.

It is also interesting to note that of the four shop weld specimens tested, the two that had the weld root and counterbores machined failed in a ductile manner, while only one of the non-machined specimens failed in a similar manner. This suggests that the inside surface condition (after welding) of a stainless steel pipe, or the corrosion films developed during exposure to the fuel pool coolant, may have an influence on the initiation of IGSCC.

The most significant result obtained during this series of tests is the partially intergranular-type fractures of the air tested field weld specimens (Figure 7). This failure occurred at a strain rate of 8.71 and 8.6 x 10^{-6} sec^{-1} and is a definite indication of a highly sensitized structure, which is substantiated by the photos of the microstructures.

ELECTROCHEMICAL POTENTIOKINETIC REACTIVATION ANALYSIS - In an attempt to quantify the degree of sensitization in the area surrounding the repaired portion of the field weld, specimens were cut, mounted and subjected to an Electrochemical Potentiokinetic Reactivation Analysis (EPR Testing). Simply stated, the EPR method is a non-destructive test being developed for determining the degree of sensitization of a stainless steel. It is believed to be correlated to provide useful information for the susceptibility of a material to intergranular stress corrosion cracking.

For ease of calculations, the faces of the samples to be tested were cut to approximately 1 cm^2 and all distances were measured from the weld centerline outward to the sample face. All samples were mounted so that the face tested was in a plane parallel to the field weld and contained the cross section area of the pipe. The susceptibility to IGSCC was evaluated electrochemically by performing a controlled potential sweep from the passive to the active region in a 0.5 M H_2SO_4 + 0.01 M KSCN solution at 30°C and recording the resultant potentiokinetic curves. The EPR tests were performed using the procedures proposed by G.E. to the ASTM for their "Round Robin" evaluation.

Each of the specimens was polarized to a passivation potential of +200 Mv (SCE) for two minutes and the current recorded during a constant potential sweep at 6 v/h in the cathodic direction until a potential of -500 mV was reached. Table #3 summarizes the results of the examinations.

The normalized charge per grain boundary area is given by the equation:

$$P_a = Q/GBA \qquad (1)$$

where: P_a= Normalized charge per grain boundary area

Q = Integrated current in coulombs

GBA = Grain boundary area, estimated from the ASTM grain size number

As can be seen on the Table, the highest P_a value obtained was approximately 8 mm, (Sample #3) away from the weld centerline (.315 in) which would indicate a significant deree of material sensitization. Sample #6 exhibited a higher P_a value than other specimens cut at an equivalent distance from the field weld probably due to its close proximity to the seam weld indicating that, even under controlled shop conditions, this material is somewhat susceptible to sensitization during welding.

Since all of the measurements were done through the cross sectional area of the pipe (through wall) it was necessary to average the grain size for the P_a calculations. The grain size varied from ASTM 1.5 near the pipe's outer surfaces to ASTM 5.5 at its center so a normalized value of 3.5 was used in the calculations. Note - the difference of using ASTM grain size 3.5 in lieu of 5.5 is approximately a multiple of 2.

The values obtained by the EPR tests are well corroborated by the photomicrographs after oxalic acid etching (ASTM A262) and clearly indicated that this particular heat of material #334165 is susceptible to sensitization by welding.

DISCUSSION

Intergranular corrosion is generally defined as a local attack on the grain boundaries of a metal by a corrosive media.

In stainless steels, susceptibility to intergranular corrosion is greatly enhanced by the sensitization process. Sensitization can be described (in austenitic stainless steels) as the formation of chromium carbide precipitates in the grain boundaries and the resultant depletion of chromium, brought about by heating the steel in the temperature range 500-800°C (1,2).

This temperature range is easily achieved during the welding process where the normal temperature of welding exceeds 1600°C during the fusion welding. Therefore, the welded base material could receive a sensitizing heat treatment in the critical range (500-800°C) at some point outward from the weld fusion time which would be maintained long enough to precipitate chromium carbide at the grain boundaries. This is not to say that the welding process

This is not to say that the welding process alone will induce a sensitizing effect on the base material, as the degree of material sensitization is a cumulation of the material's prior thermal and mechanical treatments, weld cycle history, (# passes, heat input, etc.) material chemistry, thickness and thermal conductivity, and time at temperatures in the sensitization range.

Residual stresses in a weldment are also an important factor in the stress corrosion cracking phenomenon. An overview paper (3) has cited instances of residual tensile stresses in Type 304 stainless steel heat affected zones of higher than 40 Ksi. The direction and amount of tensile stresses developed in piping seem to be closely related to the pipe's diameter, Chrenko (4) has postulated that the thinner cross sections of smaller diameter pipes provide a less efficient heat sink and a more flexible surface during welding. The heat transferred by welding is then distributed over a larger area with an increased weld shrinkage area resulting in larger axial stresses.

The effect of prior cold working can have a substantial influence on the resistance of a stainless steel to stress corrosion cracking and on its susceptibility to sensitization. Work done by Briant (5) determined that prior cold work increases the material's susceptibility to sensitization, possibly by the formation of a martensitic structure offering a more amenable crystal structure for rapid diffusion of Cr and C and providing a region where chromium depletion takes place. It is also known that cold working produces a stronger layer in the base material which is normally higher in residule stresses than the surrounding metal. It has also been observed (6) that heat treatment conducted at 1050°C for 15 minutes in order to remove the effects of prior cold work has been somewhat effective in reducing the materials susceptibility to stress corrosion cracking.

Thermal cutting of stainless steels can also affect the materials susceptibility to sensitization. Vyas and Isaacs (7) determined that prior plasma arc cutting of 304SS caused a shift in both the location and degree of sensitization after welding. The plasma cutting apparently allowed the material sufficient time at its sensitization temperature to promote chromium carbide nuclei in grain boundaries which grew during the subsequent welding operations.

Since the sensitization process depends upon the depletion of chromium in grain boundaries (1,2) it is a logical conclusion that the reduction of available carbon in the stainless steel (to react with the chromium) would enhance its resistance to sensitization and thus its susceptibility to stress corrosion cracking. This assumption has been validated by numerous investigations. Kass, Walker, and Giannuizzi (8) observed by varying carbon contents and subjecting 304 and 304L specimens to cyclic loading tests that carbon contents of less than .03% produced no failures after 2000-3000 cycles while carbon contents of .05% or higher produced failures after less than 100 cycles. Failures observed were intergranular in nature.

The effects of various alloying agents (namely S) and subsequently formed inclusions, although investigated, are somewhat less conclusive. Investigations (9,10) do indicate, however, that sulfide inclusions provide possible points for corrosion pit nucleation.

The welding process itself has a most significnt effect on the sensitization process and subsequent susceptibility of type 304 stainless steel to intergranular stress corrosion cracking. Work by Solomon (11) showed that for .05 to .08 w/o C, 304 stainless steel that the critical cooling rate for sensitization is about 5-10°C/S and that for 0.35 in plate thickness and a 30,000 V in^{-1} heat input and cooling rate of 10°C Sec^{-1} can be obtained which would be in the sensitizing range of the material.

The effects of varying chloride concentrations on stainless steel (6,12) below 135°C have produced intergranular cracking in both 304 and 304L stainless steel. Also work by Bednar (13) suggests that alloys of 301 and 305 stainless steel will crack in chloride environments at temperatures between 95 to 154°C, it is, therefore, advisable to reduce or eliminate chloride contamination in contact with stainless steel. Examples of incomplete flushing after hydrochloric acid cleaning have resulted in chloride stress corrosion cracking in 304 and 347 stainless steel (14) in the past. Since there was evidence of grain boundary etching prior cleanup as a possible example of how Cl could have entered the pipe; the results of the CERT tests do indicate, however, that even a concentration of 15 ppm Cl had no adverse affect on the pipe unless it was in an area which was previously highly sensitized; and then, the effects were similar to those found in either pure H_3O + H_3BO_3 or air.

High O_2 content can also induce stress corrosion cracking in sensitized stainless steels either singly or synergistically with Cl$^-$. van Rooyen and Kendig (15) have proposed that as the electrochemical potential in solution (O_2 concentration) is increased; the amount of Cl$^-$ required to cause stress corrosion cracking is decreased. Laboratory cracking of sensitized stainless steel has also been observed (16) in slow strain rate testing; with O_2 concentrations of 2 ppm at temperatures as low as 50°C.

In addition to O_2 and Cl$^-$, S also seems to play a role in SCC of Type 304 stainless steel. Recent studies by Brookhaven National Laboratory (17) have shown that sensitized Type 304 stainless steel can suffer stress corrosion cracking in low concentrations of thiosulfate solutions at room temperature.

The intergranular-type fracture faces observed in air is quite similar in appearance to those observed by Hippsley, Knott and Edwards (18) on 2 1/4 Cr 1 Mo steel. They attributed the dimpled appearance to the possible decohesion of grain boundary MnS/Matrix interfaces during plastic deformation causing a large amount of cavity nuclei which could induce secondary cracking.

CONCLUSIONS

1. The primary cause of the cracking appears to be intergranular stress corrosion cracking in the weld sensitized heat affected zone of a weld repair.
2. Although no definite corrosive species were identified as the cause of cracking, the various traces of Cl and S both the pipe inside surface and in the areas of the crack fracture faces detrmined by EDS is evidence of etching and possible contamination of the system by Cl and S ions.
3. The significance of the MnS stringer-like inclusions as pit nucleation sites is inconclusive.
4. The results of Electrochemical Potentiokinetic Reactivation Analysis (EPR) did show at least one area of the pipe's cross section had been sensitized significantly by the welding process.
5. It is evident by the Constant Extension Rate Test (CERT) results (corroborated by the oxalic acid etched microstructures) that the Type 304SS material HT#334165, with its high carbon content and complex thermal history (2 repairs) was severely sensitized at various locations about its girth. This degree of sensitization quantitatively was sufficient to cause intergranular-type fractures during CERT testing in air. This degree of sensitization coupled with the residual stresses from welding and the possible contamination by Cl and S ions were seemingly sufficient to crack the piping by an intergranular stress corrosion cracking mechanism.

REFERENCES

1. Pande, C.S., Suenage, M., Vyas, B., Isaacs. H.S., and Hailing, D.F., Scripta Metallurgica, 11, 681 (1977)

2. Cowan, R.L. and Tedmon, C.S., Jr., Advances in Corrosion Science and Technology, 3, 293 (1973)

3. Fox, M., Proceedings: Seminar on Countermeasures for Pipe Cracking in BWRs, EPRI, 1 Workshop Report, Paper No. 1. May (1980)

4. Cherenko, M., Proceedings: Seminar on Countermeasures for Pipe Cracking in BWRs, EPRI 2, Workshop Report, Paper No. 21 May (1980)

5. Briant, C.L., Proceedings: Seminar on Countermeasures for Pipe Cracking in BWRs, EPRI, 2, Workshop Report, Paper No. 27 May (1980)

6. Stadder, F., Duquette, D.J., Corrosion, 33, No. 2, 67 February (1977)

7. Vyas, B., Isaacs, H.S., ASTM, Special Technical Publications 656.

8. Kass, J.N., Walker, W.L., Giannuzzi, A.J., Corrosion, 36 No. 6, 299 June (1980)

9. Manning, P.E., Duquette, D.J., Savage, W.F., Corrosion, 36, No. 6, 313 June (1980)

10. Lyle F.F., Jr., Corrosion, 29, No. 3, 86 March (1973)

11. Solomon, H.D., Proceedings: Seminar on Countermeasures for Pipe Cracking in BWRs, EPRI, 2, Workshop Report, Paper No. 24 May (1980)

12. Torchio, S., Corrosion Science, 20, 555 (1980)

13. Bednar, L., Corrosion, 33, No. 9, 321 September (1977)

14. Engle, J.P., Floyd, G.L. Rosene, R.B., NACE Technical Committee Report Publication 59-5, Corrosion, 15, 69t February (1959)

15. van Rooyen, D. Kendig, M.W., BNL-NUREG-28147, Brookhaven National Laboratory, June (1980)

16. Andresen, P., Ford, F.P, Proceedings: Seminar on Countermeasures for Pipe Cracking in BWRs, EPRI, 1, Workshop Report, Paper No. 7 May (1980)

17. Isaacs, H.S., Vyas, B., Kendig, M.W., Unpubslished paper to be presented at NACE, Corrosion (1981)

18. Hippsley, C.A., Knott, J.F., Edwards, B.C., Acta Metallurgica, 28, 869 July (1980)

ACKNOWLEDGEMENTS

The author wishes to thank the USNRC (FIN A3011) for funding this work and R. Sabatini for his SEM/EDAX work, K. Sutter for his aid in the CERT experiments, Dr. B. Vyas for his help in the initial investigations, C. Schnepf and R. Graeser for the metallography and Dr. J.R. Weeks for his helpful discussions and encouragement.

Table 1 - Brookhaven Test Results

	Base Metal		Weld Metal
Fe	66.6	67.9	63.6
Cr	19.0	18.6	18.1
Ni	8.94	9.06	9.62
S	0.015	0.012	0.018
C	0.093	0.096	0.080
Ti	<.1%	<.1%	<.1%
Mn	1.73	1.77	1.83

Independent Laboratory Results

	Sample #1	Sample #2	Spec. 304	
Fe	Bal	Bal	Bal	
Cr	18.54	18.54	18-20	
Ni	9.93	9.91	8.0-10.5	
S	0.018	0.019	0.030	max
C	0.06	0.07	0.08	max
Mn	1.71	1.70	2.00	max
P	0.005	0.004	0.045	max
Si	0.53	0.53	1.0	max
Mo	<0.01	<0.01	————	

Table 2 - Constant Extension Rate Testing and Oxalic Acid Results

Sample #	Sample Condition	Where Brake	Solution for Test	Strain Rate	% Elongation	Type of Fracture	Incipient Cracks Evident	ASTM A26 Degree Sensitization
1	field weld & counterbore	HAZ	2000 ppm B as H_3BO_3	1.38×10^{-6} in/sec	15.375%	intergranular	No	ditched
2	field weld & counterbore	weld	2000 ppm B as H_3BO_3+1gMnS/$_{Lsoln}$	1.65×10^{-6} in/sec	16.875%	ductile	No	ditched
3	field weld & counterbore	HAZ	2000 ppm B as H_3BO_3+15ppmCl$^-$	1.34×10^{-6} in/sec	19.75%	70-30 intergranular	No	ditched
4	field weld & counterbore	weld	2000 ppm B as H_3BO_3+1ppm Cl$^-$	1.25×10^{-6} in/sec	31%	ductile	No	dual
5	field weld & counterbore	HAZ	2000 ppm B as H_3BO_3+1ppm Cl$^-$	1.43×10^{-6} in/sec	34.69% log	50-50 intergranular	No	ditched
6	shop weld & counterbore	adjar to weld line	2000 ppm B as H_3BO_3	1.48×10^{-6} in/sec	21.5%	partially intergranular 30-70 partially ductile	No	dual
7	shop weld machined	weld	2000 ppm B as H_3BO_3+5ppm Cl	1.46×10^{-6} in/sec	30.5%	ductile	No	dual
8	shop weld machined	weld	2000 ppm B as H_3BO_3+15ppm Cl	1.51×10^{-6} in/sec	24.5%	ductile	No	ditched
9	shop weld machined	weld	2000 ppm B as H_3BO_3+15ppm Cl$^-$	1.57×10^{-6} in/sec	31%	ductile	No	ditched
10	field weld & counterbore	HAZ	AIR	8.71×10^{-6} in/sec	49.5%	intergranular ductile 50-50	No	ditched
11	field weld & counterbore	HAZ	AIR	8.6×10^{-6} in/sec	29%	intergranular ductile 70-30	No	ditched
12	field weld & counterbore	weld	AIR	1 in/min.	approx. 32%	ductile	No	ditched/dual

Table 3 - Electrochemical Potentiokinetic Reactivation Analysis
Table of Results for Nine Tests Conducted

Sample #	Distance from weld	Area Sample	Initial Rest Potential	Integrated Current	Flade Potential	Grain Size* 100X	Pa
1	62mm	.85cm^2	-445mV	.675 $\times 10^{-3}$ C	-205mV	3.5	0.046 C/cm^2
2	61mm	.89cm^2	+ 54mV	.85 $\times 10^{-3}$ C	-190mV	3.5	0.056 C/cm^2
3	8mm	1.03cm^2	-584mV	5.54 $\times 10^{-2}$ C	-130mV	3.5	3.14 C/cm^2
4	19.5mm	1.11cm^2	-374mV	1.109 $\times 10^{-3}$ C	-187mV	3.5	0.058 C/cm^2
5	20mm	0.74cm^2	-435mV	1.55 $\times 10^{-3}$ C	-170mV	3.5	0.122 C/cm^2
6	50mm	0.89cm^2	-366mV	1.373 $\times 10^{-2}$ C	-140mV	3.5	0.892 C/cm^2
7	13mm	0.97cm^2	-384mV	2.712 $\times 10^{-3}$ C	-190mV	3.5	0.163 C/cm^2
8	base metal	1.01cm^2	-375mV	2.124 $\times 10^{-3}$ C	-120mV	3.5	0.043 C/cm^2
9	base metal	1.01cm^2	-422mV	1.055 $\times 10^{-2}$ C	-150mV	3.5	0.722 C/cm^2

*Grain size averaged to 3.5

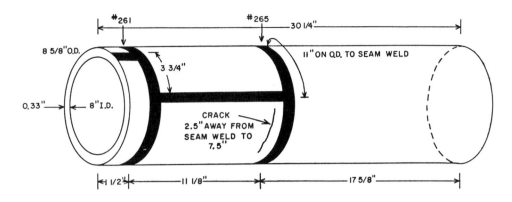

Fig. 1 – Location of welds and crack on 30 ¼" piece of 8" stainless steel pipe from Three Mile Island #1

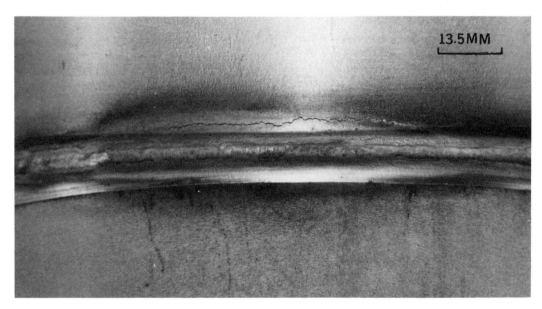

Fig. 2 – Photograph of crack on inside of 8" diameter pipe (crack approximately 5" in length)

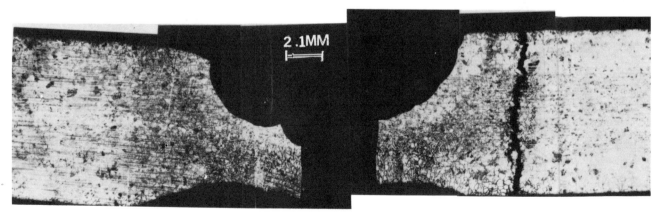

Fig. 3 – Photomicrograph of specimen showing crack counter-bore and weldment (pipe approximately 0.33" thick)

Fig. 4 - Photomicrograph of crack after ASTM
A-262 Practice A was performed

Fig. 5 - Scanning electron microscope photo
of "etched" surface on pipe inside surface

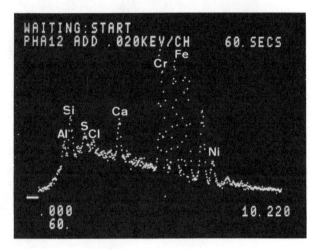

Fig. 6 - EDS scan of surface near "etched"
grain boundaries

Fig. 7 - SEM fracture face of CERT #11 tested
in air--Intergranular (approximately 70%)
type fracture

SIGNIFICANCE OF EXPERIMENTAL FACTORS
IN FRACTURE MECHANICS TEST TECHNIQUES
FOR MARINE STRESS-CORROSION CRACKING

J. A. Hauser II, R. W. Judy, T. W. Crooker
Naval Research Lab
Washington, D.C., USA

D. Tipton
Westinghouse Oceanic
Annapolis, Maryland, USA

T. Caton, A. G. S. Morton
David Taylor Naval Ship R&D
Annapolis, Maryland, USA

ABSTRACT

An experimental program is underway to evaluate the effects of relevant test variables on sea water stress-corrosion cracking (SCC) of high-strength structural alloys for marine service. Two steel alloys, AISI 4340 which is highly susceptible to SCC, and HY-130 which is highly resistant to SCC, are being tested using the cantilever beam (CB) and bolt-loaded wedge-opening-loaded (WOL) specimen configurations to evaluate various test conditions. Results indicate that cathodic protection, test duration and fatigue precracking have significant effects on SCC susceptibility. Also, that the WOL test method is not applicable for testing highly SCC resistant alloys. No significant differences have been found between sea water and 3.5 per cent salt water.

SEA WATER STRESS-CORROSION CRACKING is a significant factor in the selection and use of high-strength structural alloys for the marine environment[1]. Over the past decade, fracture mechanics test methods have become an increasingly important tool in SCC studies[2]. When considering the use of a high strength alloy the Navy normally utilizes one of two fracture mechanics test methods, the CB or the WOL, to assess the possibility of SCC. However, a major impediment in obtaining consistent and reliable data is the lack of a test standard for these methods. Although attention has been directed toward this need by Navy investigators[3] and by the American So-ciety of Testing and Materials (ASTM)[4], there is no formal document. Therefore, a Navy effort is underway to develop a standard which can be used to obtain relevant data and to decrease scatter amongst interlaboratory SCC results. Toward this end, a draft test method document has been prepared[5]. Additionally, as a necessary step in developing a standard, the Navy is conducting an experiental program to evaluate the effects of variables applicable to SCC tests conducted for naval applications. Preliminary results from these tests are presented in this paper.

FRACTURE MECHANICS CHARACTERIZATION PARAMETERS

In the past two decades, linear elastic fracture mechanics (LEFM) concepts have become an accepted method for characterizing SCC, and the plane strain stress-intensity factor, K_I, is the LEFM unit of measure. K_I, whose value depends on both stress and crack length, evaluates the driving force available at the crack tip to precipitate SCC.

The SCC process occurs in two distinct phases, incubation and propagation. After the crack tip is loaded above the level necessary for SCC to occur, there is an incubation period in which the necessary chemical and mechanical interactions occur. The rate of these reactions and length of this delay, or incubation, period depend on several variables and the complex interrelationships are not well understood at the present time. Thus, the length of the incubation period cannot be predicted, a priori, but must be based on previous tests of the same or similar material-environment combinations. This delay time establishes a lower bound on the time required for an SCC test.

Standard SCC tests are concerned with determining a K_I value below which SCC does not occur. This threshold value is defined as K_{Iscc}. Also of interest in SCC tests is the variation in crack growth rates with different levels of K_I. Studies of the propagation phase of SCC are useful when crack repair and inspection interval decisions must be made.

A generalized plot of crack growth rate as a function of K_I (Figure 1) consists of three distinct crack growth regions. Region III is of limited practical interest because crack growth rates are very high and fracture is imminent. Region II is an area where the crack growth rate is essentially constant for a large range of K_I. Region I is an area where small decreases in K_I cause large decreases in crack growth rate. The vertical asymptote approached as K_I decreases is defined as K_{Iscc} for a given material-environment system. It is possible, however, that the crack growth rate never actually becomes zero but that the rate becomes so low as to be experimentally imperceptible. Thus, the definition of K_{Iscc} may really be a function of crack length measurement capability and the investigator's patience.

FRACTURE MECHANICS TEST METHODS

SCC test methods may be classed in two categories according to the type of loading: constant displacement and constant load. For Navy applications one of two test methods are normally specified, either the bolt-loaded wedge-opening-loaded (WOL) test, a constant displacement test, or the cantilever bend (CB) test, a constant load test. The difference in loading gives rise to a major difference in crack growth kinetics: imposing constant displacement causes the stress-intensity factor, K_I, to decrease as the crack extends leading to crack arrest; whereas applying a constant load causes the stress intensity factor to increase as the crack extends, ending in fracture of the specimen. Determining an SCC threshold, K_{Iscc}, requires a different technique for each test. For the WOL test, the procedure for determining a K_{Iscc} value consists of initially loading a specimen to a stress-intensity level higher than K_{Iscc} then waiting for the crack to extend and the stress-intensity factor to concomitantly decrease until the crack arrests. The value of K_I at arrest is defined as K_{Iscc}. For the CB test method, the procedure for determining K_{Iscc} consists of loading several specimens to various levels of K_I, some above K_{Iscc} and some below, then waiting until SCC cracks initiate, grow, and fracture the specimen. K_{Iscc} is then defined as the highest value of K_I for which no fracture occurred.

TEST METHOD VARIABLES

In developing a standard test method for SCC, variables that influence the results must be identified and systematically studied so that appropriate controls and/or measurements are specified in the test standard. A study by the National Materials Advisory Board[2] has identified and prioritized variables that affect SCC tests. The following paragraphs will discuss in detail those variables listed below which were identified as being of concern and which are being evaluated by an ongoing Navy test program.

Increasing or Decreasing K
Test Duration
Environment
Cathodic Protection
Specimen Size
Load Perturbations
Fatigue Precracking

The discussions that follow are concerned only with effects on naval alloys in a marine environment.

TEST TYPE - The WOL test is a crack growth rate decreasing test and the CB test is a crack growth rate increasing test. Because of this it would not be completely unexpected to obtain different results from the two tests, and in fact, conflicting values from the two methods have been reported[6]. But there is no definitive evidence that one method consistently reports higher values of K_{Iscc} and da/dt than the other. However, results from the ongoing Navy SCC test program suggest that there is a problem with the use of the bolt-loaded WOL test for highly SCC resistant alloys. In this program two steels are being tested: AISI 4340 with a yield strength of 180 ksi (1240 MPa) which is quite susceptible to SCC, and HY-130 with a yield strength of 130 ksi (900 MPa) which is highly resistant to SCC. Table 1 lists their chemical compositions. For the 4340 steel, the CB and WOL test methods gave equal K_{Iscc} results. However, for the HY-130 steel a major discrepancy occurred. Sixteen one inch thick WOL specimens were tested and no SCC occurred. Of the sixteen specimens, ten were tested in natural seawater and six in 3.5 per cent salt water. They were tested under three levels of cathodic protection (freely corroding, zinc coupled, and potentiostated to -1000 mV vs. Ag/AgCl), at initial values of K_I from 110 ksi $\sqrt{\text{in}}$ (121 MPa $\sqrt{\text{m}}$) to 130 ksi $\sqrt{\text{in}}$ (143 MPa $\sqrt{\text{m}}$), and left in test from 7500 hours to 10,000 hours. On the other hand, CB specimens of the same one inch thickness, underwent SCC for all levels of cathodic protection.

Possible explanations for the complete absence of SCC in the HY-130 WOL specimens are being investigated. For instance, since the one inch thickness is well below the plane strain limits as given in ASTM E 399[7], additional tests using two inch thick specimens from the same plate of HY-130 are underway. Preliminary results from these thicker specimens show a lack of SCC growth also. These results strongly suggest that the bolt-loaded WOL specimen should not be used for testing materials that are highly resistant to SCC.

TEST DURATION - A lower bound on test duration is the length of the incubation period for the given material-environment combination. Because the incubation time is a complicated function of chemical reaction rates and mechanical interactions with these reactions, there is no available analytical means of predicting this delay period. Estimates must be based on previous experience with the same or a similar material-environment combination. Figure 2 presents results of apparent K_{Iscc} vs. time for CB tests of HY-130 potentiostated at -1000 mV vs. Ag/AgCl in a 3.5 percent NaCl solution. It is apparent that if the test time were too short and still on the steep part of the curve large nonconservative errors in estimating K_{Iscc} would occur.

Although the incubation time is shorter for the WOL specimen (because the initial K_I must be higher than K_{Iscc}) than the CB specimen, overall test time is normally longer because the crack is slowing at an ever decreasing rate.

ENVIRONMENT - Environmental consideration must include not only the test solution but also such parameters as pH, temperature, electrochemical potential, dissolved oxygen level, and whether the test solution is flowing or quiescent. Electrochemical potential level has such a significant effect on SCC susceptibility that it will be discussed separately.

Natural seawater is the solution of choice for Navy testing, but since it is not available at inland labs, and its composition varies from site to site, substitute solutions are often used. The two most popular choices are ASTM Substitute Ocean Water per ASTM D 1141-75[8], and 3.5 percent NaCl. There is little evidence that there is any substantive or consistent difference between these solutions, the choice is left to the discretion of the investigator.

Because of chemical reactions at the crack tip the local crack tip pH tends to remain near 4.0[9] regardless of bulk solution pH. However, because variations of solution pH could indicate experimental problems, pH should always be monitored.

Similarly, bulk solution dissolved oxygen levels may or may not be indicative of crack tip levels and the effect of small variations in dissolved oxygen is unknown. Good experimental procedure would suggest that oxygen levels be kept near saturation. Circulating systems will have high levels of dissolved oxygen naturally, where as quiescent solutions will require some means of oxygen addition. Freely corroding specimens will also deplete oxygen levels and require some action to increase the oxygen level. Dissolved oxygen levels must be monitored for all tests.

Overall temperature effects on SCC threshold behavior are not well known. However, chemical reaction rates and the dissolved oxygen saturation level both depend strongly on temperature so it is important to monitor test temperature.

Very little is known about the effect on SCC of using a quiescent test solution. It is recognized that changes in chemistry occur in stagnant solutions as reactions occur within the test cell. This can cause pH changes, ion buildup, and depletion of dissolved oxygen. The effect on SCC susceptibility is not documented but from a "good practices" standpoint it would be preferable to employ flowing natural seawater or circulating salt water.

Comparsion of interlaboratory data is complicated by environmental differences and the

effects must be evaluated carefully.

SPECIMEN SIZE - Certain size requirements must be met by SCC test specimens in order to apply LEFM and obtain a valid plain strain K_{Iscc} value. Accepted criteria are contained in ASTM E 399[7], and govern crack length, remaining ligament, and specimen thickness. These dimensions are evaluated in relation to the plastic zone size at the crack tip. In testing materials highly resistant to SCC, specimen thickness becomes the limiting criteria. ASTM E 399 requires:

$$\text{thickness} > 2.5 \times (K_I \text{ applied} / \text{yield stress})^2$$

It has been suggested [10] that this requirement is conservative for certain cases. The present Navy program is addressing this question for both WOL and CB test methods with preliminary results suggesting the CB test requires smaller section sizes than the WOL.

CATHODIC PROTECTION - It is generally accepted that for steels undergoing SCC by hydrogen embrittlement, cathodic protection potentials in the range used for general corrosion protection (-1000 to -1300 mV vs. Ag/AgCl) increase susceptibility to SCC[11]. This is attributed to electrochemical reactions which increase hydrogen availability at the crack tip. Figure 3 shows Navy results for CB testing of 4340 steel and indicates that both the incubation time and the SCC threshold are decreased by cathodic protection. Figure 4 shows similar decreases for HY-130 CB specimens where an intermittent (approximately one day in ten) potential of -1300 mV vs. Ag/AgCl was applied above a steady -1000 mV potential, causing a decreased incubation time and SCC threshold compared to a steady potential of -1000 mV vs. Ag/AgCl. This intermittent overpotential simulates possible potentiostatic control problems and demonstrates the need for absolute, reliable potential control when conducting SCC tests.

For corrosion protection, the U.S. Navy normally uses zinc anodes which generate a potential of approximately -1050 mV vs. Ag/AgCl. For SCC testing, zinc can be coupled to the test specimen, or an impressed current system, such as a potentiostat, can be used to provide an equal potential. Preliminary results from Navy testing indicate that using zinc anodes increases the incubation time for SCC of HY-130. Tests are underway to evaluate this effect.

LOAD PERTURBATIONS - Three types of load perturbations are recognized: ripple loads, step loads, and load interruptions. Preliminary research[12] indicates that ripple loads, a small cyclic load superimposed on the mean load, can have a significant, and detrimental, effect on K_{Iscc}. This phenomenon is attributed to the contribution of the cyclic load to the mechanical rupture of the oxide film which forms on the crack surface. Fracturing this film allows absortion of hydrogen ions into the metal lattice. Load interruptions should be avoided since each load cycle contributes to mechanical rupture of the protective oxide layer and also a question of low cycle corrosion fatigue can be introduced. Finally, step loading should be used only as an SCC screening process and when used, the longest possible time should be allowed between increments.

FATIGUE PRECRACKING - It is generally accepted that for precracking SCC specimens, the fatigue stress-intensity factor should be lower than the initial applied K_I. However, for materials with a low SCC threshold which may approach the fatigue threshold, ΔK_{th}, this presents the possibility of extremely long precracking times. What are the consequences if the K_I for precracking exceeds the initial applied K_I? Figures 5 and 6 depict the answer. Tests were conducted, using a 4340 steel with a yield strength of 180 ksi, for both CB and WOL specimens with the K_I used for precracking first higher, then lower, than the initial applied K_I of the test. Figure 5 shows the results for CB tests and it can be seen that both the incubation time and K_{Iscc} have been increased. Results for the WOL specimens, Figure 6, indicate an increase in incubation time for specimens that were fatigue precracked at high levels of K_I. Thus, it is important that fatigue precracking occur at levels below the initial applied K_I. This can present problems for materials with low values of K_{Iscc} which approach K_{TH}.

SUMMARY OF RECOMMENDED PRACTICE FOR SCC TESTING OF NAVAL ALLOYS

Test Type: For material environmental systems with high SCC threshold, the CB test specimen is recommended. Otherwise, either test can be used interchangably.

Test Duration: Determine empirically.

Environment: Flowing seawater or circulating 3.5 percent salt water; fully oxygenated; neutral PH; room temperature.

Cathodic Protection: Potentiostated or anode-coupled.

Specimen Size: Use ASTM E 399 criteria.

Load: Static Load.

REFERENCES

1. B. F. Brown, Ocean Engineering, Vol. 1, p. 291 (1969).

2. Characterization of Environmentally Assisted Cracking for Design, National Materials Advisory Board, Publication NMAB-386, Washington, D.C., 1982.

3. R. W. Judy, Jr., R. J. Goode, "Standard Method of Test for Plane Strain Stress-Corrosion Cracking Resistance of Metallic Materials", Naval Research Laboratory Report 7865, March 17, 1975.

4. R. P. Wei, S. R. Novak, D. P. Williams, Materials Research and Standards, Vol. 12, p. 25 (1972).

5. J. A. Hauser II, R. W. Judy, Jr., T. W. Crooker, "Draft Standard Method of Test for Plain-Strain Stress-Corrosion-Cracking Resistance of Metallic Materials in Marine Environments", Naval Research Laboratory Report 5295, March 22, 1984.

6. R. P. Wei, S. R. Novak, Interlaboratory Evaluation of K_{Iscc} Measurement Procedures for Steel, ASTM STP 821.

7. ASTM E 399-81, "Standard Test Method for Plane-Strain Fracture Toughness of Metallic Materials".

8. ASTM D 1141-75, "Standard Specification for Substitute Ocean Water".

9. B. F. Brown, C. T. Fujii, E. P. Dahlberg, Journal of the Electrochemical Society, Vol. 116, No. 2, p. 218 (1969).

10. C. T. Fujii, "Effect of Specimen Dimensions on K_{Iscc} Determination by the Cantilever Method", Naval Research Laboratory Report 8236, May 31, 1978.

11. C. T. Fujii, and E. A. Metzbower, "The SCC Properties of HY-130 Steel Plate and Welds", Naval Research Laboratory Report 3463, March 1977.

12. R. R. Fessler and T. J. Barlo, Threshold-Stress Determination Using Tapered Specimens and Cyclic Stresses, ASTM STP 821, 1984.

Table 1 - Chemical Composition (Weight percent)

	C	Mn	P	S	Si	Cu	Ni	Cr	Mo	V	Fe
4340	0.41	0.74	0.01	0.016	0.21		2.00	0.74	0.26	0.05	Bal
HY-130	0.13	0.82	-	0.002	0.24	0.05	5.20	0.44	0.52	0.05	Bal

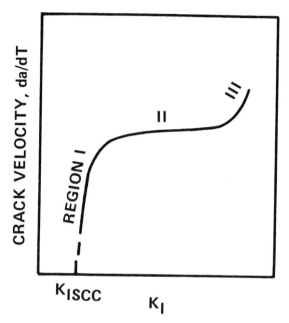

FIGURE 1 - Relationship between crack growth rate and stress intensity factor for SCC testing.

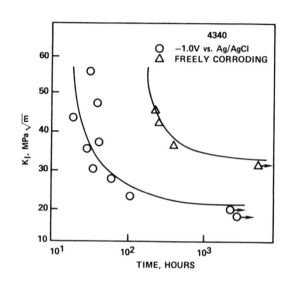

FIGURE 3 - Effect of cathodic protection on 4340 CB specimens.

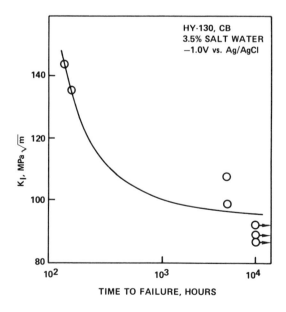

FIGURE 2 - Time to failure for HY-130 CB specimens.

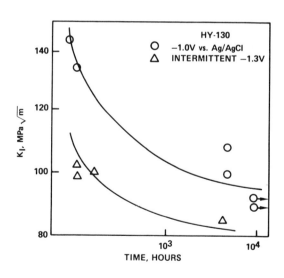

FIGURE 4 - Effect of intermittent overpotential on HY-130 CB specimens.

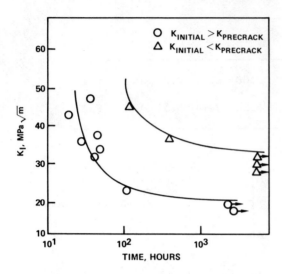

FIGURE 5 - Effect of fatigue precracking above applied K_I for 4340 CB specimens.

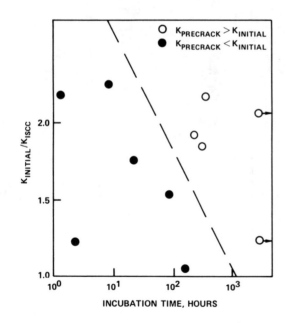

FIGURE 6 - Incubation time increase for 4340 WOL specimens when the pre-cracking K_I is greater than applied K_I.

INFLUENCE OF CYCLIC CONDENSATION ENVIRONMENT ON THE FATIGUE PROPERTIES OF AA 7075 IN T651 AND T651 RRA CONDITION

Sten Å. H. Johansson
Department of Mechanical Engineering
Linköping University
Linköping, Sweden

ABSTRACT

Constant amplitude fatigue testing was performed using unnotched specimens of AA 7075 aluminium alloy in T651 and T651RRA condition in dry air and cyclic condensation of water vapor environment.
The fatigue fracture surfaces were studied in SEM both qualitatively and quantitatively.
The results showed a decrease in fatigue life caused by environment.
The fracture surface displayed special beachmarks in the initiation region caused by the environments. From measurements of the beachmark spacing the crack growth rate for very short cracks was calculated. The results indicate that short cracks seem to grow faster than cracks measured by using fracture mechanics type of specimens.

INTRODUCTION

Materials selection and structural design have, when considering fatigue, been based on testing of undamaged flaw free material (safe life design) or pre-cracked material (fail safe design). From the testing of flaw free material at various stress levels information about total fatigue life, including the crack incubation and crack growth period, can be represented by an S-N-curve and used by the designer. The crack growth in a precracked specimen can be measured, either visually or by some other method, as a function of elapsed cycles and expressed as the growth rate versus the stress intensity factor range. The da/dN ΔK-curves are very useful for the designer when determining the necessary inspection intervals.

In order to improve the strength of a material treated by a double ageing treatment (T7x), Cina (1,2) recently introduced the Retrogression and Reaging treatment. This treatment (RRA) consists of a short time retrogression of the alloy in the T6 temper in the annealing temperature range, followed by water quenching and a reaging treatment identical to the original T6 temper. Some investigations have already been done and it has been confirmed that the RRA treatment gives about the same strength as T6, improvement in resistance to SCC and better exfoliation corrosion resistance (2,3,4).

The influence of environment on fatigue has been known for a long time (5) but it is only during the last decades that the deleterious influence of mild environments, such as humid air, has been realized. The environmental effects can be seen as changes in the total fatigue life or as displacement of the curve in the da/dN ΔK-plot. During the last year considerable work has been done in studying the low crack growth regime of this curve (6). The reduction in fatigue life is probably taking place during the crack initiation as well as during the crack propagation period (5). Unfortunately it has not until recently been wholly realized that crack growth rates obtained by standard fracture mechanics type of specimens are not representative for small cracks in bulk specimens (7,8). Studies of short cracks have so far been done by observing surface flaws by replica technique (9) optical microscope or potential drop techniques (7,8,10). For short cracks there are normally no markings present like the fatigue striations observed in the Paris region of crack propagation. A variation in the environmental parameter during fatigue causes variations in crack growth rate (mode) and thereby introduces markings which can be used to monitor crack growth (14). The crack initiation in Al-Zn-Mg-alloys is often considered to be of slipless type in the high cycle fatigue region (11) and occurs when a critical dislocation loop density is reached in highly localized slip bands. The early fatigue crack growth rate is observed to be higher for short cracks than for equivalent large ones and show a deceleration period at the beginning of growth for a surface crack of about 40 μm in length (7).

The environmental effect is most pronounced at the shortest crack lengths (12) for "long" cracks. The crack propagation is often considered cleavage-like until the region of striation formation is reached (12) for higher stress intensities. In this region the influence of environment, for example humid air, can be clearly seen due to formation of brittle striations with bigger spacing than the ductile ones described by Laird (13). The brittle striations are formed on (100) planes and are brittle in appearance (12,14). The crack extension has been extensively studied by Bowles (15) and it has been proposed that the ductile crack propagation is initially brittle during the rising part of a load cycle. Near maximum stress, however, the crack blunts due to plastic deformation and the subsequent unloading results in a blunted elliptical crack tip profile. In this model the crack propagation mode is dependent on whether the tensile fracture stress is reached before the shear process has started on favourably oriented slip planes or not. The cyclic stress strain response of an Al-Zn-Mg-alloy (16) shows that cyclic hardening takes place. This hardening is probably also present in the crack tip region since the dislocation structure has been found to develop into subgrains in the vicinity of the crack tip (17,14) thereby giving rise to reduced slip activity. The embrittling effect of environment on crack propagation can probably be explained by the reaction of water vapour with freshly exposed reactive metal at the crack tip causing hydrogen evolution. The hydrogen is known to cause embrittlement (18) measured as a reduction of area loss. This means that the crack tip can be embrittled either by reduced slip activity caused by cyclic hardening and hydrogen or by reducing the cohesive strength of the lattice in some way.

The purpose of the present investigation is to determine the influence of environment, in the form of cyclic condensation, on the fatigue properties of 7075T651 and 7075T651 RRA using dry air as a reference environment. Before fatigue testing the state of the surface has been determined by residual stress measurements (14). To correlate the fatigue crack propagation to the observations made during fatigue and to markings caused by the cyclic condensation environment, the fracture surfaces are subjected to both qualitative and quantitative fractographic analyses.

EXPERIMENTAL DETAILS

ALLOY-The commercial alloy AA7075T651 in the form of 6mm thick sheet is used in this investigation. The tensile and yield strength is 577MPa and 524MPa respectively. The material was heat treated to T651RRA condition in the form of slabs (45·150 mm). The retrogression treatment (14) was performed in a salt bath with agitation at 240°C for 35 s (heating up time 18 s). The slabs were stored at room temperature for 24 h before reaging at 120°C took place in order to regain T651 hardness.

SPECIMENS-The specimens were machined from slabs with the tensile axis in the rolling direction and with a theoretical stress concentration factor K_t=1.1. From the results of residual

stress measurements (14) the depth of detectable damage caused by machining was estimated and material to a depth of 210 μm was removed by means of grinding, diamond polishing and electrolytic polishing.

APPARATUS-The fatigue testing machine used was a 5 ton load frame equipped with an MTS 40kN Servohydraulic actuator in closed loop with a 436 and 406 MTS controller. For environmental control a climate chamber was mounted around the specimen and the gaseous environment was fed through tubes from a climate cabinet modified in order to make instantaneous changes between humid air and dry air environment possible. Two types of environments were used namely dray air (20-25% rel humidity) and cyclic condensation environment produced by changing from 100% rel humidity to room air every 30 s. Fracture surfaces were examined in a SEM (JEOL JSM-25 S3). The specimen stage movement and magnification was calibrated using TEM grids.

RESULTS

FATIGUE TESTING-Constant amplitude fatigue testing was performed at two stress levels (200 and 250 MPa, zero min stress) in dry air and cyclic condensation environment. The result is plotted in an S-N-curve (fig. 1).

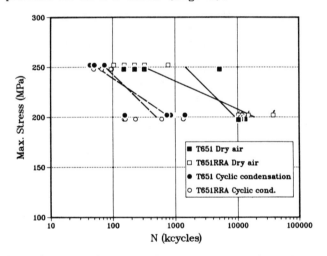

Fig. 1 - S-N diagram showing number of cycles to failure at two levels of MAX stress (200, 250 MPa R=0, points displaced for clarity). Environment: dry air and cyclic condensation. Heat treatment: T651 and T651RRA. (straight lines connecting mean values.

The highest stress level shows fatigue lives between 10^5 cycles and $5 \cdot 10^6$ cycles in dry air environment. The large scatter is caused by one data point for each temper around $225 \cdot 10^3$ cycles. The cyclic condensation of water vapour causes a decrease in fatigue life for the two types (Mean values: T651 $6 \cdot 10^4$, RRA $8 \cdot 10^4$). At the lower stress level the test in dry air was interrupted at about 10^6 cycles. In cyclic condensation environment the fatigue life for the two heat treatments lies between $5 \cdot 10^5$ cycles and 10^6 cycles (Mean values: T651 $8 \cdot 10^5$, RRA $5 \cdot 10^5$).

FRACTOGRAPHIC EXAMINATION-The specimen surface was studied by using plastic replicas during fatigue and the results show that cracks seem to start at inclusions (Fe-rich) and the specimens cycled in humid air display corrosion products around the inclusions. The fracture surfaces of the specimens were studied in SEM. The initiation zone of the specimens fatigued in dry air shows a small flat area at the initiation point and a crack propagation in a cleavage like manner. The specimens cycled in cyclic condensation environment show a more flat initiation point (fig. 2a). There are also certain markings present in the form of semicircular rings centered around the initiation point. The rings are probably due to the change in crack propagation mode caused by the cyclic condensation. The profile of the markings is not very sharp although there seems to be a small angular difference between the planes of crack propagation (fig. 2b) forming a saw tooth profile. The surface structure seem to be somewhat smoother on one of the markings and irregular on the other one. Those irregularities seem to consist of cleavage like pattern. The ratio of the distance between the markings is about 2. The crack front looks almost penny shaped and the crack retardation often reported (7) can not be clearly seen in this case though there are sometimes some irregularities in the crack front. The markings can be observed to a distance of about 100 to 200 μm from the initiation point. The markings then disappear and the crack surface displays the usual cleavage like appearance until the striated region is reached at a stress intensity of 7 MPa√m. This region contains areas covered with ductile striations, in the case of dry air separated with tear fracture dimples caused by primary phases (fig. 3a and b). The environmental crack propagation in this region takes place on elongated flat areas containing brittle striations (fig. 4a and b) sometimes linked together by ductile striations. The ductile striations, when accompanying brittle ones, often have a direction of crack propagation of 45° to the brittle ones.

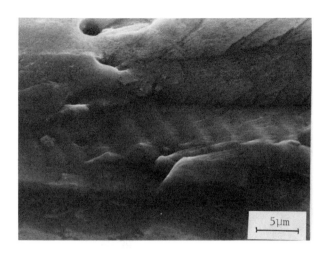

Fig. 2b - Markings in higher magnification showing that the beachmarks can be divided into two parts.

Fig. 2c - Primary electron image of the initiation zone.

Fig. 2d - Optical micrograph of initiation zone showing the markings in black and white contrast.

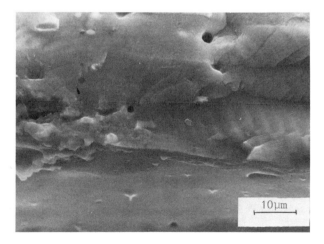

Fig. 2a - A colony of primary phase particles is seen together with corrosion products and a cavity.

Fig. 2 - Fractographs of the initiation region. Max stress 250 MPa, cyclic condensation.

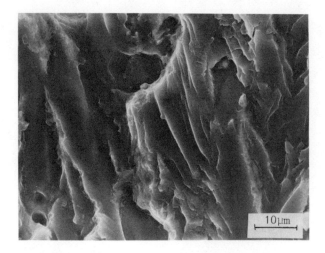

Fig. 3a

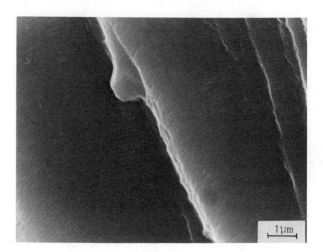

Fig. 3b

Fig. 3 - SEM micrograph showing fracture surface of area with ductile striations. $\Delta K=7.6$ MPa√m, dry air.
a) Ridges of ductile striations in steps, sometimes separated by areas of tearing.
b) High magnification of center part in a).

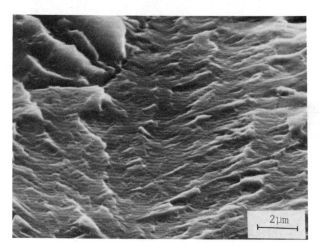

Fig. 4a

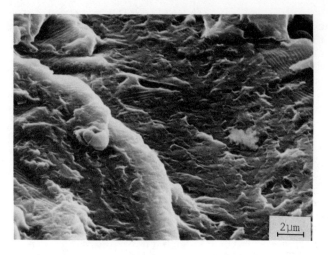

Fig. 4b

Fig. 4 - SEM micrographs of a fracture surface of a specimen tested in cyclic condensation environment at 250 MPa max stress.
a) Environmental propagation by brittle striation formation. $\Delta K=7.2$ MPa√m.
b) Brittle striations are sometimes accompanied by parallel ductile ones at some angle to the grain boundary. $\Delta K=13.5$ MPa√m.

For higher values of stress intensity the propagation in dry air shows more accentuated striation profile, more tearing, microvoid coalescence and dimples caused by primary phases (fig. 5). The brittle striations show a similar trend (fig. 5b) and the accompanying ductile striations show slip lines and have a saw tooth profile. The brittle striations seem to be more dependent on microstructure since they tend to be localised to certain grains. For higher stress intensities the whole environmental cycles can easily be recognized on the fracture surface as bands with higher contrast.

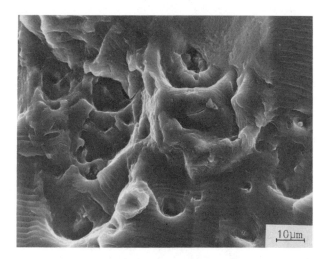

Fig. 5a

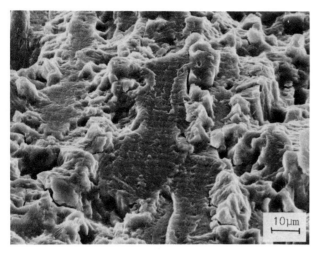

Fig. 5b

Fig. 5 - SEM micrograph of a specimen cycled in air and cyclic condensation, 250 MPa max stress.
a) Ductile striations and dimples caused by microvoid coalescence and primary phases, ΔK=17 MPa√m (dry air)
b) Areas of brittle striations, ΔK=17 MPa√m.

Quantitative Study of Crack Growth in SEM - The crack advance per humidity cycle in the initiation zone has been shown to display markings. From good contrast micrographs it is possible to determine two different stages of propagation probably a wet and a dry part. The ratios between the length of wet part and the dry part is assumed to be 2. Measurements of those markings were performed on photographic prints from light optical micrographs magnified to a total magnification of 1750x (fig. 2d). Since the environmental cycle is composed by one wet and one dry part the wet part was chosen by taking 67% of the total crack advance per environmental cycle divided by the number of load cycles elapsed during the wet part in order to obtain the mean crack advance per fatigue cycle.

Assuming that the crack has a semicircular shape, the cyclic stress intensity was calculated by the formula:

$$\Delta K = 0.69 \ \Delta\sigma\sqrt{\pi \cdot a} \qquad (1)$$

a = crack radius
Δσ = stress range

The cyclic stress intensity calculated by formula 1 is plotted versus crack advance per cycle and is plotted in a logarithmic diagram (fig. 6). Measurements of striation spacing in the region of higher stress intensity were performed for both ductile and brittle striations. The one to one correspondence between the macroscopic growth rate and striation spacing has been investigated and reported (19,20). In this study special care was taken to avoid measurements of striation spacing near inclusions and tear ridges. The whole crack front was inspected during the measurements and large flat representative areas were chosen. As a consequence of this measurements in the lower

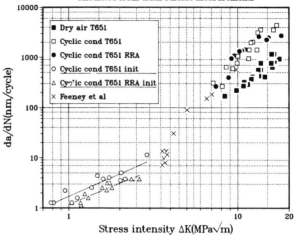

Crack growth rate versus cyclic stress intensity in dry air and cyclic condensation environment.
Calculated from crack surface measurements

Fig. 6 - Growth rate data for 7075 in T651 and T651RRA condition calculated from measurements on fatigue fractographs.

ΔK-region were avoided which is an accordance with findings by Nix et al (20). The brittle striation spacings were also measured as they appear on large flat areas (fig. 3a and b) (fig. 5b). The results are plotted in fig. 6 together with crack growth data of the same alloy produced in measurements performed in humid air by Feeney et al (12).

DISCUSSION

The negative influence of cyclic condensation environment can be seen in the fatigue life curves for the two heat treatments (fig. 1). The scatter in fatigue life makes it impossible to rank the two heat treatments.

The crack propagation in cyclic condensation environment gives rise to markings. The markings are obviously caused by the cyclic nature of the environment but the exact behaviour of the crack during propagation is impossible to describe. The crack can either propagate in a continuous way and change direction of crack propagation i e crack plane or slow down propagation to zero and thereby cause arrest marks. The crack growth can also be completely (21) noncontinuous i e a type of brittle striation were every growth increment is triggered by the presence of water. The lower limit for observations of the markings is a crack radius of about 10 μm. The reason is probably that the primary phase or the colony of primary phases present influences the growth. It might be possible to observe markings in smaller cracks provided they not start on inclusions. Above a crack radius of 100 μm the markings can no longer be observed. The reason for this is probably that the crack growth turns over to a more discontinuous mode more dependent on microstructure.

The noncontinuous crack growth mentioned above seem to proceed in cleavage like manner until striations appear. The crack thereby seems to start to propagate in a continuous manner and then goes over to discontinuous propagation.

If crack growth is looked upon as a process caused by accumulation of damage the production of one striation is caused by extra damage added to damage caused by previous cycles. If more than one cycle is needed to make the crack propagate and the crack extension caused by those cycles is depending on microstructure (orientation of grains) the growth could be noncontinuous and cleavage like. But why is the growth continuous when the cracks are very short in this investigation. The answer is either that the growth is only apparently continuous or that the growth is continuous because the residual stresses causing crack closure are not present for short cracks.

CONCLUSIONS

Fatigue testing in air and cyclic condensation of water vapour has been performed on alloy AA7075 in T651 and T651RRA conditions. The fracture surfaces were studied in order to understand the nature of crack growth.

- The fatigue properties of 7075 in T651 and T651 RRA condition are decreased by cyclic condensation. The difference in fatigue life between the two heat treatments is negligible.
- The growth of very short cracks is reflected by markings on the fracture surface caused by the cyclic nature of the environment. The markings were used to estimate crack growth rate.
- Crack growth rate in the Paris region has been estimated by measurement on ductile and brittle striations.

ACKNOWLEDGMENT

This investigation was part of an internordic project sponsored by the Swedish Board for Technical Development (STU) with contributions from A/S Raufoss and Saab-Scania AB. These important contributions are greatfully acknowledged.

REFERENCES

1. Cina, B. M., U.S. Patent No. 3, 856, 584, Dec. 24, 1974
2. Cina, B. M. and B. Ranish, "New Technique for Reducing Susceptibility to Stress Corrosion of High Strength Aluminium Alloys" in Proceedings of International Conference on Aluminium Industrial Products, Pittsburg, Chapter of the American Soc. of Metal, Oct. 1974
3. Islam, M. U. and W. Wallace, Metals Technology 10, 3986-91 (1983)
4. Nisancioglu, K. and R. Tusvik, The Effect of RRA Treatment on High-Strength Al Alloys of 7000 Series. SINTEF Project Memo 34/243/KN/b1
5. Laird, C. and D. J. Duquette, "Mechanisms of Fatigue Crack Nucleation", p. 88, NACE-2, Houston, Texas (1972)
6. Bäcklund, J., A. F. Blom and C. J. Beevers, ed. "Fatigue Thresholds", EMAS, West Midlands, UK, (1982)
7. Lankford, J., Fat. of Eng. Mat. and Str. 5, 233 (1982)
8. Pearson, S., Eng. Fr. Mech. 7, 235-46 (1975)
9. L. Magnusson, "Cyclic behaviour of Carburized Steel", Linköping Studies in Science and Technology. Dissertations, Linköping 1980
10. R. P. Gangloff, "Fatigue Crack Growth Measurements and Data Analysis", ed. Hudak, Bucci STP 738, Philadelphia (1979)
11. Duquette, D. J. and P. R. Swann, Acta Met. 24, p. 241
12. Feeney, J. A., J. C. McMillan and R. P. Wei Met. Trans. 1, p. 1741 (1970)
13. C. Laird, "Fatigue Crack Prop.", p. 131, ASTM STP, Philadelphia (1966)
14. S. Johansson, "Corrosion Fatigue and Microstructure Studies of the Agehardening Al-Alloy AA7075", Linköping (1984)ISBN 91-7372-810-1
15. C. Q. Bowles, "The Role of Env. Freq. and Waveshape During Fat. Crack Growth in Al Allloys", Report LR-270, Delft (1978)
16. C. Laird, "The General Cyclic Stress Strain Response of Aluminium Alloys", ASTM STP 637 (1976)
17. Grosskreutz, J. C. and G. G. Shaw, Acta Met. 20, p. 523 (1972)
18. Albrecht, J., A. W. Thompson and I. M. Bernstein, Met. Trans. A., 10, p. 1759 (1979)
19. Cina, B. and T. Kaatz, Fat. of Eng. Mat. and Str., 5, p. 233-248 (1982)
20. Nix, K. J. and H. M. Flower, "Mat. Exper. and Design in Fat.", Fatigue 81, Warwick Univ., England (1981)
21. Davidson, D. L. and L. Lankford, Fat. of Eng. Mat. and Str., 6, pp. 241 (1983)

STRESS-CORROSION CRACKING FAILURES OF HIGHER NICKEL-CHROMIUM ALLOY STEELS IN THE PULP AND PAPER INDUSTRY

D. G. Chakrapani
MEI-Charlton, Inc.
Portland, Oregon, USA

ABSTRACT

Use of higher nickel-chromium alloy steels has made significant inroads in pulp and paper industrial machinery in recent years. In general, these alloys out perform carbon steels during corrosion service; however, they are susceptible to stress-corrosion cracking (SCC) under specific environmental conditions. Several case histories of SCC failures of pulp and paper industrial machinery are presented. The equipment includes pulp digesters, liquor feed pipes, black liquor evaporators, desuperheater diffusers, and suction rolls. The materials include several grades of austenitic stainless steel and Inconel. Specific environmental/service conditions that led to the SCC failures are presented and appropriate corrective procedures are discussed.

CORROSION IS ONE OF THE MAJOR PROBLEMS facing operation and maintenance personnel in the pulp and paper industry. The cost of corrosion is high in terms of detrimental and degrading effects on the machinery. The economic incentive for the mill manager to combat corrosion is high, considering that the cost of corrosion is valued at 4.4 percent of the value of goods produced by the pulp and paper industry.[1] Many times service failures of critical components results in expensive downtime and loss in production.

In recent years, to minimize corrosion-related maintenance and equipment downtime, the tendency in the pulp and paper mills is to use corrosion-resistant alloys, in particular, higher nickel-chromium alloys, e.g., various grades of stainless steels, Inconel, etc. These higher nickel-chromium alloys also are increasingly used as replacement materials for traditionally employed carbon steels and nonferrous alloys. Many times such replacements are necessitated by aggressive environments due to increased use of recycled water (e.g., higher chloride ion concentrations) or use of higher strength corrosion-resistant materials to meet the design requirements of modern high speed machinery.

The majority of higher chromium-nickel alloys used in the pulp and paper industry are austenitic stainless steels type 304 and molybdenum-containing types 316 and 317, both containing normal carbon content (maximum 0.08 percent) and low carbon content (maximum 0.03 percent). With the increasing severity of corrosion conditions, special grades of stainless steels and Inconel claddings have been used in recent years for specific applications.[2]

Higher nickel-chromium alloys, in general, offer excellent protection against corrosion. However, they are susceptible to varying degrees of SCC under specific environmental/service conditions. In this paper, SCC of higher nickel-chromium alloys in pulp and paper machinery is presented through case histories. Specific environmental/service conditions responsible for the SCC failures are identified and optional available corrective measures are presented.

CASE HISTORIES

PANDIA DIGESTER INLET CONE - The Pandia digester is a long cylindrical vessel which uses alkaline sulfite liquor to cook sawdust for pulping. The cylindrical vessel is generally fabricated of carbon steel. However, the inlet cone, because of severe service erosion/corrosion conditions where sawdust, cooking liquor, and steam enter, was fabricated of American Iron and Steel Institute (AISI) 304L stainless steel with E308 welds. Typical liquor concentration was approximately 80 percent sodium hydroxide, 20 percent sodium sulfite with chloride concentrations at 2 grams per liter. The operating pressures in the inlet cone were up to 1.2 MN/m^2 (170 psig).

The inlet cone had developed leaks within a year of service. Liquid penetrant inspection showed significant through-wall cracking near the fillet welds joining the bottom flange and side wall (Fig. 1) and the butt welds. Metallographic specimens were prepared from the welds to examine the microstructure and nature of the cracks.

Figures 2 and 3 show the specimens from a butt weld. The cracks had initiated in the base metal, weld heat-affected zone and weld metal and penetrated through the wall thickness. The cracks were transgranular and distinct with significant crack branching which is characteristic of SCC. A few cracks also initiated from the outside surface of the vessel, probably from the liquor leaking onto the outside surface.

The cooking liquor at the inlet cone contained over 2000 parts per million (ppm) chlorides and was aggressive to 304 stainless steel. The cracking was identified as chloride-induced SCC. Austenitic steels of 300-series, while offering good resistance to general corrosion, are highly susceptible to SCC, particularly in the presence of chloride-bearing high temperature liquors. The inlet cone was replaced with an Inconel clad carbon steel inlet cone to combat the SCC. Although the SCC was resolved, this led to a different type of problem described in the next case history.

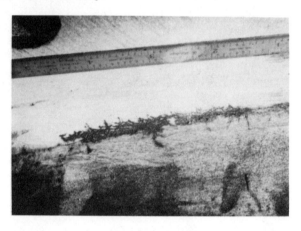

Fig. 1 - Crack indications from liquid penetrant inspection at the weld between the flange and cone.

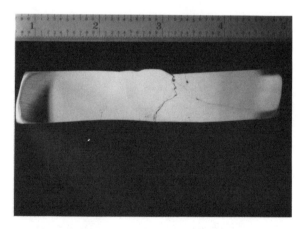

Fig. 2 - Cracks penetrated through the wall thickness and resulted in leaks.

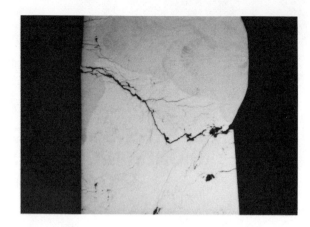

Fig. 3 - SCC initiated on the liquor side in the type 304L base metal and E308 weldment.

Magnification 3X

INCONEL CLAD CARBON STEEL INLET CONE OF THE PANDIA DIGESTER - The Inconel clad carbon steel inlet cone failed in a localized area after 7 years of service. Although the service life of the Inconel clad inlet cone was substantially greater than the 304L austenitic stainless steel inlet cone, it was still less than the anticipated life of 25 years.

The inlet cone was fabricated of 1/2-inch thick type SA-212 Grade B carbon steel with 0.075-inch thick Inconel cladding alloy 600. The failure was caused by an erosion/corrosion leak at the midsection (Fig. 4). Erosion/corrosion was confined to a localized area directly facing the steam inlet nozzle. The erosion/corrosion cavities were smooth-bottomed pits (Figs. 5 and 6). Limited intergranular corrosion was found at the erosion/corrosion pits. The microstructure also had carbide precipitation at the grain boundaries, hence, susceptibility to intergranular corrosion (Fig. 7).

The erosion/corrosion penetrated the Inconel cladding and created crevices at the interface between the carbon steel and Inconel cladding. At some locations, this led to debonding of the Inconel cladding and SCC in the Inconel cladding and carbon steel (Fig. 8). The Inconel cladding was susceptible to SCC, however, under the severe conditions introduced by the crevices between the cladding and carbon steel.

The Inconel cladding was intact elsewhere in the inlet cone with insignificant corrosion-related degradation. In the absence of the conditions that led to erosion/corrosion, the Inconel clad carbon steel was considered adequate for the intended service. As a corrective measure, a solid Inconel liner was recommended in the areas of direct steam impingement. The liners are periodically inspected and replaced. Once the Inconel cladding is protected from erosion/corrosion, the life expectancy of the inlet cone should meet the design criterion.

Fig. 4 - Inconel clad carbon steel inlet cone from the Pandia digester.

Fig. 5 - Erosion/corrosion cavities in the Inconel cladding at the steam impingement area.

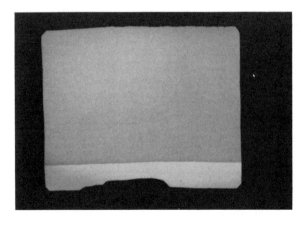

Fig. 6 - Metallographic cross section of erosion/corrosion cavity in the Inconel cladding.
Magnification 4X

Fig. 7 - Intergranular corrosion at the cavities in the Inconel cladding.   Magnification 100X

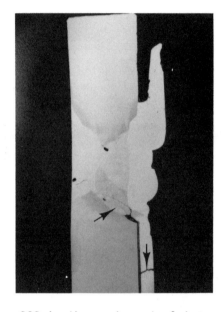

Fig. 8 - SCC in the carbon steel base metal and Inconel cladding (arrows).   Magnification 2X

STAINLESS STEEL BLACK LIQUOR FEED PIPE - An 8-inch diameter stainless steel black liquor feed pipe to a carbon steel digester had failed within 1 year of service. The material was type 316 molybdenum containing austenitic stainless steel. The service environment was alkaline black liquor at 175° C (350° F). The pipe had developed cracks on the inside surface coincident with an external support gusset.

The cracks had initiated at wide corrosion grooves. The early stages were corrosion-assisted fatigue cracks (Fig. 9). The cracks initiated at the corrosion grooves and propagated as transgranular SCC with characteristic branching (Figs. 10 and 11).

Our evaluation indicated that the cracks were localized in an area of high cyclic stresses as a consequence of geometrical constraints on the piping and unsupported cantilever loads. No cracks were found elsewhere in the pipe. In the absence of highly localized service stresses (exceeding yield strength of the material), the corrosion grooving and subsequent SCC would not have occurred in this service environment. The pipe support system was modified with additional gussets to reduce the magnitude of cyclic stresses at the critical areas. The modification was apparently successful.

WHITE LIQUOR TRANSFER PIPELINE – A 4 1/2-inch diameter stainless steel pipe transferring hot white liquor solution of sodium hydroxide and sodium sulfite, developed leaks adjacent to the welds within 4 years of service. The stainless steel pipe was AISI type 304 and welded with E308 weld electrodes. The service temperature was 190° C (375° F) and the solution contained approximately 700 ppm chlorides.

Liquid penetrant inspection of the pipeline showed that the leaks were numerous and were confined adjacent to the welds (Fig. 12). A metallographic specimen from the circumferential weld showed that the cracks initiated at the inside surface (Figs. 13 and 14). In addition to the base metal, SCC also had initiated at a notch at the weld root due to improper welding procedures (Fig. 15).

The failure was attributed to chloride-induced SCC with secondary contributory factors, including improper welding procedures. We recommended that the pipeline be replaced with a material more resistant to SCC. The candidate materials are commercial grade unalloyed titanium or Inconel-600 which have superior resistance to SCC compared to austenitic stainless steels.

Fig. 9 – Cracks at the inside surface of the black liquor pipe were coincident with an external support gusset.

Fig. 10 – SCC initiated at the corrosion grooves in the type 316 stainless steel.

Magnification 4X

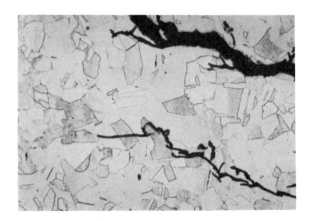

Fig. 11 – Transgranular SCC in type 316 stainless steel.　　Magnification 100X

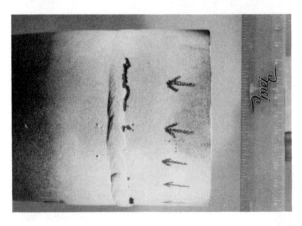

Fig. 12 – Crack indications from the liquid penetrant inspection in type 304 stainless steel white liquor transfer line.

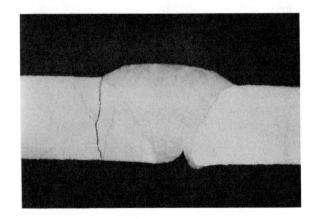

Fig. 13 – Cracks initiated on the liquor side and penetrated the wall thickness.

Magnification 2X

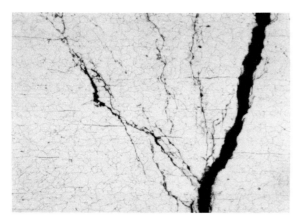

Fig. 14 – Transgranular and branching SCC in the type 304 stainless steel white liquor line.
Magnification 100X

Fig. 15 – SCC in the weld initiated from the weld defect.
Magnification 100X

POWER BOILER STEAM DESUPERHEATER – The desuperheater diffuser nozzle in the steam supply line failed within 9 months of service in a 8.25 MN/m^2 (1200 psig) steam line. The nozzle was an austenitic stainless steel casting in conformance to the American Society for Testing and Materials (ASTM) Specification A 296, Grade CF8 material.

The nozzle had numerous cracks on the inside and outside surfaces, and the cracks had penetrated through the wall thickness in several areas (Fig. 16). Energy dispersive spectrometry (EDS) analysis of deposits on the fracture surface and nozzle surfaces revealed calcium, silicon, sulfur, and sodium with traces of chlorides. The presence of these constituents in the desuperheater assembly was attributed to occasional boiler water carry-over during service. The fracture surfaces had distinct beach markings delineating the crack front, representative of crack propagation stages (Figs. 17 and 18). The cracks were transgranular and unlike classical

al corrosion-fatigue cracks exhibited branching (Fig 19), characteristic of chloride-induced SCC in austenitic stainless steels.

The failure resulted from chloride-induced SCC, possibly assisted by cyclic stress. The sources of cyclic stress are likely intrinsic, probably from temperature fluctuations in the desuperheater assembly. A temperature fluctuation as low as 94° C (200° F) in the assembly, about the mean operating temperature of 480° C (900° F), could induce thermal strains in excess of the yield of the material. The beach markings on the fracture surface were attributed to crack propagation under the influence of cyclic thermal strains.

Our recommendation for alternate material for the desuperheater nozzle included nickel base alloys per ASTM B 564, Grades 600 or 800 titanium alloy per ASTM B 367, Grades C3/C4, or ferritic stainless steel alloy per ASTM 182, Grade FXM27. These materials, in addition to having a superior resistance to chloride-induced SCC, exhibit low thermal strains as compared to austenitic stainless steels. We recommended identifying and eliminating the service-related factors which contributed to carry-over and subsequent chloride contamination.

Fig. 16 – Desuperheater nozzle failed after 9 months service in a high pressure steam line.

Fig. 17 – Fracture surface showing beach marks representative of crack fronts.

Fig. 18 - Beach marks and cleavage-like river markings on the fracture surface.

weld into the base metal (Fig. 22). Metallographic examination revealed that the cracks were transgranular and branching, characteristic of SCC in austenitic stainless steels (Fig. 23). The fracture surfaces had a brittle cleavage-like appearance, typical of SCC in austenitic stainless steels (Fig. 24). Chlorides in the service environment were a contributory factor.

Our evaluation showed that the primary factor causing SCC localized at the electric resistant welds was substantial residual stresses as a result of fabrication procedures. In the absence of large residual stresses, the possibility of SCC in the material in the service environment would be significantly reduced. We recommended that the heat exchanger plates be subjected to stress-relief heat treatment following fabrication and welding.

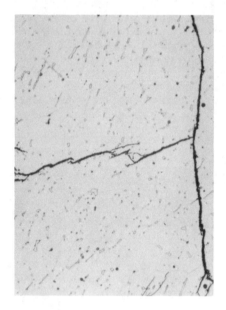

Fig. 19 - Transgranular branching SCC in the CF8 austenitic stainless steel casting.

Magnification 100X

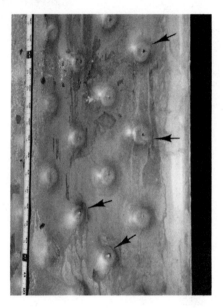

Fig. 20 - Type 316L stainless steel plate heat exchangers developed leaks at the spot welds (arrows).

FALLING FILM BLACK LIQUOR EVAPORATOR - The falling film black liquor evaporator consisted of flat twin plate heat exchangers and were employed in several effects to increase the black liquor solids content prior to its burning in the recovery boiler. Several plate heat exchangers were fabricated of AISI type 316L stainless steel by electric resistance welding.

The plate heat exchangers were formed by two parallel plates, electric-resistant stitch welded at the formed dimples to provide space between the plates. Cracks initiated at the inside surface of the welded areas and penetrated through the wall thickness (Fig. 20). In several locations, the weld fractured and the plates separated with significant spring back, indicative of high residual stresses attributed to fabrication and weld procedures (Fig. 21). The cracks had extended radially from the electric resistant

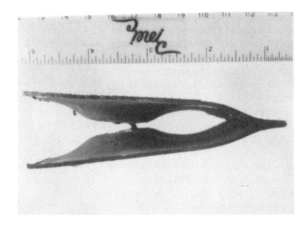

Fig. 21 - Large spring back of the plates indicates significant residual stresses.

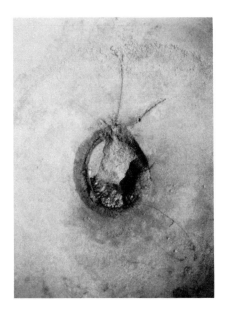

Fig. 22 - Cracking initiated at the spot welds and progressed into the base metal.

Magnification 2X

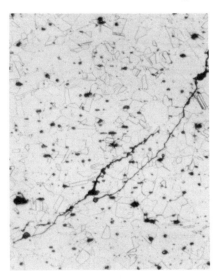

Fig. 23 - Transgranular, branching type SCC in the 316L heat exchangers.　Magnification 100X

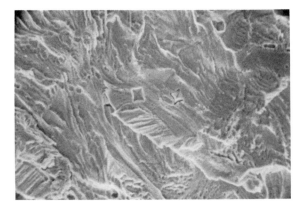

Fig. 24 - SCC fracture surface with cleavage-like river markings.　Magnification 100X

STAINLESS STEEL SUCTION ROLL IN A PAPER MACHINE - Traditionally, suction rolls were made of bronze which provided the advantage of castability, machineability, and corrosion resistance. However, with high speeds and wider paper machines in recent years, stainless steels have been replacing bronze as suction roll materials because of their higher modulus of elasticity. Service performance of the stainless steel alloys are highly dependent upon the severity of the service environment.[3]

Two suction rolls at the first press section of a 25-foot wide paper machine developed cracks within 2 years of service. The rolls were austenitic stainless steel castings made of ASTM A 351 Grade CF8M alloy containing molybdenum. The rolls were exposed to slightly acidic white water (pH approximately 4.7) containing chlorides (45 ppm).

Visual and liquid penetrant inspections of the rolls revealed extensive cracking at the roll inside surface (Figs. 25 and 26). The cracks penetrated more than 30 percent of the wall thickness and a few cracks were several inches long. The cracks were preferentially oriented along the roll length and primarily at the roll inside surface. Field metallographic examination showed significant grain boundary chromium-carbide precipitation and intergranular corrosion (Fig. 27 and 28).

The roll failures were attributed to chromium depletion along the grain boundaries (sensitization) resulting from slow cooling of the casting to avoid large residual stresses. The service environment consisted of acidic chloride solutions which induced intergranular corrosion. Tensile stresses (residual plus service) promoted the propagation of intergranular SCC. Considering the severity of the service environment, the roll manufacturer recommended a proprietary ferritic/austenitic stainless steel as the replacement material for the rolls. This duplex stainless steel had exhibited[3] superior corrosion-fatigue resistance and low residual stress in laboratory tests. Limited service experience thus far seems to confirm these findings.

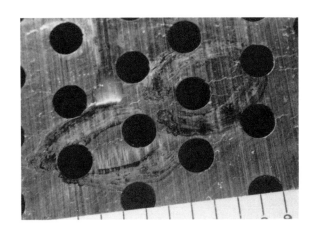

Fig. 25 - Cracks at the inside surface of CF8M austenitic stainless steel suction roll.

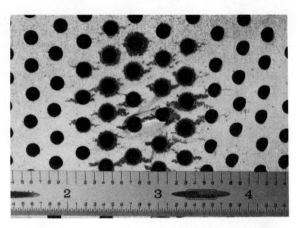

Fig. 26 - Crack indications from liquid penetrant inspection.

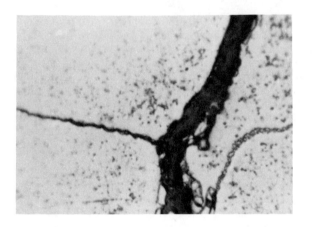

Fig. 27 - Intergranular carbides and corrosion in CF8M suction roll.　　　　Magnification 100X

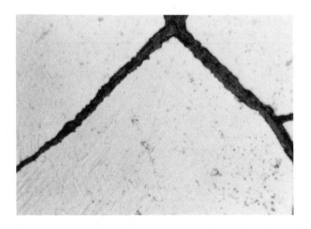

Fig. 28 - Intergranular corrosion in the CF8M suction roll.　　　　Magnification 100X

SUMMARY

The purpose of this paper is to illustrate SCC failures in pulp and paper making machinery. Although higher nickel-chromium alloys offer excellent resistance to corrosion, they are susceptible to SCC under specific environmental/ service conditions. While the failures may bear common characteristics of failures in other industries, the environmental/service stress conditions differ significantly in pulp and paper machinery and requires a closer study to identify appropriate corrective measures. A systematic analyses of the failure can provide valuable information to develop corrective procedures whether design, material, or environmentally related. These case histories provide information about potential problem areas and should be used with caution in predicting performance without considering other variables.

1　"Economic Effects of Metallic Corrosion in the United States." A Report to Congress by the National Bureau of Standards, March 1978.

2　Karl-Erik Johnson, Pulp & Paper Industry Corrosion Problems, National Association of Corrosion Engineers, Houston, Texas, 1980, Vol. 3, p. 2.

3　C.B. Dahl and C.W. Ranger, "Stainless Steel Suction Roll Performance Design and Materials," Pulp & Paper Industry Corrosion Problems, National Association of Corrosion Engineers, 1977, Vol. 2, p. 105.

FRACTURE MECHANICS BEHAVIOUR OF STRESS CORROSION CRACKS

Christina Berger
Kraftwerk Union AG
Muelheim-Ruhr, West Germany

Abstract

Stress corrosion cracks may be initiated and grow in certain components during service under corrosive media. In quenched and tempered steels stress corrosion cracks are intercrystalline, always multiple, branched and have different depths and configurations at different cross sections. Due to their complex crack configuration it is very difficult to calculate the stress intensity factor at the tip of these cracks. Experimental fracture mechanics investigations of such service induced cracks show that their stress intensity factor is 30 to 70 % smaller than the stress intensity calculated for single straight cracks. Theoretical calculations arrive at the same results. Therefore, the critical crack size is at least two times larger than the standard fracture mechanics calculation would predict for a single straight crack.

IF A STRESS CORROSION CRACK is found during in-service inspection of an engineering component it is important to predict its subsequent growth rate under service loading conditions and its critical size to unstable propagation for estimating the residual life time. To calculate the critical crack size by fracture mechanics methodology it is essential to know the stress intensity factor of the stress corrosion cracks.

In quenched and tempered steels stress corrosion cracks have an intercrystalline path and are always multiple, branched and zig-zag formed. Along the length of a major branched crack secondary branching and changes in branch angles can occur. This complex configuration influences the real stress intensity factor and needs more clarification.

The aim of this work is to compare different theoretical calculations of idealized crack configurations with practical investigations of service induced stress corrosion cracks of complex configurations, to quantify the stress concentration factor of such cracks and to look for the consequence to the residual life time of components with such stress corrosion cracks.

CRACK CONFIGURATION OF SERVICE INDUCED STRESS CORROSION CRACKS

Metallographic examinations were performed on stress corrosion cracks (SCC) which were, for example, emanating from the keyways of steam turbine discs due to corrosive environments at high stress concentrations [1-4]. The materials are quenched and tempered steels. All of these cracks have a very complex configuration with regard to number, shape, depth and length. Figures 1 and 2 show the crack initiation path at a keyway and the fracture surfaces of such service induced stress corrosion cracks. Typical cross sections of these cracks are seen in Figures 3 to 5. Very often more than one crack is initiated at the surface. In Figure 3a e.g. four cracks are to be seen, but one of these cracks has no contact to the surface. Sometimes only one or two cracks appear to be nucleated from the surface (Figs. 3f, 4d, 4e). However, parallel metallographic sections always reveal several cracks and crack branches just below. The crack path is always intercrystalline, i.e. along the primary austenitic grain boundaries.

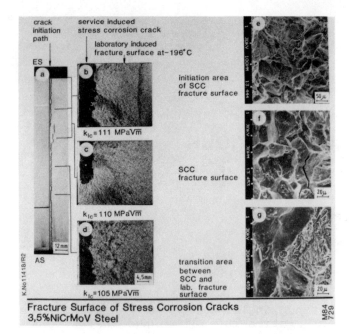

Fracture Surface of Stress Corrosion Cracks 3,5%NiCrMoV Steel

Figure 1

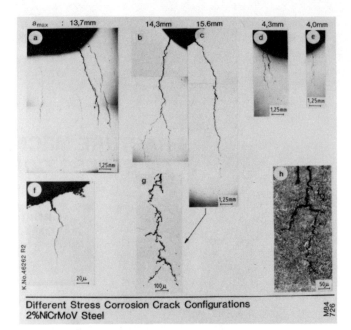

Different Stress Corrosion Crack Configurations 2%NiCrMoV Steel

Figure 3

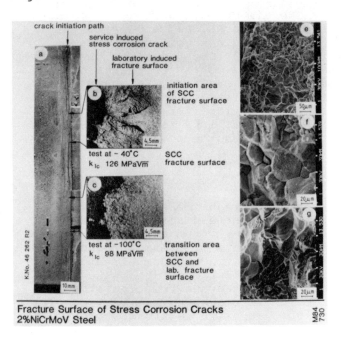

Fracture Surface of Stress Corrosion Cracks 2%NiCrMoV Steel

Figure 2

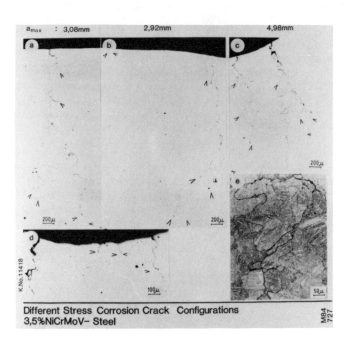

Different Stress Corrosion Crack Configurations 3,5%NiCrMoV- Steel

Figure 4

Therefore, the orientations of the crack tips have to change at every grain boundary (Figs. 3f, 4d, 4e) and they result in a zig-zag crack path. Also crack branching occurs over more than one grain (Figs. 4c, 4e). Due to this, typical macroscopic crack tip branching is induced (Figs. 3b, 3e, 4, 5). In most cases at least two crack tips occur but sometimes many more crack tips are observed (Fig. 5). Each of these macroscopic cracks have further microscopic crack branching (Figs. 3g, 3h).

Figures 3 and 4 can only show the crack configuration in one specific metallographic section. If one grinds in parallel sections only several tenths of a millimeter below one metallographic section one can find very different configurations of the cracks with regard to number, branching and depth of the cracks (Fig. 5). The very different crack configurations are seen much better in Figure 3, where the metallographic sections are at a distance up to 25 mm from each other. Stress corrosion cracks sometimes appear to start from a single crack

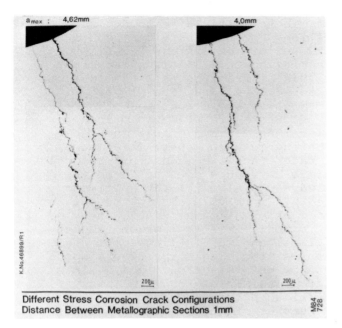

Different Stress Corrosion Crack Configurations
Distance Between Metallographic Sections 1mm

Figure 5

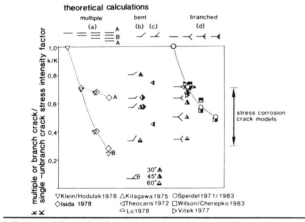

Influence of Crack Configuration on Stress Intensity Factor   M85 846e

Figure 6

but inevitably such cracks branch
to form multiple cracks with very
different angles and shapes.

Summarizing these multiple cracks,
macroscopic and microscopic crack
branching, bending from the straight
crack path, different depths and configurations at different cross sections
all lead to a very complex crack path
which is typical for these stress
corrosion cracks [5].

INFLUENCE OF CRACK CONFIGURATION ON STRESS INTENSITY FACTOR

The value of the stress intensity
factor (SIF) for a given loading condition and given crack length is the
most important point for a fracture
mechanics calculation. Beyond it, the
SIF strongly depends on the crack
configuration.

A lot of publications about this
subject are available [e.g. 6-16]. They
assume crack models for their theoretical as well as for their finite element
calculations and also carry out photoelastic examinations to determine the
SIF of such crack tips. These calculations of SIF $\underline{k}$ for multiple (a), bent
(b, c) and branched (d) cracks are
related to single straight (unbranched)
cracks $\underline{K}$. The resulting relations k/K
are shown in Figure 6. The different
crack models are schematically depicted
in the upper part of this diagram.
By more than one parallel orientated
crack (a) the k-value at the specific
crack tip is reduced by at least 30 %
in comparison to a single crack for
the considered crack distances. Therefore, the ratio is k/K = 0.7. With
more than two cracks, the ratio decreases further, while the outer crack
tips (A) have higher k-values than the
inner ones (B).

Concerning the determination of
SIF of bent (b, c) or branched (d)
cracks, the ratio of straight to bent
or branched crack length and also the
bending angle ⊘ play an important role.
With increasing number of branched
cracks and bending angles the ratio k/K
is decreasing drastically down to 0.35.

Comparing these crack models
in Figure 6 with the metallographically
examined SCC (Figs. 3 to 5) a correlation is possible. From there one can
establish that also parallel cracks
have no straight path due to the intercrystalline crack growth. Therefore,
an additional reduction of stress
intensity has to be taken into account
for this kind of crack path. This is
also true for macroscopic branched
cracks in addition to microscopic
branching with angles larger than 30°.
Also very short cracks change their
orientation at the first primary austenite grain in a typical branching
angle of about 60° (see Figs. 3f, 4d).

Summarizing, one can say that
every change of the crack path means a
decrease of SIF and leads, for the
present SCC configuration models,
to k/K-ratios between 0.3 to 0.7.
However, these crack configuration
models can only consider one specific
section and not the strong variations
of the crack configurations in different
sections. Therefore, these theoretical
crack models lead to an overestimation
of the real k-value.

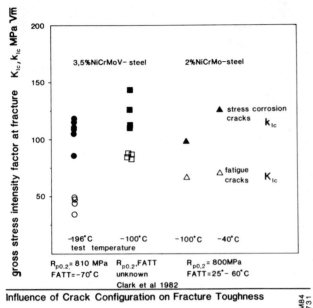

Figure 7

EXPERIMENTAL INVESTIGATIONS OF SPECIMENS WITH STRESS CORROSION CRACKS

To establish the influence of service induced stress corrosion cracks on the stress intensity factor, fracture mechanics tests were carried out on specimens with service induced cracks and with cracks initiated in the laboratory under high cycle fatigue conditions.

25 mm thick three point bend specimens with service induced SCC were taken from two turbine discs of 2 % NiCrMo and 3.5 % NiCrMoV quenched and tempered steels. In addition to that, specimens of the same material and the same size were precracked according to ASTM E 399 in air as mentioned before.

The fracture mechanics tests were carried out at temperatures which guaranteed linear elastic fracture behaviour of the specimens. Therefore, one can state that the constraint of these specimens simulates the stress condition in a large component.

The fracture toughness test results are shown in Figure 7. The specimens with fatigue precracks (unshaded points) are real K_{Ic}-values with a definite stress intensity factor at the crack tip. For the specimens with service induced cracks a so-called gross SIF at fracture k_{Ic} could be determined with the average crack length of each specimen (shaded points).

The test results of 3.5 % NiCrMoV steel at -196°C belong to the service induced cracks according to Figures 1

and 4. These multiple tests for each condition show that service induced cracks sustain on average 2.5 times and a minimum of 1.75 times larger gross SIF at fracture than the fatigue precracked specimens. The plastic zone sizes at the crack tips of service induced cracks are at maximum load smaller than 0.5 mm so that they do not influence each other. This means that each crack tip leads to a reduction of gross SIF. This does not seem to be always decisive for the specimens tested by Clark [17]. Due to the higher plastic deformation at -100°C the plastic zone sizes are larger (about two times) and may influence the reactions at the crack tips. These specimens show on average only a 1.6 times larger gross SIF at fracture than the fatigue precracked specimens.

Further fracture mechanics tests were carried out on 2 % NiCrMo steel at -100°C and -40°C. Crack configurations and fracture surfaces are according to Figures 2 and 3. These two tests show that the specimens with stress corrosion cracks sustain 1.5 and 1.8 higher gross SIF k_{Ic} than specimens with fatigue cracks.

The reason for the scatter of test results for specimens with SCC is the severe changing of crack configuration along the crack length in one specimen (Figs. 1 to 4). The depth and length variations as well as the multiple, zig-zag path and crack tip branching influence the resulting SIF. Furthermore, it is seen from Figure 1 that parallel initiated cracks at the surface do not always have connections to each other so that a mutual relief of the SIF's takes place.

These experimental investigations show clearly that specimens with such stress corrosion cracks have to be loaded up to fracture at least 30 % higher than specimens with normal fatigue cracks.

DISCUSSION

These experimental results on quenched and tempered steels can correlate with further experimental results on Maraging steel and PMMA [18] and with theoretical calculations. For this the SIF's of specimens with stress corrosion cracks and fatigue cracks are compared at the same loading condition. This comparison is shown in Figure 8. The points are the mean values of each test series and the scattering is the result of a comparison of the lowest K_{Ic}-value with the largest gross SIF k_{Ic} and the largest K_{Ic} with

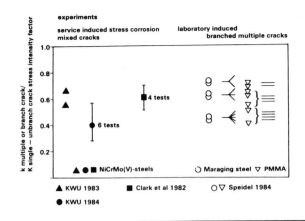

Influence of Crack Configuration on Stress Intensity Factor   M85 845e

Figure 8

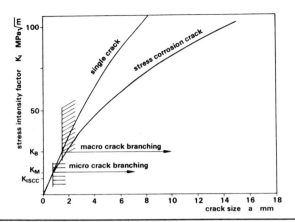

Influence of Stress Intensity Factor on Crack Configuration   M85 706e

Figure 10

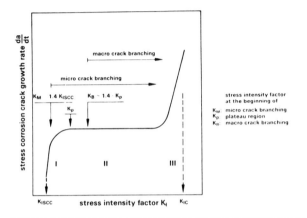

Stress Corrosion Crack Growth Behavior (Schematically after Speidel)   M85 705e b

Figure 9

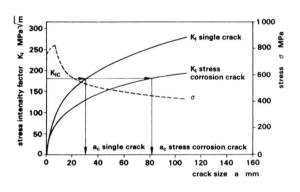

Influence of Stress Intensity Factor on Critical Crack Size in a Notch Stress Field   M85 704e

Figure 11

the lowest k_{IC}-value respectively.

Considering the total effect of the stress corrosion cracks, they lead to k/K-ratios from 0.3 to 0.7. These experimental results are in good agreement with theoretically calculated k/K-ratios for the observed SCC configuration. These results allow a reduction of the stress intensity factor calculated for a single crack by a factor of 0.7 in minimum if there is a multiple and branched intercrystalline stress corrosion crack.

What does this mean for life time prediction of components with such stress corrosion cracks? The stress corrosion crack behaviour and its configuration depends according to Speidel [12] on the K_{ISCC}-value and the stress intensity factor (Fig. 9). If one assumes a value of K_{ISCC} larger than 10 MPa$\sqrt{m}$ [19] then a stress corrosion crack which is induced in a notch stress field shows already micro crack branching after crossing the first austenite grain sizes. That means a

crack length of about 1 mm (Fig. 10). The crack branches macroscopically at a length smaller than 2 mm. In that case the stress intensity factor calculated for a single crack has to be reduced by the factor of 0.7 in minimum.

This reduction of SIF of stress corrosion cracks in comparison to single cracks means that a component with stress corrosion cracks may be loaded at least 1.4 times higher than a similar component with a single straight (fatigue) crack. Therefore, the critical crack size is at least two times larger than standard fracture mechanics calculations would predict for a single straight crack (Fig. 11). Furthermore, if there is a decreasing stress in a notch stress field this yields to a further increase of the crack size before it becomes critical.

This increase of critical crack sizes of components with stress corrosion cracks allows an important prolongation of their tolerable life time.

SUMMARY

Service induced stress corrosion cracks in quenched and tempered steels show a complex configuration. They have always an intercrystalline path, are multiple, branched and have different depths, lengths and shapes. Experimental fracture mechanics investigations on different test materials with service induced stress corrosion cracks show that the stress intensity factors of these stress corrosion cracks are 30-70 % smaller than those of single (fatigue) cracks. Theoretical calculations arrive at the same results. This increases the critical crack sizes in minimum by the factor of two and prolongs the life time of a component with stress corrosion cracks.

REFERENCES

1. Engelke, W., Jestrich, H. A., Schleithoff, K. and H. Termuehlen, American Power Conference, Chicago, Illinois (1983)

2. McMinn, A., NACE, AIME-ANS-Conf., Myrtle Beach (1983)

3. Lyle Jr., F. F. and G. A. Lamping, ASME/IEEE Power Generating Conf., St. Louis, Missouri (1981)

4. Eiselstein, L. E. and R. D. Caligiuri, EPRI Seminar on Low Pressure Turbine Disc Integrity, San Antonio, Texas (1983)

5. Berger, C. and G. Mundt, 5th European Conf. on Fracture, Lisboa (1984)

6. Klein, G. and L. Hodulak, Z. Werkstofftechn. $\underline{9}$, 86-92 (1978)

7. Isida, M., Proc. of the First Int. Conf. on numerical methods in fracture mechanics, Univ. College Swansea, 81-94 (1978)

8. Kitagawa, H., Eng. Fracture Mech. $\underline{5}$, 515-529 (1975)

9. Theocaris, P. S., J. Mech. Phys. Solids $\underline{20}$, 265-279 (1972)

10. Theocaris, P. S., Eng. Fracture Mech. $\underline{17}$, 361-366 (1983)

11. Lo, K. K., J. of Applied Mech. $\underline{45}$, 797-802 (1978)

12. Speidel, M. O., NATO Scientific Affairs Division, Brussels, 345-354 (1971)

13. Speidel, M. O., EPRI Seminar on Low Pressure Turbine Disc Integrity, San Antonio, Texas (1983)

14. Wilson, W. K. and J. Cherepko, Int. J. of Fracture $\underline{22}$, 303-315, (1983)

15. Vitek, V., Int. J. of Fracture $\underline{13}$, 481-501 (1977)

16. Keller, H. P., TÜV-Reports 129 (1981) and 135 (1982), Germany

17. Clark Jr., W. G., Seth, B. B. and D. H. Shaffer, ASME/IEEE Power Generating Conf., St. Louis, Missouri (1981)

18. Magdowski, R. M., Uggowitzer, P. J. and M. O. Speidel, Conf. on Environment Sensitive Cracking Problems, Munich (1984)

19. Speidel, M. O. and R. M. Magdowski, Proc. of Second Int. Symp. on Environmental Degradation of Materials in NPWR, Monterey, Cal. (1985)

STRESS CORROSION CRACKING
OF A TITANIUM ALLOY
IN A HYDROGEN-FREE ENVIRONMENT

R. A. Bayles, D. A. Meyn
Code 6312, Naval Research Lab.
Washington, DC, USA

Stress corrosion cracking experiments performed using the slow strain rate technique with precracked cantilever bend specimens of a titanium alloy are described. Carbon tetrachloride and salt water environments were compared. A computer system monitored the time and load. A compliance calibration was used to calculate the crack depth throughout the experiment permitting calculation of crack growth rate and stress intensity factor. Scanning electron microscopy permitted study of the fractography. Crack growth and fractographic results are discussed with respect to proposed mechanisms of stress corrosion cracking. Future work is described.

THE MECHANISMS of stress corrosion cracking (SCC) in titanium alloys are controversial, particularly regarding the role of hydrogen. Several mechanisms for SCC in titanium alloys in aqueous chloride solutions have been proposed. Bomberger, Meyn, and Fraker [1], Burale and Pugh [2], and Blackburn and Smyrl [3] have written reviews of the subject. The most commonly accepted mechanism for SCC is that the strain due to applied or residual stresses breaks the passive film allowing reaction of water molecules on the bare metal surface, generating hydrogen which may enter the metal [4]. The hydrogen may then form brittle hydrides [5] or facilitate localized slip [6], either of which may promote subcritical cracking. The chloride ion is presumed to inhibit repassivation and, possibly, increase electrochemical activity by increasing solution conductivity [7]. An alternative mechanism, that of stress-sorption, involves the action of chlorine directly weakening interatomic bonds by adsorbing to the metal surface. Although some researchers believe that the hydrogen mechanism is also responsible for cracking in carbon tetrachloride (CCl$_4$) due to residual water [8], the stress-sorption mechanism is viewed more favorably for the CCl$_4$ environment [9,10]. Dissolution mechanisms are disfavored because of the large electrochemical currents required to sustain the observed crack growth rates [11], although the currents would be much less if the dissolution was only required to trigger an increment of mechanical fracture [12]. Much work has been done on surface reactions of chloride ions on metal surfaces [13] including titanium alloys [14]. Some studies of reactions of chlorinated solvents on titanium alloy surfaces have been reported [15] and many more, possibly relevant, studies with aluminum alloys are reported [16,17,18].

In this paper which describes ongoing research, comparisons between crack growth in CCl$_4$ which is a hydrogen-free environment and salt water which contains hydrogen will be made. The crack growth kinetics and subsequent fractographic appearance will be examined to highlight similarities and differences which may provide insight into the cracking mechanisms.

RESEARCH MATERIAL

An alloy known to be very susceptible to stress corrosion cracking was chosen for this work: Ti-8Al-1Mo-1V. The susceptibility was confirmed by comparing experiments (described later) in air (the control environment) with those in the environments of interest. The threshold for cracking and the deformation absorbed by the specimens were much greater in air than in the test environments. The hydrogen content of this material is relatively low (14ppm) which precludes sustained load intergranular or alpha-beta interface cracking due to internal hydrogen from significantly interfering with the environmentally induced transgranular cracking which is the subject of this paper. The microstructure, shown in Fig. 1, consists of elongated alpha phase surrounded by beta phase resulting from the mill anneal. Specimens were cut from the 7mm thick plate

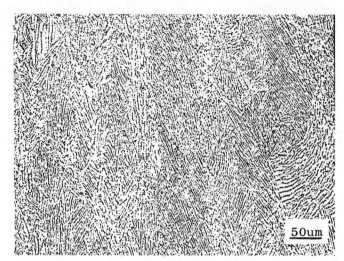

Fig. 1. Microstructure of Ti-8Al-1Mo-1V alloy showing elongated alpha phase (light) surrounded by beta phase (dark). Section normal to long transverse direction.

using a water-cooled abrasive cutoff wheel. A single edge notch and side grooves were put into each specimen with a water-cooled low speed abrasive saw. The relevant dimensions were measured to permit calculation of the stress intensity factor.

The CCl4 used in these experiments is A.C.S. Certified carbon tetrachloride which contains 40ppm water by lot analysis. A Karl-Fischer titration confirmed this water content. The salt water used was prepared by dissolving A.C.S. Reagent grade sodium chloride in distilled water to a concentration of 3.5 weight percent.

TEST METHODS

Initially the slow strain rate (SSR) technique was primarily used for rapid screening of SCC susceptibility, but with more laboratories developing experience with SSR, it is gaining respectability as a sophisticated research tool. The value of the technique is both that the slow straining simulates in a severe manner the creep processes that often occur in practice, and that experiments may be performed in a predictable period of time [19,20]. The specimens in the present work were tested in cantilever bending with the notch on the tension side. The test machine was designed to permit fatiguing the specimens (load ratio, R = 0.1, frequency 15Hz) to produce a precrack of the desired depth, followed by a slow strain rate experiment. Either an adjustable, motor-driven eccentric for fatigue or a reduction gear-driven micrometer for SSR can be attached to the load train which consists of linear bearings for guidance, a 1.1kN capacity load cell, and a pointed rod which deflects one end of the specimen. The other end of the specimen is rigidly clamped. Fig. 2 shows the experimental

apparatus. The deflection rate of the SSR system is 53nm/s producing a rate of increase of stress intensity of 1MPa√m/ks prior to crack propagation. The SSR experiments typically run for about 50ks which is a similar duration to the more usual tensile SSR experiments. The deflection rate was the same throughout this work. Performing SSR experiments over a range of rates can identify a rate at which susceptibility is greatest, which may be different for different environments. The length of the moment arm is 80mm, the specimens are approximately 14mm wide (in the direction of crack propagation) and they are approximately 7mm thick. In the plane of the crack, the side grooves reduce the thickness by about 15%. In order to exclude moisture from the CCl4 environment and to protect workers from CCl4 vapors, the specimen is enclosed in a glove box purged with flowing nitrogen gas. Purging is begun at least 12 hours before the SSR experiment and continues throughout the experiment. The box is open to lab air during the fatigue precracking and the salt water SSR experiments. The liquid environments are contained in a PTFE cup which is press-fit onto the specimens. No adhesive sealants which may contaminate the environment were used. The orientation of the material is TL; the fracture surface is in a plane perpendicular to the long transverse direction and the crack grows in the longitudinal (primary rolling) direction.

Fig. 2. Apparatus for fatigue precracking or slowly deflecting cantilever bend specimen. Glove box permits CCl4 experiments to be performed in N2 to exclude moisture.

MEASUREMENT OF LOAD AND CRACK DEPTH

A microcomputer system [21] using software written by the authors at the Naval Research Laboratory (NRL) is used to monitor the time and the load on the specimen, and to control the motors which power the fatigue and SSR machinery. The load cell is powered and monitored by a conventional load cell conditioner/amplifier, and the amplified signal is sent to the computer via an analog-to-digital converter. The load signal is filtered by a noise rejection routine in the data acquisition software. In addition, during the fatigue experiment, a large capacitor filters the load cell output so that what is monitored is the mean load. The essential function of the computer system is to determine whether a change in load (up or down) has occurred which is greater than a predetermined tolerance. If so, the load and the time are stored as a data point. This approach allows following a relatively fast load drop while not storing a large number of data points associated with constant load as is the case with time interval (chart) recording.

Both the fatigue and the SSR experiments are displacement controlled. Therefore, because the displacement rate is very slow, the load will drop as crack growth reduces the stiffness of the specmen. This load drop has been used with a compliance calibration to provide a means of determining the crack depth at intermediate points in the experiment. In the fatigue precracking portion of the experiment, experience indicated that the desired precrack depth could be obtained simply by having the computer stop the machine after a 20% load drop. In the SSR experiment a more sophisticated calibration was required since intermediate values for crack depth were desired. SSR experiments were performed in air on a series of calibration specimens having saw-cut notches of various depths. Due to the constant deflection rate, the deflection may be calculated from the time recorded by the computer. Up to the point of severe yielding the slope of the load-deflection curve was constant for each specimen with the slopes being lower for the specimens with deeper notches. This series of curves is shown in Fig. 3. In the case of a/W = 0.2 the slope

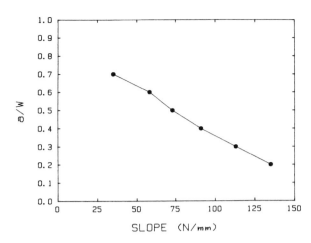

Fig. 4. Linear relationship between crack depths and slopes of load versus deflection curves for wide range of a/W.

decreased to a lower value after about 1mm deflection. For calibration purposes the initial higher slope was used. In practice, after precracking, most specimens had a depth such that a/W = 0.3. The accuracy of calibrations based on such slopes is affected by plasticity in the specimens. The specimens used in this calibration are not as wide as those used in later experiments where more distance for the crack to grow was desired. When the depth of the notch is plotted versus the slope of the load-deflection curve, a straight line is obtained over a wide range of the depth, Fig. 4. This linearity of the depth versus slope permits using the slope, calibrated by the depth at the beginning and

end of the SSR experiment, to calculate the crack depth for each load-deflection data point. An example of this calculation is performed in the discussion section of this paper. Knowledge of the crack depth permits calculation of the stress intensity factor and crack growth rate. The stress intensity factor calculation assumes a single crack front; crack branching will cause the stress intensity factor to be overestimated. Figs. 5a and 5b show typical load-deflection curves for both environments from which crack growth kinetics may be inferred from the steepness of the slopes of the load drops. Fig. 6 shows the corresponding crack growth rate curves calculated using the compliance calibration. Although the calibration is less precise than one based on a crack mouth opening displacement guage, it quantifies the trends observed in the load-deflection curves and is used in this paper to describe the crack growth kinetics. The slow strain rate system and associated data analysis methods described here have been developed over a period of time. All of the specimens tested through this development have contributed to the general characterizations made of the SCC process, while only recent experiments were used to quantify crack growth rates.

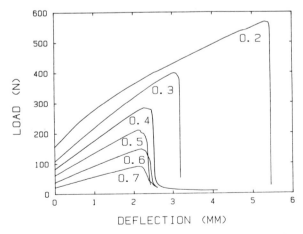

Fig. 3. Saw-notched calibration specimens produced these plots of load versus deflection for several values of a/W. W = 9.2mm. Initial slopes are used in crack length calibration.

FRACTOGRAPHY TECHNIQUES

Scanning electron microscopy was used to study the fracture surfaces, both at normal incidence (sometimes in stereo) to determine the fracture mode, and for viewing the cross sections to determine the amount of secondary cracking. A high-gain backscattered electron detector provided good low-magnificaton images of the fracture surfaces and clearly indicated the microstructural features underlying the fracture surfaces in un-etched cross-sections by utilizing the atomic number contrast due to enrichment of the beta phase with the molybdenum in the alloy. Cross-sections were also polished and etched for examination by conventional metallographic techniques.

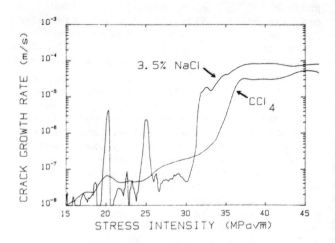

Fig. 6. Crack growth rate as function of stress intensity factor calculated using compliance calibration. Note gradual increase in rate for CCl4 and spurts for salt water.

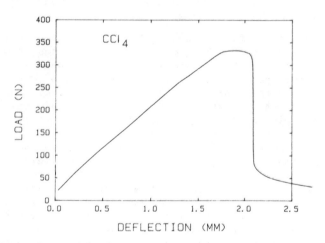

Fig. 5a. Load versus deflection curve for CCl4 environment. Initial ramp is due to slow strain rate deflection. Roll off and steep drop are due to crack growth.

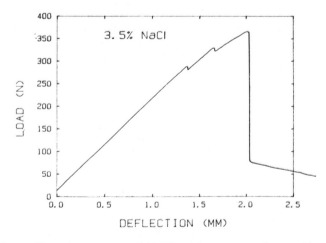

Fig. 5b. Load versus deflection curve for salt water environment. Note two small load drops associated with spurts of arrested crack growth.

CRACK GROWTH RESULTS

In Fig. 5b it is seen that in the salt water case the increasing load due to deflection of the specimen is interrupted by small load drops. At least one of these small load drops was observed in each slow strain rate experiment in the salt water environment whereas no such drop was ever observed in the CCl4 environment. These load drops are a result of spurts of crack growth which were arrested. The small load drops in Fig. 5b correspond to the crack growth rate spikes in Fig. 6 which show that the apparent stress intensity factor at which crack growth begins is between 20 and 25MPa√m in the salt water environment. In the CCl4 environment, crack growth begins at about the same point but the rate smoothly increases with increasing stress intensity factor. In Figs. 5a and 5b the major load drop at about 2mm deflection is associated with the steady state growth above 35MPa√m in Fig. 6. In some cases this drop is steeper for the salt water environment indicating slightly faster crack growth. Although the initial crack growth is steady in CCl4 and intermittent in salt water, the steady state rates are about the same. In the salt water environment the major load drop is often not complete, with the load dropping to an intermediate value and then decreasing at a much more gradual rate as the specimen continues to be deflected. Presumably this interruption of the load drop is due to secondary cracking producing several crack fronts with the stress intensity produced by the applied load divided among them. This effect is less pronounced for the CCl4.

FRACTOGRAPHIC RESULTS

The fractographic appearance of the specimens tested in both salt water and CCl4 is the same. The fracture surfaces consist of transgranular alpha cleavage facets at different altitudes connected by fluted cliffs, Figs. 7a and 7b, with little integranular or alpha-beta interface cracking (IFC). Due to the texture and banded grain structure of the material, the cleavage facets are sometimes linked into narrow longitudinal bands which stretch through the fatigue zone and the SCC zone. These bands seem to occur more in some samples than in others, but their prevalence does not appear to be a function of environment.

Metallographic cross sections show somewhat more transgranular secondary cracking in the salt water environment than in CCl4, Fig. 8a and 8b. Little transverse splitting (longitudinal secondary cracks perpendicular to the main crack caused by transverse stresses) is observed in either environment.

Fig. 7a. Fracture surface from CCl4 environment showing cleavage facets and fluted shearing.

Fig. 7b. Fracture surface from salt water environment showing similar features as in CCl4.

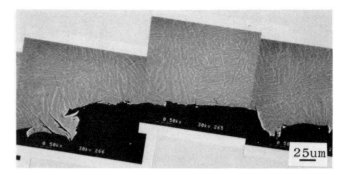

Fig. 8a. Cross section of SSR specimen from CCl4 environment; unetched, microstructure revealed by backscatter SEM.

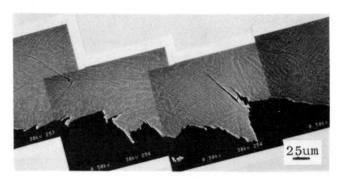

Fig. 8b. Cross section from salt water environment. Somewhat more longutudinal secondary cracking observed.

STAINING IN CCl4

Every specimen tested in CCl4 experienced brown and blue staining which was not observed in any of the specimens tested in salt water. The stain appeared on the SCC fracture surface and as an arc on the fatigue zone shown in Fig. 9. The fatigue precracking was done in air, not in CCl4, so the stain is believed to be a residue produced during subcritical cracking in the SSR experiment. The residue may be soluble in the CCl4 only in the special environment of the crack tip, precipitating out of solution farther away. Alternatively, it may precipitate out during evaporation of the CCl4 at the end of the experiment in the nitrogen gas environment in the glove box with the arc produced by some surface tension effect. An energy dispersive x-ray spectrometer (EDS) on an SEM was used to examine the stain on the relatively flat fatigue fracture surface. The resulting spectrum was compared with one from the shiny area of the fatigue zone inside the arc. As shown in Fig. 10, the stain contains chlorine which is not observed at the shiny area a millimeter away where the stain is absent.

Weber, et al. [22] have found similar corrosion products for SCC experiments with Ti-8Al-1Mo-1V done in both CCl4 and salt water. Possibly the circumstances of the present experiments causes the residue to be deposited only in the CCl4 case.

Fig. 9. Comparison of fracture surfaces showing stain arc on fatigue zone of CCl4 specimen and not on adjacent salt water specimen. Color from stain also visible in subcritical crack growth region of CCl4 specimen.

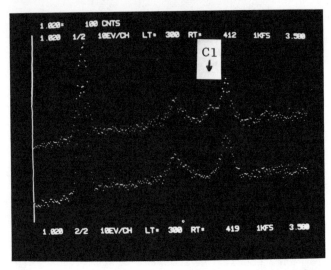

Fig. 10. Energy dispersive x-ray spectra showing chlorine peak for CCl4 environment. Peak is not present for salt water environment.

DISCUSSION

A compliance calibration was used to determine the crack length, allowing calculation of crack growth rate and stress intensity factor producing curves of the type shown in Fig. 6. Concerns regarding the low precision of the compliance calibration are mitigated by the fact that the specimens tested in CCl4 and salt water have nearly the same dimensions and experience crack initiation at about the same load. The effects of plasticity in the specimens and flexibility in the load train and its support structures should be identical in both environments.

The cantilever bend configuration and system compliance used here mandates that the stress intensity factor will increase as the crack grows, despite load drops (shown in the following calculation). This increase should cause continued crack growth possibly at an even higher rate, contrary to what is seen in the initial stages of crack growth in the salt water environment where crack growth arrests temporarily. The following calculation briefly illustrates the algorithm used to calculate the compliance calibration parameters and then applies that result to determine the crack growth and stress intensity increase associated with the first small load drop in Fig. 5b. The crack lengths at the beginning and end of the experiment are measured to be a_O = 4.9mm and a_f = 13.3mm. The corresponding slopes of the load-deflection curves are measured from Fig. 5b to be m_O = 206N/mm and m_f = 7.1N/mm. The slope of the function of the crack growth versus load-deflection-slope is:

$$s = \frac{a_f - a_O}{m_f - m_O} = \frac{13.3 - 4.9}{7.1 - 206} = -0.042 \qquad (1)$$

Since at zero deflection the load is zero, each of the load-deflecton curves passes through the same origin. Therefore, the intercept on the crack length axis of Fig. 4 may also be calculated. In order to calculate the crack growth associated with the small load drop, Eq. (1) may be rearranged to solve for the change in crack length:

$$\Delta a = s \times \Delta m \qquad (2)$$

Before the load drop the slope of the load deflection curve is the initial slope of 206N/mm. After the load drop the slope measured from Fig. 5b is 197N/mm, so Δm = -9N/mm. From Eq. (2) one calculates that Δa = 0.37mm or about 400um, which is much larger than features in the microstructure. Calculation of the increase in stress intensity for this cantilever bend configuraton during the first small load drop makes use of an adaptation of the 3-point bend formula of Kies, et al. [23]. They derived a function, f, of the crack depth, a, and the specimen width, W:

$$R = \frac{a}{W} \qquad (3)$$

$$f(R) = 4.12 \sqrt{\frac{1}{(1-R)^3} - (1-R)^3} \qquad (4)$$

Their formula for the stress intensity factor, K, in terms of the load, P, the length of the moment arm, L, and the thickness and net thickness, B and B_N is:

$$K = \frac{PLf(R)}{W\sqrt{BB_N W}} \qquad (5)$$

A value for $f(R)$ may be calculated for the original crack depth and for the crack depth after the small load drop. The stress intensity factor may be calculated for both cases using the appropriate values of P and $f(R)$ in Eq. (5). During the small load drop the stress intensity factor increased from an initial value of $18.6 MPa\sqrt{m}$ to $19.4 MPa\sqrt{m}$.

In the slow strain rate experiments described here the likely first step in the cracking process is breaking of the passive film due to strain at the tip of the fatigue crack, thereby providing the CCl_4 or salt water with access to the reactive metal under the film. The following mechanisms may be involved in crack propagation:

1. Strong film rutures allowing slip to exit the surface.
2. Alloy dissolves producing stress concentration.
3. Hydrides form and then fracture.
4. Hydrogen hardens the alloy causing embrittlement.
5. Hydrogen softens the alloy allowing slip.
6. Work hardened surface layer dissolves triggering increment of mechanical fracture.
7. Some species, possibly H or Cl, adsorbs to the surface weakening interatomic bonding as in liquid metal embrittlement (LME).

The first mechanism may be disregarded since, by itself, it does not involve an environmental effect. The second mechanism would require too much current flow to be plausible [11]. The next three mechanisms require diffusion of hydrogen which should slow the cracking process; also hydrogen is in extremely short supply in the CCl_4 case. At least in the case of salt water, diffusion may not be such a limitation if a small amount of hydrogen embrittlement in the surface at the crack tip is sufficient to trigger an increment of purely mechanical crack growth. On the other hand, each of the hydrogen effects probably requires a significant depth of absorption to induce cracking. The sixth mechanism is appealing for both environments if dissolution is only neccessary to trigger mechanical crack growth. The seventh, stress-sorption, has been shown to produce very high crack growth rates in LME and so does not require mechanical fracture to occur in order to produce the observed high crack growth rates. Detailed analysis of the kinetics involved in these mechanisms is beyond the scope of the present paper. Particularly if crack advance is primarily due to mechanical fracture triggered by an environmental effect, the similarity of the fractographic appearance does not guarantee that the same mechanism is responsible for cracking in both environments.

Scully [24] argues that such similarity in fractographic appearance is not unreasonable even when the cracking mechanism is different.

The primary difference between SCC in the two environments which the present work reveals is in the initial stages of cracking where, in salt water, spurts of crack growth are arrested. The high rate of crack growth may be due to abrupt release of elastic energy or a very active cracking mechanism. The crack arrest is due to repassivation or secondary cracking, both active chemical processes. A possible explanation for the observed behavior is that in the salt water, a strong passive film exists which protects the metal until enough strain accumulates to fracture the film.

Then the appropriate cracking mechanism(s) advance the crack. Repassivation and/or secondary cracking begin to stifle the cracking process; the crack arrests and waits for more strain to accumulate. In the CCl_4 environment, conditions may not be suitable for formation of a strong film. In that case, the environment has more steady, although still restricted, access to the metal causing cracking to advance in much smaller steps appearing to respond more immediately to the stress intensity. A lower propensity for secondary cracking would also encourage steady crack growth.

CONCLUDING REMARKS

Several mechanisms of cracking are consistent with the results of the present work. The staining in CCl_4 indicates that some dissolution occurs, but no evidence was observed which links such dissolution with crack advance. Also, dissolution may have occurred in salt water without leaving a visible residue. The experimental technique described highlights the early stages of subcritical cracking and may permit future work to eliminate some possible cracking mechanisms from contention. Future experiments will be performed using a new high-capacity, variable speed, slow strain rate test machine (also computer controlled and monitored). Larger specimens of the WOL type will be used and a crack mouth opening displacement guage will permit more accurate analysis of crack growth rates. Determining the crack growth kinetics as a function of strain rate, permitted by the new machine will aid in determining the details of the cracking process. Corrosion inhibitors of different types [25] will studied to determine what effect an inhibitor's particular function has on the SCC process. Also, further work must be done to identify the nature of the residue which produced the stain on the specimens tested in CCl_4. A Karl-Fischer titrimeter will be used to monitor the water content of the carbon tetrachloride during each experiment. This instrument will permit determining the effect of small water additions to the CCl_4. Electron diffraction will be used to look for hydrides on the fracture surfaces. The presence of hydrides would be strong evidence for a hydrogen absorption mechanism. These results will then be integrated with the results and hypotheses of others to clarify the role of various environmental agents in SCC of titanium alloys.

ACKNOWLEDGEMENTS

The authors thank Dr. A. John Sedriks and Dr. E. Neville Pugh for encouraging the initiation of this work. We thank Dr. Mitchell Jolles for his discussions on the applicability and limitations of the compliance calibration. This project was supported by ONR/NRL basic research funding. Support was also provided by the Naval Air Systems Command.

1. Bomberger, H. B., Jr., Meyn, D. A., and Fraker, A. C., in "Titanium Science and Technology", pp. 2435-2454, G. Lutjering, U. Zwicker, and W. Bunk, Eds., Deutsche Gesellschaft fur Metallkunde e. V., Oberusel, West Germany (1985)

2. Bursle, A. J., and Pugh, E. N., in "Environment-Sensitive Fracture of Engineering Materials", pp. 18-47, Z. A. Foroulis, Ed., The Metallurgical Society of the American Institute of Mining, Metallurgical, and Petroleum Engineers, Warrendale, PA, USA (1979)

3. Blackburn, M. J., and Smyrl, W. H., in "Titanium Science and Technology", pp. 2577-2609, R. I. Jaffee and H. M. Burte, Eds., The Metallurgical Society of the American Institute of Mining, Metallurgical, and Petroleum Engineers, Warrendale, PA, USA (1973)

4. Smith, J. A., Peterson, M. H., Brown, B. F., Corrosion, 26, 12, 539-542 (1970)

5. Sanderson, G., Powell, D. T., Scully, J. C., in "Fundamental Aspects of Stress Corrosion Cracking", pp. 638-649, R. W. Staehle, A. J. Forty, and D. Van Rooyen, Eds., National Association of Corrosion Engineers, Houston, TX, USA (1969)

6. Beachem, C. D., Met Trans, 3, 437-451 (1972)

7. Green, J. A. S. and Sedriks, A. J., Corrosion, 28, 6, 226-230 (1972)

8. Scully, J. C. and Powell, D. T., Corrosion Science, 10, 719-733 (1970)

9. Vassel, A., in Ref. 2, pp. 277-283

10. Hoffmann, C. L., Judy, R. W., Jr., and Rath, B. B., in Ref. 1, pp. 2495-2502

11. Scully, J. C., in "Stress Corrosion Research", pp.209-224, H. Arup and R. N. Parkins, Eds., NATO, Sijthoff & Noordhoff, The Netherlands (1979)

12. Jones, D., Met Trans, 16A, 1133-1141 (1985)

13. Hoar, T. P. and Jacob, W. R., Nature, 216, 1299-1301 (1967)

14. Beck, T. R. and Grens, E. A., II, J. Electrochem. Soc., 116, 2, 177-184 (1969)

15. Leith, I. R., Hightower, J. W., Harkins, C. G., Corrosion, 26, 9, 377-380 (1970)

16. Foroulis, Z. A. and Thubrikar, M. J., J. Electrochem. Soc., 122, 10, 1296-1301 (1975)

17. Archer, W. L. and Simpson, E. L., Industrial & Engineering Chemistry Product Research and Development, 16, 158-162 (1977)

18. Archer, W. L., Industrial & Engineering Chemistry Product Research and Development, 21, 670-672 (1982)

19. Parkins, R. N., in "Stress Corrosion Cracking - The Slow Strain-Rate Technique, ASTM STP 665", pp. 5-25, G. M. Ugiansky and J> H. Payer, Eds., American Society for Testing and Materials, Philadelphia, PA, USA (1979)

20. Payer, J. H., Berry, W. E., and Boyd, W. K., in Ref. 19, pp.61-77

21. Meyn, D. A., Moore, P. G., Bayles, R. A., and Denney, P. E., in "Automated Test Methods for Fracture and Fatigue Crack Growth, ASTM STP 877", pp. 27-43, W. H. Cullen, R. W. Landgraf, L. R. Kaisand, and J. H. Underwood, Eds., American Society for Testing and Materials, Philadelphia, PA, USA (1985)

22. Weber, K. E., Fritzen, J. S., Cowgill, D. S., and Gilchriest, W. C., in "Accelerated Crack Propagation of Titanium by Methanol, Halogenated Hydrocarbons, and Other Solutions, DMIC Memorandum 228", pp. 39-42, Battelle Memorial Institute, Columbus, OH, USA (1967)

23. Kies, J. A., Smith, H. H., Romine, H. E., and Bernstein, H., in "Fracture Toughness Testing and Its Applications, ASTM STP 381", pp.328-356, American Society for Testing and Materials, Philadelphia, PA, USA (1965)

24. in Ref. 11

25. Sedriks, A. J. and Green, J. A. S., Journal of Metals, 48-54, (April 1971)

ONLINE MONITORING OF HYDROGEN IN STEELS AT AMBIENT AND ELEVATED TEMPERATURES

Koji Yamakawa, Harushige Tsubakino
Department of Metallurgical Engineering
College of Engineering
University of Osaka
Prefecture, Sakai, Osaka, Japan

Abstract

Two new methods for monitoring hydrogen content in steel at ambient and elevated temperatures by using an electrochemical permeation method are proposed. One is to monitor hydrogen content in pipe line continuously by using a nickel-plating method. This probe is based on the formation of NiOOH which contains a mobile hydrogen. The field test was performed with high reliability over three months. The detection limit of this hydrogen probe is less than 0.01 ppm as a hydrogen content. The other is to monitor hydrogen content in steel at elevated temperatures (673-773 K) in the range of practical interest for hydrogen attack. Molten sodium hydroxide as an electrolyte and a stabilized zirconia reference electrode are used. The detection limit is quite enough to monitor hydrogen attack in steel.

THREE BASIC METHODS for measuring hydrogen content at ambient temperatures in steels are in current use[1], i.e., a pressure method, a vacuum method, and an electrochemical method. However the electrochemical method has numerous remarkable features[2-7]; (a) good sensitivity, (b) a simpler measuring apparatus and (c) suitability for successive measurement. The Petrolite Corporation developed a hydrogen probe using an electrochemical method, in which a palladium foil was placed on the extraction surface of steels[8]. Such a probe can not apply for a thin or annealed iron specimen because the background (residual) current is considerably large (smaller residual current gives higher sensitivity for measuring hydrogen content in steels). Yamakawa Et al.[4] have found that a nickel-plating method has better sensitivity than the palladium method. This is because thin and uniform nickel-plating is possible and the nickel plated metal is changed

to NiOOH, whose hydrogen is mobile, in a alkaline solution under a certain anodic polarization.

On the other hand, there is no monitoring probe at elevated temperatures in the range of practical interest of hydrogen attack. All of the studies on hydrogen attack have been performed by a usual gaseous method using pressure vessel or autoclave[9] but not performed by an electrochemical method because an aqueous solution has been used as an electrolyte.

In this study, a monitoring apparatus and some techniques to monitor the hydrogen content in pipe line over a long term at ambient temperature and, also, in a steel at elevated temperatures(673-773 K), by an electrochemical method are presented.

BASIC CONCEPT TO MEASURE HYDROGEN CONTENT

An electrochemical permeation method is based on the following concept (Fig.1): hydrogen is introduced into one side of a metallic specimen during cathodic polarization, while the other side is held at an anodic potential sufficient to ionize any hydrogen arriving at the surface after passage through the metal:

$$H \rightarrow H^+ + e^-. \tag{1}$$

The resulting ionization current is a direct measure of the instantaneous rate of the hydrogen permeated through the metal[2]. At the steady state, the hydrogen content C_o (ppm) in a planar specimen can be calculated from eq.(2):

$$C_o = 1.318 \times 10^{-6} \frac{J_\infty L}{s\, D} \tag{2}$$

and for hollow cylindrical specimen, when the ratio of outer to inner radii is larger than 1.1,

$$C_o = 1.318 \times 10^{-6} \frac{J_\infty a \ln(b/a)}{s\, D} \tag{3}$$

where, J_∞(A) is the permeation current at the steady state, L(m) the thickness of specimen, D (m²/s) the diffusion coefficient of hydrogen in the specimen, s(m²) the area, and a and b are the inner and outer radii, respectively.

The time-dependent equalization of the hydrogen content in the specimen is described by Fick's second law, and by applying Fick's second law with the appropriate initial and boundary conditions, theoretical permeation curves during the build-up and decay processes were obtained. D was calculated from the fitting of the theoretical curves to the build-up and decay transient curves obtained by changing the cathodic current (Fig.2).

MONITORING HYDROGEN CONTENT IN PIPE LINE

PROCEDURE - The sample used was a pipe which was equivalent to API grade 5LX-X60 (outer diameter of 0.5 m and 10 mm in thickness). The chemical composition of the pipe is shown in Table 1. The pipe was cut into 0.15x0.08 m² for laboratory test and two steel plates were welded on both sides of the pipe of 0.3 m in length for field test.

Before the test, the surface of the pipe was prepared, as shown in Table 2. The thickness of nickel-plating layer was about 50 nm, which was determined from the relationship between the plated thickness and diffusivities of hydrogen in a mild steel[4].

The hydrogen permeation experiment was performed by using the apparatus shown in Fig.3. The specimen was put into two cells, the entry and extraction cells. The extraction cell was filled with 1 or 3 N NaOH solution, which was selected depending on a test temperature, and the specimen was polarized at +150 mV vs Hg/HgO potentiostatically. Then a high anodic current was observed due to the passive reaction of eq.(4):

$$Ni + 3OH^- \rightarrow NiOOH + H_2O + 3e^- \qquad (4)$$

After the anodic or residual current density

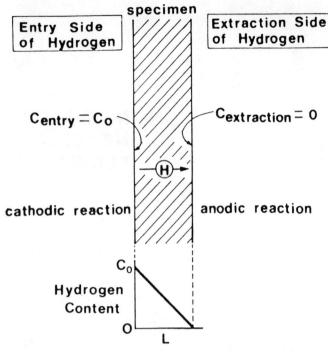

Fig.1 Principle of electrochemical permeation method.

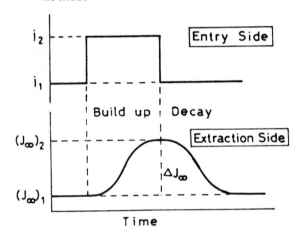

Fig.2 Build up and decay permeation transients accompanying with different cathodic currents.

Table 1 Chemical compositions of steels (wt%).

	C	Si	Mn	P	S	Cu	Cr	Ni	Mo	Nb	Ti	Al	Ca
5LX-X60	0.07	0.03	1.23	0.016	0.001	0.03	0.03	0.03	0.06	0.032	0.017	0.041	0.023
Carbon Steel	0.19	0.22	1.00	0.018	0.012	--	--	--	--	--	--	0.015	--

Table 2 Surface preparation and monitoring of pipe line.

1. Mechanical Polishing --- Emery Paper (up to #600)

   (Electropolishing ------ $H_3PO_4 - CrO_3$, 3 kA/m² x 12 min)

2. Acid Pickling ---------- 3N HCl

3. Nickel-Plating --------- Watt's Bath, 30 A/m² x 3 min

4. Passivating ------------ 1 or 3N NaOH, +150 mV vs Hg/HgO x 24 hr

5. Monitoring ------------- +150 mV vs Hg/HgO

decreased less than 0.3 mA/m², the extraction cell was filled with a acidified and buffered solution (0.2 M CH₃COOH + 0.017 M CH₃COONa). The cathodic hydrogen charging was then begun to obtain diffusivities at various temperatures between 263 and 323 K.

The hydrogen monitoring system for a field type consisted of cells for nickel-plating and monitoring, which were made from acrylic resin, pumping equipment and electronic equipment. Photo.1(a) and Fig.4 show the pumping equipment used for supplying electrolyte to the cell and degassing the electrolyte. Fig.5 shows the electrolytic cells for nickel-plating and for monitoring the hydrogen content in pipe line. The monitoring cell has a auxiliary chamber containing a reference electrode (Hg/HgO) and a counter electrode connected to the cell by a universal joint. Because of this connecting method, the cell could be set at any position on the external surface of pipe line, as shown in Fig.6.

RESULTS AND DISCUSSION - Laboratory Test -
The diffusivities of hydrogen obtained from both specimens with and without electropolishing are shown in Fig.7. These data can be expressed by eq.(5), obtained from the method of least squares.

$$\log D = -1.26 \times 10^3 \frac{1}{T} - 5.14 \qquad (5)$$

Furthermore, the steady state permeation current densities J_∞ in both specimens with and without electropolishing were very close when the same cathodic current was applied.

Thus, the electropolishing could be avoided in this monitoring.

Fig.8 shows the time-dependency of the passive current density of the nickel-plating layer. The anodic current density decreases gradually with time and attains about 0.25 mA/m² in 24 h, which is independent of temperature and concentration of NaOH. From this current density, when we select 0.4 mA/m², the hydrogen content is 0.01 ppm, i.e., the detection limit of this monitoring system. Ikeda[10,11]) tried to elucidate the relationship between the hydrogen content and hydrogen induced cracking in three different solutions of NACE bath (5% NaCl + 0.5% CH₃COOH), AS bath (0.5% CH₃COOH) and SS bath (artificial sea water). These solutions were saturated with H₂S gas of 1 atm. He obtained the hydrogen content in a similar steel as in this study by a glycerin method. From his data, the critical permeation current density for the hydrogen induced cracking in the steel can be calculated (Fig.9), assuming that the steel is homogeneous.

It was concluded that the detection limit is quite enough to be used for a hydrogen induced cracking test.

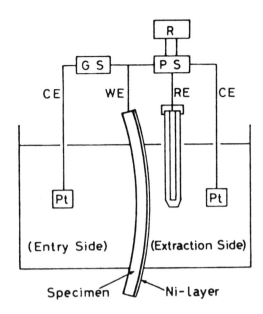

Fig.3 Electronic circuit and cells for laboratory test.

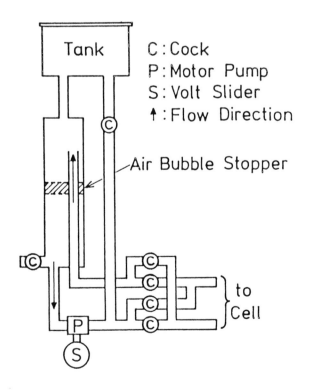

Fig.4 Schematic representation of pumping equipment for supply of electrolyte.

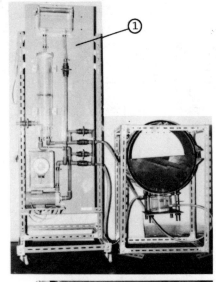

Photo.1 (a) Hydrogen monitoring system and
          pumping equipment,
          (b) fixing of cell.
          1 :pumping equipment, 2 :fixing belt,
          3 : rubber plate, 4 :pipe, 5 :cell

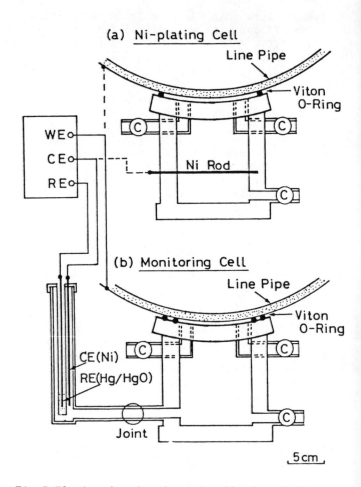

Fig.5 Electronic circuit and cells for field
      test.

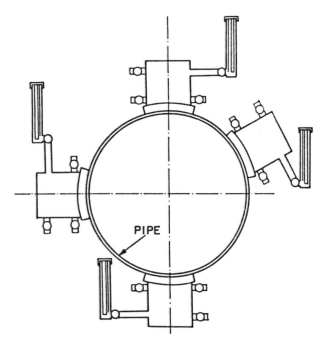

Fig.6 Setting of cell.

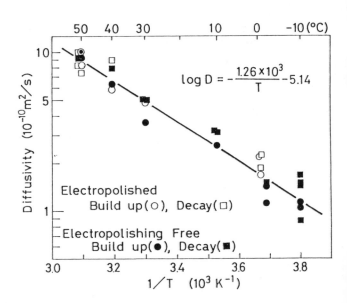

$$\log D = -\frac{1.26 \times 10^3}{T} - 5.14$$

Fig.7 Arrhenius plots of diffusivities of
      hydrogen in line pipe.

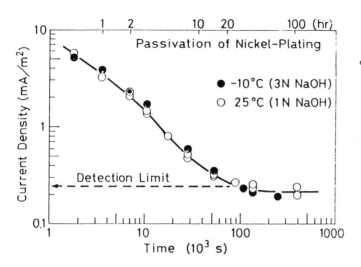

(a)

Fig.8 Time- dependency of passive current
density of nickel-plating.

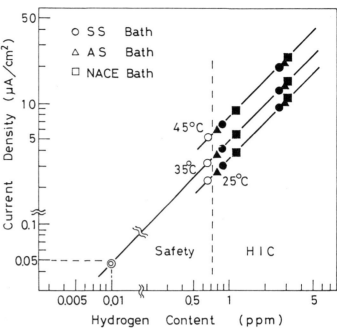

Fig.9 The critical hydrogen content in line
pipe for HIC.

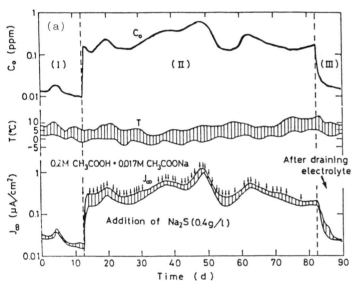

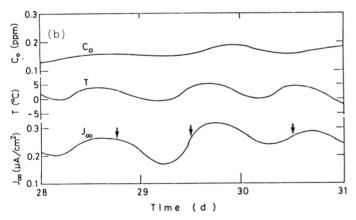

Fig.10 Monitoring of hydrogen content in pipe
line. (a) long term measurement, (b)
details of (a)

Field Test - In field test, the interior
of the pipe was filled with the buffer
solution at first, and then $Na_2S(7g/l)$ was
added to this solution which corresponded
to 1000 ppm as H_2S. The monitored
permeation current density, temperature and
hydrogen content calculated from eqs.(2)
and (5) are shown in Fig.10(a). After
addition of Na_2S, the hydrogen content
increased suddenly from about 0.015 to
0.2-0.5 ppm. The arrows show the suppli-
mental addition of $Na_2S(0.7g/l)$. Details of
the region II (Fig.10(b)) show that the
variation of the permeation current density
corresponds well to the variation of
temperatures but the hydrogen content
remains almost constant. After monitoring,
the electrolyte was drawn out from the pipe
(the stage III) and the anodic current was
measured again at the same polarization
(+150 mV vs Hg/HgO). Anodic current density
decreased drastically to about 0.2 mA/m^2
within the experimental time. This current
density corresponds well to the residual
current density shown in Fig.8.

Therefore, this monitoring has been
performed with high reliability.

MONITORING HYDROGEN CONTENT
AT ELEVATED TEMPERATURES

PROCEDURES - A schematic diagram of experimental setup for the hydrogen permeation at elevated temperatures is shown in Fig.11. The specimen had a cylindrical shape and its bottom was about three times thicker than the thickness of the side. The thickness (b-a) was $2-6.5 \times 10^{-3}$·m and its inner radius (a) was 12.5×10^{-3} m. The chemical composition of the carbon steel used was shown in Table 1. The specimen, into which a stainless steel tube was screwed, separated the entry side (A) and extraction side (B) of the hydrogen. Both sides were filled with sodium hydroxide (melting point = 595 K).

In side (A), argon gas was bubbled through a water bath at 303 K with a gas flow rate of 1.7×10^{-6} m^3/s. The cathodic charging of hydrogen was carried out galvanostatically, using four graphite counter electrodes, located on the same circumference. The extraction surface of the specimen was plated with a gold-coating of about 0.15 μm to prevent the dissolution of the steel. Platinum-coated stabilized zirconia tube was used as the reference electrode. Side (B) was kept in a good airtight condition and filled with argon gas. Under these conditions, the cathodic current above -1.5 V vs $Air/O^{2-}(ZrO_2)$ decreased gradually as the arrow shown in Fig.12.

Then the anodic potential from -1.0 to -0.4 V vs $Air/O^{2-}(ZrO_2)$ was applied to the surface (Side(A)) to extract the hydrogen permeated passage through the steel, and the cathodic current density corresponded to the range between -2.0 and -1.4 V vs $Air/O^{2-}(ZrO_2)$ was used for the introduction of hydrogen to the steel, i.e., 50-200 A/m^2 [12]).

RESULTS AND DISCUSSION - From the relatioship between the permeation current and the extraction potential at a constant cathodic current (Fig.13), the most suitable potential 1o extract the hydrogen was the range from -0.8 to -0.6 V vs $Air/O^{2-}(ZrO_2)$.

Typical build up and decay transients at 673 K are shown in Fig.14. Fig.15 shows the comparison of these transients with theoretical curves. The data agreed fairly well with the theoretical curves. This result indicated that the extraction surface was held at an sufficient anodic potential to ionize any hydrogen arriving at the surface which passed through the metal. But the data tended to deviate from the theoretical curves as the specimen thickness decreased. This would be due to the effect of the gold-plating on the hydrogen permeation rate. However, the wall thickness in actual pressure vessels and actual pipes is usually thicker than 5×10^{-3} m.

The diffusivities obtained from several build up and decay runs at each temperature for thicker thickness specimen were in good agreement with the values obtained from the gaseous method (Fig.16), i.e., gaseous permeation [13-16]) and evolution techniques [16-18]).

The detection limit in this electro-chemical method could be evaluated from Fig.12 as less than 0.1 ppm, which was much less than the critical hydrogen contents for hydrogen attack in carbon steel and in Cr-Mo steel, as shown in Fig.17 [19]).

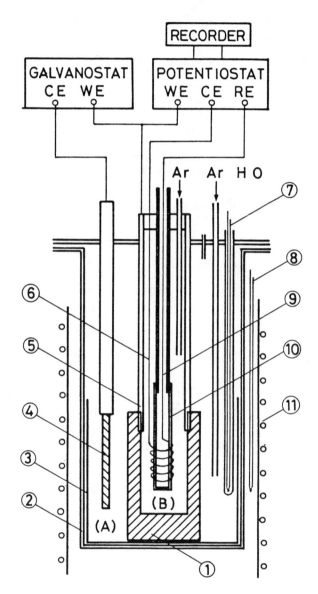

Fig.11 Experimental setup.
(A):Entry side, (B):Extraction side
1:specimen, 2:stainless steel tube,
3:alumina crucible, 4:graphite
rod, 5:stainless steel tube,
6:nickel wire, 7,8:thermocouple,
9:stainless steel wire,
10:stabilized zirconia tube,
11:furnace

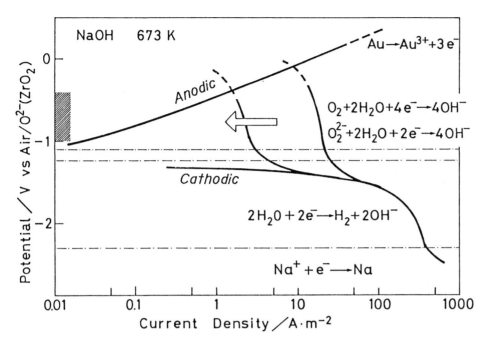

Fig.12 Polarization curves
of the steel in
molten sodium hy-
oxide at 673 K.

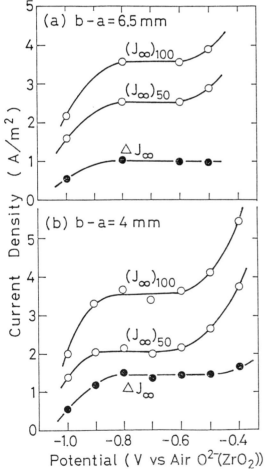

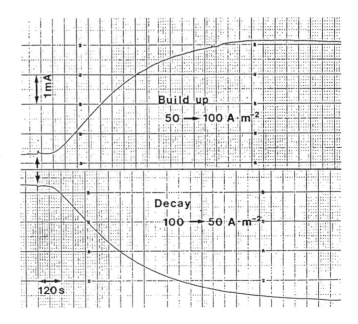

Fig.13 Relationship between permeation current density
and extraction potential for two cathodic
current densities, 50 and 100 A/m^2 at 673 K.

Fig.14 Typical permeation transients at 673 K.
Extraction potential is −0.6 V.

255

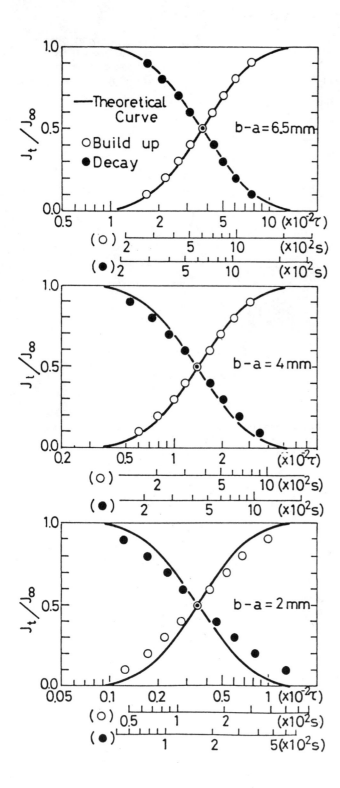

Fi. 15 Comparison between theoretical curves
and experimental transients at various
specimen thickness. (673 K)

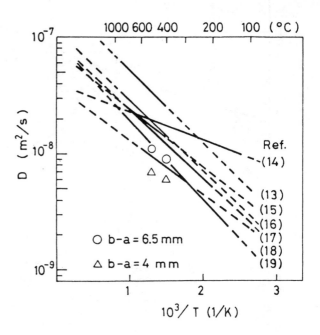

Fig.16 Diffusivities obtained from this
electrochemical method.

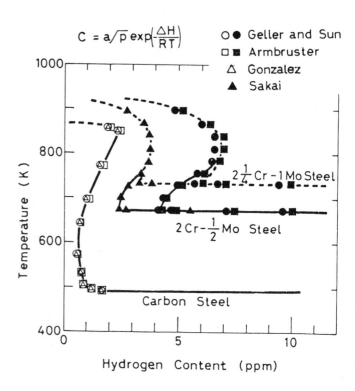

Fig.17 The critical hydrogen content in steels
for hydrogen attack, evaluated from
the Nelson Curve.

CONCLUSIONS

(1) A new electrochemical hydrogen probe, in which thin nickel layer was deposited at the extraction side to improve the detection limit of the hydrogen content, was developed and was useful for application to an actual pipe line.

(2) An electrochemical permeation method at elevated temperatures by using the electrochemical method was developed for the first time, and was useful for predicting the hydrogen attack in steels.

ACKNOWLEDGEMENTS

The part of this work was supported by a Grant-in-Aid for Developmental Scientific Research from the Ministry of Education, Science and Culture. The authors wish to thank Mr.T.Mizuno and Mr.A.Ando for the cooperation of the work. Thanks are also due to Sumitomo Metal Industries, LTD and Kobe Steel, LTD for supplying the line pipe steel and the carbon steel, respectively.

REFERENCES

1) W. H. Thomason, Corrosion/84, Paper No.237
2) M. A. V. Devanathan and Z. Stachurski, Proc. Roy. Soc. London A270, (1962) 90.
3) J. McBreen, L. Nanis and W. Beck, J. Electrochem. Soc. 113, (1966) 1222.
4) S. Yoshizawa, K. Tsuruta and K. Yamakawa, Boshoku Gijutsu (Corrosion Engineering) 24, (1975) 511.
5) N. Boes and H. Zuchner, J. Less-Common Metals 49, (1976) 223.
6) P. K. Subramayan, "Comprehensive Treatise of Electrochemistry", Vol.4, p.411, Plenum Press, New York, (1982)
7) I. M. Bernstein and A. W. Thompson, "Advanced Techniques for Characterizing Hydrogen in Metals", p.89, TMS-AIME, (1982)
8) C. G. Arnold, "Corrosion and Corrosion Protection Handbook", ed. by P. A. Schweitzer, p.435, Marcel Deckker Inc., New York (1983)
9) H.Tsubakino and K.Yamakawa, Tetsu-to-Hagane (J. of ISIJ) 71, (1985) 1070.
10) A. Ikeda, T. Kaneko and F. Terasaki, Corrosion/80, Paper No. 8.
11) A.Ikeda, Dr. Thesis to Kyoto Univ., (1981)
12) H. Tsubakino, A. Ando, T. Masuda and K. Yamakawa, Trans. ISIJ 25, (1985) 999.
13) W.Geller and T-H. Sun, Arch. Eisenhuttenw. 21, (1950) 423
14) W. L. Bryan and B. F. Dodge, A.I.Ch.E.J. 9, (1963), 223
15) C. Sykes, H. H. Burton and C. C. Gegg, JISI 156, (1947) 173
16) H. Schunck and H.Taxhet, Arch Eisenhuttenw. 30, (1959), 661
17) W. Eichenauer, H. Kunzig and A. Pebler, Z.Metallk. 49, (1958) 220
18) F. R. Coe and J. Moreton, JISI 204, (1966) 366
19) T. M. Stross and F. C. Tompkins, J.Chem. Soc. London (1956) 230
20) H. Tsubakino and K. Yamakawa, Boshoku Gijutsu (Corrosion Engineering) 24, (1984) 159

AUTOMATION IN CORROSION FATIGUE CRACK GROWTH RATE MEASUREMENTS

J. B. Boodey, V. S. Agarwala
Naval Air Development Center
Warminster, Pennsylvania, USA

ABSTRACT

The constant amplitude corrosion fatigue behavior of 300M steel (UTS 280-300 ksi) was studied with an automated mechanical testing system. The computer program utilized an elastic compliance method with crack opening displacement (COD) measurements to compute crack length. The crack growth rate and stress intensity factor were determined from the methods described in ASTM standards E 647 and E 399 respectively. The corrosion fatigue tests were conducted at frequencies of 0.1, 0.5 and 1.0 Hz and at a load ratio of 0.1. The test environments were dry air, salt-moist and salt-moist containing a corrosion inhibitor. The effect of environment on crack growth rate of 300M steel was most pronounced at low frequencies while becoming insignificant at high i.e. greater than 1 Hz.

THE DEVELOPMENT OF A STANDARD test method for constant amplitude corrosion fatigue crack growth behavior of high strength aerospace alloys in marine environments has been a concern of the navy for many years(1). The corrosion fatigue properties of these materials are usually determined by optical crack length measurements. This method has proven to be expensive and time consuming. In addition, test results may lack reproducibility. This inconsistancy may be due to operator error and transient effects such as test interruptions, and loading and frequency changes(1). In an attempt to reduce the disadvantages of the current method, various crack length measurement techniques have been integrated into automated testing systems. The techniques include the use of electrical potential(2), crack opening displacement (COD) gages (3-7) and linear voltage differential transducers (LVDT)(8).

The electrical potential techniques, both ac and dc, have been sucessfully used for measuring crack lengths of fatigue and fracture specimens. However, these may introduce additional variables with specimens tested in a corrosive environment. One possible effect is that the applied constant current may polarize the specimen in the test medium thereby altering the electrochemical reactions occuring at the crack tip. Thus, to avoid any unknown effects, the use of non-interferring methods such as COD gages or LVDT can be considered.
In these techniques, crack lengths are correlated with elastic compliance relationships as monitored by the COD gage and LVDT. These non-interferring methods have been studied extensively, and their compliance has been verified for both compact tension specimens and single edge notched bend bars in corrosive environments(3-8).

An automated testing system was utilized in this investigation to study the effects of environment, frequency and a corrosion inhibitor on the fatigue crack growth behavior of 300M steel. This testing system was designed to comply with the ASTM Test Method for Plane-Strain Fracture Toughness of Metallic Materials(E 399) and the ASTM Test Method for Constant-Load-Amplitude Fatigue Crack Growth Rates Above 10E-8 m/cycle(E 647). For this study, crack length was calculated by an elastic compliance method using COD and load measurements.

AUTOMATED TEST APPARATUS, MATERIAL AND TEST CONDITIONS

HARDWARE- The test apparatus, a MTS Alpha System, is comprised of a Hewlet-Packard model 9845B microcomputer and a ultrastiff MTS servohydraulic test stand. The two systems when interfaced, were operated by closed loop control. This

Table 1 - Elemental Composition (Weight Percent) of 300M Steel

Ni	Si	Cr	C	Mo	V	Fe
1.80	1.60	0.80	0.43	0.40	0.07	balance

technique allows the system to control the testing process by feedback from the load, strain, stroke and auxiliary channels. A unique feature of the servohydraulic system is the linear voltage differential transducer in the stroke actuator. This LVDT, when in the precise mode (18 bits), is capable of a resolution of 3×10^{-5} in (8×10^{-4} mm) and an accuracy of 2.4×10^{-4} in (6×10^{-3} mm) over a travel of 1.77 in (45mm). With this precision, combined with the excellent lateral and axial stiffness, and grip alignment, the stroke displacement measurements can be used for crack length calculations and closed loop control. The crack opening displacement was measured with an MTS model 632.02 clip gage designed specifically for ASTM E 399 testing. The load cell, an MTS model 661.16, has a full scale capacity of 2000 lbf (8.9kN) and due to the nature of constant amplitude testing was the primary device for closed loop control.

SOFTWARE- The computer program controlled the acquisition of data and monitored the cyclic loading (load ratio,frequency and waveform) during both the precrack and fatigue phases. The software also reduced the raw data with various mathematical routines. A brief overview of these routines was as follows. A flow diagram is shown in the Appendix A.

Modulus of Elasticity- The modulus of elasticity was computed from corresponding elastic compliances, COD/load(δ/P) and optical crack length measurements collected during the precrack phase. The elastic compliance was calculated from the slope of 200 pairs of COD and load over one loading cycle. A linear regression analysis was used to calculate the slope. This method is believed to reduce problems asssociated with crack closure and crack tip plasticity(5). The algorithm used for calculating Young's modulus, E', was the Tada-Paris compliance relationship for three point bend specimens(9):

$$E' = \frac{6*P*S*a*V(a/W)}{B*\delta*(W^2)} \qquad Eq.(1)$$

where
- P = load
- S = load span
- a = crack length
- B = depth
- W = width
- δ = COD
- V(a/W) = 0.76-2.28(a/W)+3.87(a/W)^2 -2.04(a/W)+0.66/(1-(a/W))^2
- E'= Young's modulus

The average modulus of elasticity was calculated from the precrack data, crack length range of 0.15in(3.75mm) to 0.20in(5.00mm), and was used for the crack length computations during the fatigue phase.

Crack Length Determination- The crack length during the fatigue phase was calculated from the elastic compliance slope, the average Young's modulus of elasticity, E', and the Tada-Paris compliance relationships given in Equation(1). A unique feature of this method is the use of Newton's iterative convergence for solving two simultaneous equations. This method replaces the old method of solving for a/W with a polynomial algorithm. Newton's method uses Equation(1) as a function, f(a/W), and it's first derivitive, df(a/W), to solve for the crack length. If the estimated crack length does not converge after 100 iterations, the test is terminated.

Crack Growth Rate- Whenever a change in crack length of 0.01 in (0.25mm) occured the crack growth rate (CGR), da/dN, was calculated. The values for the da/dN were determined from both the secant method and the seven point incremental polynomial technique according to ASTM standard (E647).

Table 2 - Mechanical Properties of 300M Steel

Tensile strength, ksi(MPa)	Yield strength, ksi(MPa)	Fracture Toughness, ksi$\sqrt{in}$(MPa$\sqrt{m}$)	Elongation, %(in 2in)
295(2030)	245(1690)	60(66)	8

Stress Intensity Factor- The stress intensity factor range , ΔK, was also calculated after an increase in crack length of 0.01 in (0.25mm) or greater. The stress intensity factor range was computed as described in ASTM standard (E399) using:

$$\Delta K = (\Delta P*S/B*W^1.5)*f(a/W) \qquad \text{Eq.(2)}$$

where
ΔP = load range
S = load span
B = depth
W = width
a = crack+notch length
f(a/W)=3*((a/W)^0.5)*[1.99-(a/W)*(1-a/W)*2.15
-3.93*(a/W)+2.7*(a/W)^2]
/[2*(1+2*(a/W))*(1-a/W)^1.5]

This algorithm was used for both the raw crack length data and crack length data computed from the seven point incremental polynomial method as described in ASTM standard (E 647 para. x1.2).

MATERIALS AND TEST CONDITIONS- The material tested for this investigation was 300M high strength steel. The elemental composition and the mechanical properties are presented in Tables 1 and 2 respectively. The thermal treatment to the steel plate was to austenitize at 1600°F + 25°F, oil quench, and double temper at 575°F. Three point bend test specimens were utilized in compliance with ASTM E 399. The specimen dimensions were as shown in Figure 1. The tests were conducted at frequencies of 0.1, 0.5 and 1.0 Hz in dry air(< 10% R.H.), salt-moist air (1% NaCl + >90% R.H.) and in these environments containing a corrosion inhibitor called DNBM. The inhibitor which contained phase transferred salts of dichromate(D), nitrite(N), borate(B) and molybdate(M) in a non-aqueous solvent was chosen because of it's effective corrosion fatigue inhibition in 4340 high strength steel(10). The cyclic loads applied were 150/1500lbs (R=0.1) and the pre-crack length was set to produce an initial stress intensity of 25 ksi√in (27.5 MPa√m). The waveform was sinosuidal and computations used haversine function.

RESULTS AND DISCUSSION

The automation developed to perform constant amplitude fatigue testing for the MTS-Alpha System showed excellent results using a COD gage. The CGR data generated from the elastic compliance method agreed well with the optical measurements. The agreements between the optical and COD methods ranged from 0.2 in. (5 mm) to approximately 0.45 in. (11 mm) in the measurements of crack length. The stroke displacement measurements, when used to compute the CGR, lacked in precision and resolution compared to optical and COD measurements. It was noted that the lack of compliance was due to a shift of reference point on the stroke position (or LVDT output) after every load-cycle. A correction factor is being developed in the algorithm (software) to compensate for this change. However, discussions in this paper deal with the COD data in determining the feasibility of this technique to be used in the corrosion fatigue studies.

The plots in Figures 2-5 show the (da/dn) vs. ΔK relationships for the 300M steel studied under various test conditions. The effects of frequency on the fatigue crack growth rate in a salt-moist environment were quite significant as shown in Figure 2. At

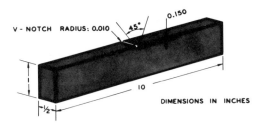

Fig.1 - Three point bend specimen dimensions.

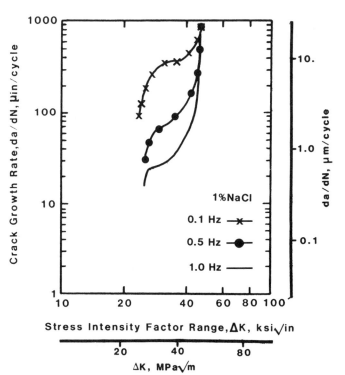

Fig.2 - Effect of frequency on crack growth rate in a salt-moist environment.

low frequency of 0.1 Hz the CGR was found to be 4.5 and 9 times greater than those at 0.5 and 1.0 Hz, respectively (Table 3). At extremely low frequencies, i.e. less than 0.1 Hz, the CGR is probably most sensitive to even slightest of environment. Conversely, at higher frequencies, e.g. 1.0 Hz, the effects of environment seem to disappear even in the most corrosive environments. Figure 5 shows only small displacements in between the curves for dry air, salt-moist air and the one inhibited by DNBM. In other words, the fatigue behavior of steel is not significantly affected by the environment when cyclic frequency is high. This is in agreement with the observations of Wei and Shim(11). However, the threshold frequency, where the effects of environment are insignificant on the CGR, will be different for different materials and strengths. In high strength alloys, the effects of environment may extend to some higher frequencies as well.

The effects of environment on high strength steels (e.g. 300M steel) at low frequencies are mostly the manifestations of corrosion reactions. The CGR is predominantly controlled by the active path dissolution (anodic reaction) kinetics which occur at the crack tip during fatigue. Thus, as the cyclic frequency decreases, the time for anodic reaction to occur at the newly

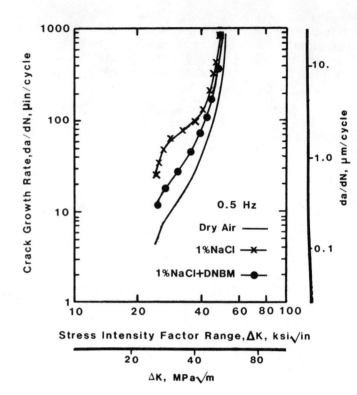

Fig.4 - Effect of environment on crack growth rate at 0.5 Hz.

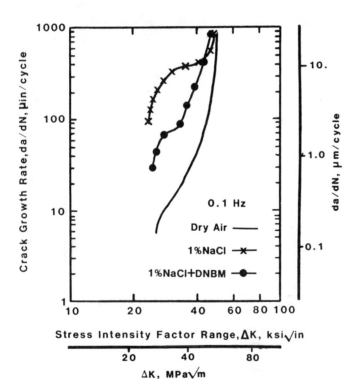

Fig.3 - Effect of environment on crack growth rate at 0.1 Hz.

generated surface increases. Additionally, hydrogen evolution reactions taking place will further increase the CGR by hydrogen embrittlement. At high frequencies, the reaction times become insignificantly small for the corrosion reactions to occur, hence no increase in the CGR. However, if the surface reactions can be controlled even at low frequencies, such as with the use of corrosion inhibitor, DNBM, a reduction in the CGR should occur. The experiments with DNBM, as shown in Figures 3 and 4, support this theory. The mechanisms by which DNBM modify the crack-tip chemistry have been discussed elsewhere(10).

The SEM fractographs (Figure 6) of the failure surface taken in the plateau region of the crack growth show a significant change in the morphology of the surface as the test environment was changed: from ductile (pure fatigue) failure in dry air to highly intergranular brittle fracture in salt-moist environment. The presence of inhibitor changes the intergranular-brittle feature to a partly, ductile-failure feature resembling the dry air morphology.

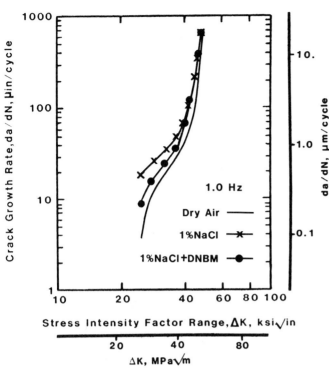

Fig. 5 - Effect of environment on crack growth rate at 1.0 Hz.

An automation in the performance of constant amplitude corrosion fatigue tests using an MTS-Alpha System was achieved. The CGR data generated from the elastic compliance method and using the COD gage showed excellent agreement with previous studies done by optical methods. The crack growth rate verses stress intensity factor range studies were made for high strength 300M steel under various test conditions. The effect of environment showed a most predominant effect at low frequecies. At high frequencies, the contribution of corrosion on CGR was insignificant. The control of surface reactions at the crack tip due to the presence of an inhibitor at low frequencies showed significant reduction in the CGR.

Table 3 - Corrosion Fatigue Test Data for 300M Steel

Environment-	Dry Air			1% NaCl			1% NaCl+DNBM		
Frequency-	0.1	0.5	1.0	0.1	0.5	1.0	0.1	0.5	1.0
* Crack Growth Rate, μin/cycle (μm/cycle)	22 (.6)	20 (.5)	22 (.6)	375 (10.)	90 (2.3)	45 (1.1)	120 (3.1)	40 (1.)	33 (.8)
** Stress Intensity Factor Range, ksi√in (MPa√m)	43 (47)	44 (48)	45 (49)	24 (26)	38 (42)	41 (45)	35 (38)	41 (45)	42 (46)

* Plateau velocity at 35 ksi√in (38 MPa√m).
** At a crack growth rate of 100 μin/cycle (2.5 μm/cycle).

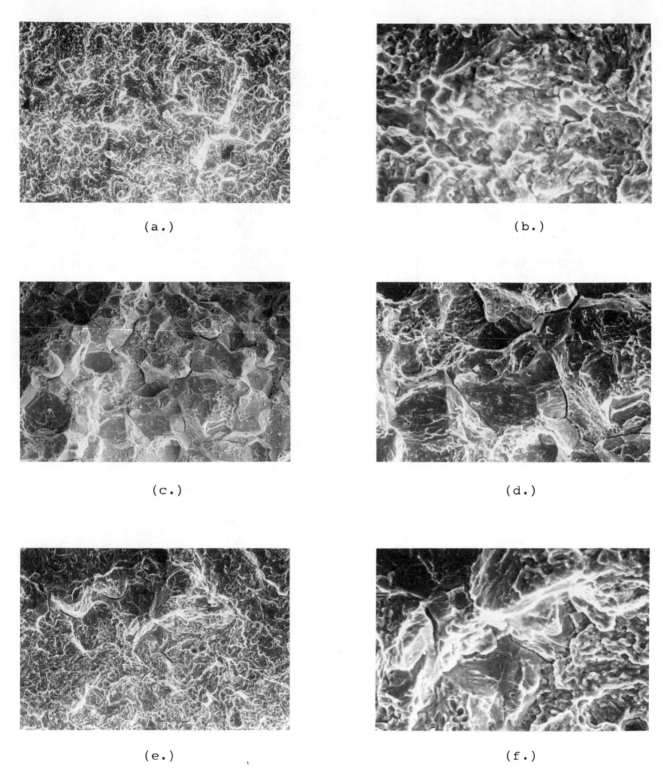

Fig.6 - Effect of environment on fracture surface morphology at 0.1 Hz and approx. 35 ksi√in (38 MPa√m). Environments were: Dry Air (a and b); Salt-Moist (c and d); and Salt-Moist + DNBM (e and f). Micrographs taken at (500 and 2000X) respectively.

References

1. Crooker,T.W.,Gill,S.J.,Yoder,G.R.,and Bogar,F.D., "Development of a Navy Standard Test Method for Fatigue Crack Growth Rates in Marine Environments," Environment - Sensitive Fracture: Evaluation and Comparison of Test Methods, ASTM STP 821,S.W. Dean,E.N. Pugh and G.M. Ugiansky, Eds., ASTM, Phila.,1984, pp415-425.

2. Wei,R.P: and Brazill,R.L., "An Assessment of A-C and D-C Potential Systems for Monitoring Fatigue Crack Growth," Fatigue Crack Growth Measurement and Data Analysis, ASTM STP 738, S.J.Hudak,Jr., and R.J. Bucci,Eds., ASTM, 1981, pp. 103-119.

3. Yoder,G.R.,Cooley,L.A., and Crooker,T.W., "Procedures for Precision Measurement of Fatigue Crack Growth Rate Using Crack-Opening Displacement Techniques, "Fatigue Crack Growth Measurement and Data Analysis, ASTM STP 738, S.J. Hudak, Jr., and R.J. Bucci, Eds., ASTM, 1981,pp. 85-102.

4. Van Der Sluys,W.A. and DeMiglio,D.S., "Use of a Constant delta-K Test Method in the Investigation of Fatigue Crack Growth in 288 C Water Environments, "Environment - Sensitive Fracture: Evaluation and Comparison of Test Methods, ASTM STP 821, S.W.Dean, E.N.Pugh and G.M.Ugiansky, Eds., ASTM, Phila, 1984, pp. 443-469.

5. Fabis,T.R., Liaw,P.K., Ceschini, L.J., Leax,T.R., and Landes, J.D.,"Computer Controlled Fatigue Crack Growth Rate Testing on Bend Bars in a Corrosive Environment, "Environment - Sensitive Fracture: Evaluation and Comparison of Test Methods, ASTM STP 821, S.W.Dean, E.N.Pugh and G.M.Ugiansky, Eds.,ASTM, Phila., 1984, pp. 470-483.

6. Tipton,D.G., "A Low Cost Microcomputer Data Acquisition System for Fatigue Crack Growth Testing," Environment-Sensitive Fracture: Evaluation and Comparison of Test Methods, ASTM STP 821, S.W.Dean, E.N.Pugh and G.W.Ugiansky,Eds., ASTM, Phila, 1984, pp.484-496.

7. Yoder,G.R., Cooley,L.A. and Crooker,T.W., "Effects of Microstructure and Frequency on Corrosion Fatigue Crack Growth in Ti-8Al-1Mo-1V and Ti-6Al-4V, "Corrosion Fatigue: Mechanics, Metallurgy, Electrochemistry and Engineering, ASTM STP 801, T.W. Crooker and B.N. Leis, Eds., ASTM, 1983, pp. 159-174.

8. Ceschini,L.J., Liaw,P.K., Rudd,G.E., Logsdon,W.A., "Automated Corrosion Fatigue Crack Growth Testing in Pressurized Water Environments, "Environment - Sensitive Fracture: Evaluation and Comparison of Testing Methods, ASTM STP 821, Dean,Pugh and Ugiansky, Eds., ASTM, Phila, 1984, pp. 426-442.

9. Tada, H., "The Stress Analysis of Cracks Handbook, "Del Research Corp, Hellertown, Pa., 1973, pp.2.16-2.17.

10.Agarwala,V.S., DeLuccia,J.J., "Modification of Crack Tip Chemistry to Inhibit Corrosion and Stress Corrosion Cracking in High Strength Alloys," Embrittlement by the Localized Crack Environment, R.P. Gangloff, Eds., AIME Pub., 1984, pp.405-419.

11.Wei,R.P. and Shim,G.,"Fracture Mechanics and Corrosion Fatigue," Corrosion Fatigue: Mechanics, Metallurgy, Electrochemistry and Engineering, ASTM STP 801, T.W. Crooker and B.N.Leis, Eds., ASTM, 1983, pp. 5-25.

Computer Program Flow Diagram

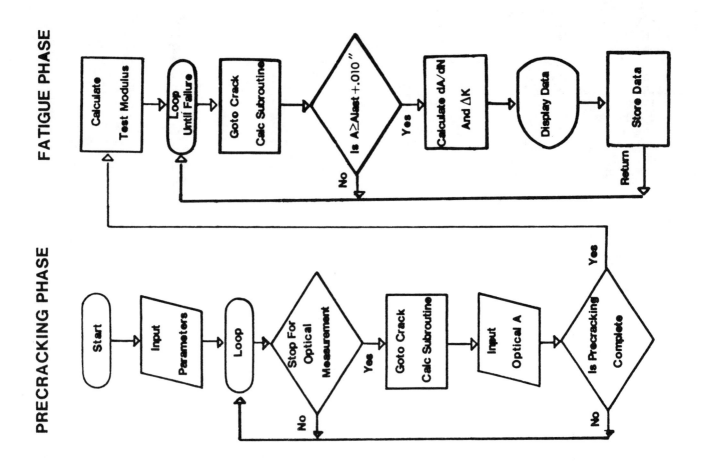

THE BREAKING LOAD METHOD: A NEW APPROACH FOR ASSESSING RESISTANCE TO GROWTH OF EARLY STAGE STRESS CORROSION CRACKS

R. J. Bucci, R. L. Brazill, D. O. Sprowls*,
B. M. Ponchel*, P. E. Bretz
Alcoa Laboratories
Alcoa Center, Pennsylvania, USA
*retired

Abstract

The breaking load test method - a much improved accelerated technique for assessing stress corrosion cracking (SCC) behavior has been developed at Alcoa Laboratories. In comparison with present industry standards, the breaking load method provides more information with fewer specimens and shorter exposure times; additionally, it is more discriminating of SCC performance among relatively resistant materials. The method involves breaking strength measurements determined from tension tests of cylindrical specimens that have been exposed to corrosive environment in the presence of a sustained tensile stress. The degree of degradation due to the stress corrosion attack is measured by comparing the specimen post-exposure strength with the original tensile strength (no exposure). The breaking load method provides SCC description in terms of numerical quantities that are amenable to statistics and fracture mechanics interpretations. This enables probability of SCC initiation and growth to a given flaw size to be used as a criterion for rating materials. An added feature of this interpretive approach is that the effects of alloy strength and toughness and specimen size on measured performance can be normalized.

INTRODUCTION - THE BREAKING LOAD METHOD

An extensive review of state-of-the-art stress corrosion (SCC) test procedures (1) identified the need for improved accelerated methods capable of discriminating degrees of susceptibility among materials in meaningful terms. Under contract to NASA Langley, Alcoa Laboratories examined various SCC test procedures and their effectiveness for evaluating aluminum alloys. One result of this four year program has been the development of the "breaking load method" - a much improved test technique for evaluating SCC (2). In comparison to present day methods, this new approach is capable of providing more quantitative information, with fewer specimens and shorter exposure times, while removing many of the recognized deficiencies of conventional lab accelerated procedures.

The breaking load approach utilizes data from tension tests performed on replicate groups of smooth specimens after exposure to static stress and corrosive environment for various lengths of time. In studies involving high strength 7075 aluminum alloys, an exposure period of 4 to 10 days was found to be optimum (2). The breaking loads from the tension tests are converted to "apparent" tensile strength values which can then be compared against the material original tensile properties (no exposure). Analyzed in its simplest form, the breaking load test senses SCC damage as a loss in strength (failure load/original area). In general, the larger the strength decrease, the greater the degree of SCC attack; and by making area of the specimen cross section small, the method's sensitivity to detect cracking at an early stage is enhanced.

Mean trends in breaking load test results that fit the above description are shown in Figure 1. The data shown are for three temper variants of 7075 aluminum alloy plate possessing high (T651), medium (T7X1) and low (T7X2) degrees of SCC susceptibility when subjected to short-transverse tension stress. All three temper variants were obtained from as-received, 2.5-in. thickness, 7075-T651 commercial plate. Sample preparation in this manner enabled evaluation of controlled materials having varying degrees of SCC susceptibility, but with identical compositions and grain structure. Subsequent aging practices for the T7X1 and T7X2 conditions were chosen respectively to obtain mechanical and corrosion properties typical of T7651 and T7351 commercial plate tempers. Short-transverse mechanical properties for the three 7075 temper conditions are in given in Table 1. The experimental procedure employed smooth 0.125 in. diameter by 2-in. long

TABLE 1

Short-Transverse Mechanical Properties
of 7075 Aluminum Alloys
Subjected to Breaking Load Experiments

Temper	UTS, ksi	TYS, ksi	Kic, ksi√in.
T651	76.5	63.8	19.3
T7X1	70.6	62.0	19.4
T7X2	69.4	60.4	20.2

threaded-end tension specimens per ASTM standard G47 (3). These specimens were mounted in a self stressing frame, and then exposed to 3.5% NaCl solution by alternate immersion in accordance with ASTM standard G44 (4). The specimens were solvent cleaned prior to exposure, and inspected daily for possible fractures while exposed. Replicate specimens were exposed in sets of five, with each set scheduled to be removed from the alternate immersion tank after various periods. Upon termination of the exposure period, the surviving specimens were rinsed in water, dried and tension tested within one hour per ASTM method B557 (5). Each data point in Figure 1 represents the averaged breaking strength of the five-specimen group scheduled to identical exposure conditions. In those cases where individual specimens failed prior to completion of the intended exposure interval, the breaking stress was equated to the exposure stress for the purpose of calculating average values. The raw

data from these experiments are reported in Appendix C of Reference (2).

The results in Figure 1 clearly show that the breaking load procedure is able to discriminate degrees of SCC resistance among the three 7075 temper variants with relative ease. Moreover, this is done without waiting for the specimens to fail in the environment. In contrast, traditional pass-fail methods require higher replication and longer test times to screen differences among more SCC resistant materials. Additionally, by comparing breaking strengths of specimens exposed with and without stress, a means of partitioning SCC response from that attributed to general corrosion is provided.

STATISTICAL ANALYSIS OF BREAKING LOAD DATA

The statistical characterization of fracture data has been studied for many years (6,7) with much effort devoted to the statistics of extreme values (8 to 14). The rationale behind this approach is that materials contain weakening flaws and that, although there may be a wide spectrum of flaw sizes in a specimen, the fracture seeks out the dominant flaw (weakest link). Thus, the distribution of all flaw sizes within a specimen group is not as important as the distribution of largest flaws (the extreme values). Assuming fracture strength to be inversely related to size of the largest SCC flaw, then the breaking strengths of replicate specimens subjected to identical exposure conditions should follow an extreme value

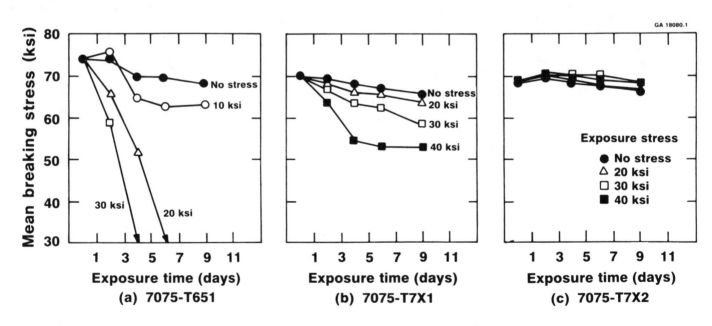

Mean Breaking Stress vs Exposure Times for Short Transverse 0.125 in. Diameter Tension Specimens Exposed to 3.5% NaCl Solution by Alternate Immersion (ASTM, G44) at Various Exposure Stress Levels. Each Point Represents an Average of Five Specimens

Figure 1

distribution of smallest values (2). This distribution is bell-shaped and skewed such that the population is likely to produce a wider range of fracture stresses below the average than above. Under these assumptions the probability for survival (no fracture at the stress of interest) can be calculated from breaking strength data according to Equation 1, attributed to Gompertz (14);

$$P = \exp\left[-e^{Z}\right] \qquad [1]$$

where P is the probability of survival and Z is the reduced variate. The reduced variate is of the form

$$Z = (S - \zeta)/\eta \qquad [2]$$

where S is the stress of interest (e.g., exposure stress) and ζ and η are estimates of the distribution location and scale parameters for the sample size, N. Determining the parameters ζ and η involves probability plotting. This graphical technique has the ability to handle truncated data sets, such as test groups containing specimens that failed in the environment prior to tensile testing. Briefly, values of ζ and η can be estimated from a straight line fit to plotted data that order ranks the replicate specimen breaking strengths against their expected values according to the distribution of interest (15). Specimens failing in the environment are counted in the sample size, but have breaking strengths below the exposure stress and can not be plotted to determine the straight line fit.

Survival probabilities calculated in the above manner are shown in Table 2 for the intermediate T7X1 temper alloy tested at 40 ksi exposure stress. Calculated survival probabilities for all conditions shown in Figure 1 are given in Appendix C of Reference (2), along with a detailed description of the probability plotting procedure. Although the sample size was small for each group of test conditions analyzed in this work (N = 5), all breaking load data showed nearly linear relationships when plotted on extreme value probability paper. Hence, the assumed extreme value distribution of breaking strengths appears to be reasonable for the test

conditions considered. The results of Table 2 and Reference (2) show that survival probabilities estimated from the extreme distribution of specimen breaking strengths agree favorably with those obtained by traditional pass-fail analysis (no. survivors/no. tested).

Comparative studies involving more SCC resistant materials require large, and often prohibitive, amounts of testing if traditional pass-fail sampling techniques are to be utilized (2). More quantitative use of breaking load data can be made by calculating the tensile stress that 99 percent of the specimens could be expected to sustain after being subjected to a period of exposure at a given stress. This 99 percent survival stress, S_{99}, is calculated by substituting 0.99 for P in Equation 1 and solving Equations 1 and 2 for S as follows:

$$S_{99} = \zeta + \eta \cdot \ln[\ln(0.99)^{-1}] \qquad [3]$$

$$= \zeta - 4.60\,\eta$$

A practical advantage of the 99 percent survival stress is illustrated by the case where five replicate specimens of each of the three 7075 tempers of Figure 1 are exposed in test, and all survive the designated exposure conditions. The estimated survival probability, P, determined from the pass-fail data would equal 1.0 for all three tempers (i.e., five survivors/five specimens tested). For a sample size of five and probability of survival estimated at 1.0, the use of pass-fail (binomial) statistics estimates the actual probability of survival to lie between 0.55 and 1.0 with 95 percent confidence. Thus, not only is the confidence interval large, but the test has shown no differences between tempers. In contrast, the table below illustrates that more discriminating information can be obtained via the 99 percent survival stress.

Plate Temper	Exposure Conditions	No. Survivors out of Five Tested	Breaking Load Data Extreme Value Analysis	
			Prob. of Survival, %	99% Survival Stress, ksi
T651	30 ksi, 2 days	5	98	22
T7X1	30 ksi, 9 days	5	99	31
T7X2	30 ksi, 9 days	5	100	63

A threshold stress can also be estimated from breaking load test results. This estimation requires considerably fewer specimens and shorter test times than traditional pass-fail threshold determinations. Consider a practical threshold stress definition to be that exposure stress for which it can be demonstrated with 95 percent confidence that the probability of specimen failure under specified testing conditions is less than one percent. Following a procedure described in Appendix B of Reference (2), a threshold stress can be estimated by plotting the 99 percent survival stresses versus the exposure stress for each exposure period as in Figure 2. The threshold stress estimate is then taken where

TABLE 2

Breaking Load Test Results on Alloy 7075-T7X1 Short Transverse, 0.125-in. Round Tensile Bars Exposed to 3.5% NaCl Solution by Alternate Immersion (ASTM G44) at 40 ksi Exposure Stress

Expos. Days	No. Survivors/ No. Tested	Breaking Strength Distribution Parameters, ksi				Survival Prob.(a) @ 40 ksi
		Mean	Std. Dev.	ζ	η	
2	5/5	63.9	4.20	66.6	6.21	0.99
4	5/5	54.7	3.35	56.7	4.66	0.97
6	4/5	54.7(b)	8.86(b)	57.2	15.66	0.72
9	4/5	54.5(b)	7.96(b)	56.5	12.12	0.77

(a) Calculated using extreme value distribution parameters.
(b) Calculated using 40 ksi as the breaking stress for the specimen that failed during exposure.

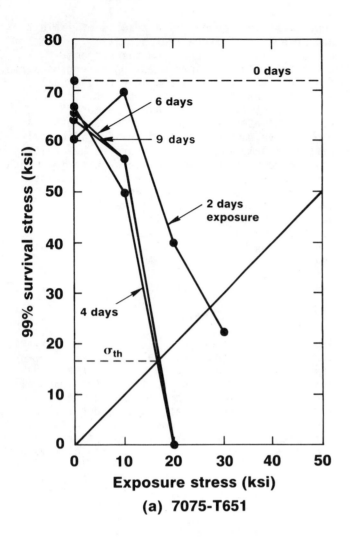

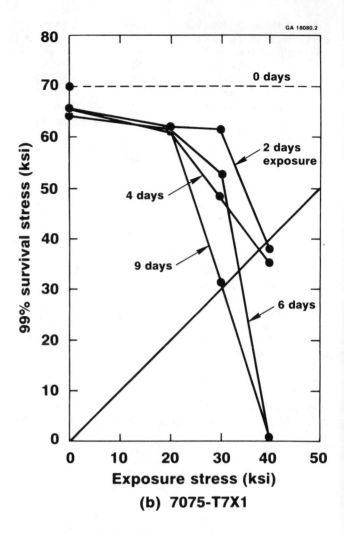

(a) 7075-T651

(b) 7075-T7X1

**Calculated 99% Survival Stress vs. Exposure Stress
for 0.125-in. Round Tensile Bars Exposed to 3.5% NaCl
Solution by Alternate Immersion**
Figure 2

the trend in the data intersects the 1:1 line indicating equal exposure stresses and survival stresses. The intersection point estimates the exposure stress that would produce a 99 percent survival stress at the test time considered. The estimated alloy short-transverse threshold stress, from Figure 2, is about 17 ksi and 30 to 35 ksi respectively for alloys 7075-T651 and 7075-T7X1 after 4, 6 and 9 day exposures. For the example shown in Figure 2(a), premature failures at 20 ksi for the scheduled 6 and 9 day exposures precluded statistical computation of the 99 percent survival stress, and a conservative estimate of zero was taken.

Table 3 compares traditional pass-fail and breaking load SCC ratings of the three 7075 alloy tempers. Both approaches produce consistent performance rankings, and each readily distinguish between the high susceptibility (T651) and low susceptibility (T7X2) temper

conditions. The comparisons of greater interest, however, are the intermediate T7X1 temper with the two extremes. The 99 percent survival stress from the breaking load analysis gives a more definitive comparison of the T7X1 and T7X2 materials than either of the survival probability computations. Conventionally estimated thresholds are in good agreement with the thresholds determined from breaking load data. Each of the above conclusions is consistent with ratings under the broader range of test conditions covered in Reference (2).

METALLOGRAPHIC AND FRACTOGRAPHIC EXAMINATIONS

Metallographic profiles from short-transverse tensile bars exposed and then tension tested are shown in Figure 3 for the three 7075 alloy tempers. These photos of localized corrosion sites were taken from

TABLE 3

SCC COMPARISONS OF THREE TEMPER VARIANTS OF 7075 ALLOY PLATE BY
DIFFERENT ACCELERATED SCC TEST METHODS - SHORT TRANSVERSE STRESS

Measure of SCC Susceptibility	Plate Temper		
	T651	T7X1	T7X2
Smooth 0.125-in. Round Tensile Bars Exposed to 3.5% NaCl Alternate Immersion (ASTM G44) :			
o Breaking Load Test Method:			
- Probability of Survival, % (a)	0	100	100
- 99% Survival Stress, ksi (a)	0	53	69
- Threshold Stress, ksi (b)	17	30-34	56-66
- Calculated Mean Depth of SCC, in. (a)	>0.052	0.018	0.001
- 99% Penetration Limit, in. (a)	∞	0.028	0.006
- Avg. SCC Growth Rate, 10^{-5} in./hr.	54(c)	6.9(c)	3.3(d)
o Traditional Pass-Fail Analysis:			
- Prob. of Survival, 95% conf. interval (a)	0-31	74-100	74-100
- Approx. Threshold Stress, ksi (e)	10-20	20- 40	>40
Traditional Precracked Specimens Exposed to 3.5% NaCl Dropwise, 3 Times Daily:			
o Ring Load WOL Specimen, Fatigue Precracked:			
- K_{Iscc}, ksi $\sqrt{in.}$	5	13	19
o Bolt Load DCB Specimen, Tension Pop-in:			
- Plateau Velocity, 10^{-5} in./hr.	45	11	1.5

(a) 6 days at 30 ksi
(b) Range, 4-9 day exposure; estimated per method of Figure 2.
(c),(d) Over 2 to 6 and 2 to 9 days, respectively, at 30 ksi.
(e) 20 days per ASTM G47

longitudinal sections on the specimen diameter and normal to the plane of tensile failure. In the sections examined, general appearance of the attack for each temper condition was similar to that shown in the figure, except that penetration typically was deeper in sections taken from specimens exposed for longer intervals at higher stress. In all cases the corrosion was directional, primarily following grain boundaries in the rolling plane of the plate. This is best shown by the network of fine features observed in the T651 temper material, with directional pitting also present. The propensity for intergranular attack is normally less in overaged T7-type tempers (16) and, as expected, was reduced in specimens of the T7X1 temper, and practically non-existent in the T7X2 temper. In the T7X2 temper specimens, corrosion took the form of relatively shallow directional pitting perpendicular to the specimen axis with only traces of grain boundary attack associated with some of the pits.

The fracture surfaces of selected breaking load specimens were examined in the scanning electron microscope (SEM); one such typical observation is shown in Figure 4. A primary objective of this exercise was to obtain flaw size measurements as benchmarks for subsequent fracture mechanics characterization of SCC behavior. These measurements are given in Table 4. The examinations were performed on specimens with the highest and lowest breaking strengths from selected groups of five replicate specimens. For the groups of specimens involved in these measurements, occurrences of premature failures during exposure are also noted in the table. Two measurements of flaw size taken from the SEM photos were: the maximum depth of penetration along a diameter, denoted a_{max} in Figure 4; and the cracked area volume fraction (Ac/Ao), defined

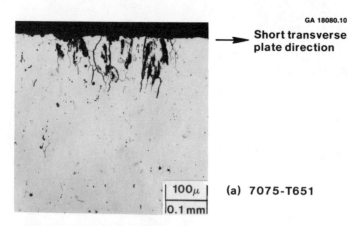

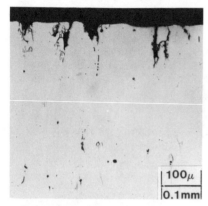

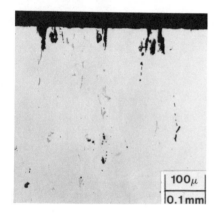

(a) 7075-T651

(b) 7075-T7X1

(c) 7075-T7X2

Typical Surface Attack in the Three Temper Variants of Alloy 7075 Plate Exposed to 3.5% NaCl Solution by Alternate Immersion
Figure 3

as the SCC fracture area (Ac) divided by original area (Ao) of the specimen test section.

The flaw size data of Table 4 illustrate several expected trends. The first is that, for each material and applied stress condition, the measured flaw sizes generally increase with length of exposure. Secondly, the 2-day, 0, 20 and 30 ksi exposure data on the highly SCC-susceptible 7075-T651 alloy illustrate that

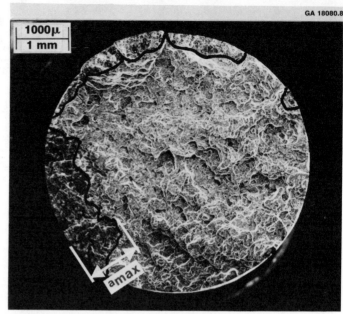

7075-T7X1   0.225 in. diameter specimen
Exposed 9 days at 40 ksi, fracture stress 47.0 ksi

Magnified View of Specimen Fracture Surface with Border of Stress Corrosion Flaw Outlined for Clarity
Figure 4

increasing exposure stress progressively increases SCC flaw size. In contrast, the highly SCC resistant T7X2 temper shows no effect of exposure stress on flaw size, suggesting that the observed degradation at the 9-day exposure is more attributable to generalized corrosion than SCC. A third observation is that under comparable exposure conditions the two specimen diameters (0.125 and 0.225-inch) of T7X1 temper material have similar SCC penetrations (a_{max}).

CORRELATION OF FRACTURE STRESS WITH FLAW SIZE

An alternative interpretation of the breaking load test results is to quantify SCC in terms of an "effective flaw size", linked to tensile strength degradation in the corroded specimen. To establish this link, mathematical equations were formulated to relate breaking stress and flaw size for the specimen configuration and materials of interest (2). Effectiveness of the analytical approach was initially verified by good agreement of predicted flaw sizes with actual sizes measured from tensile fractures of specimens containing partial-thickness fatigue cracks (2). The technique and its application to breaking load data from corroded 7075 alloy tensile bars are described below.

The idealized flaw configuration shown in Figure 5 is supported by fractographic evidence

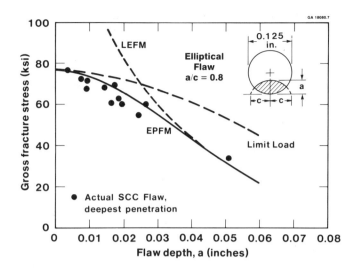

Comparison of Actual and Predicted Fracture Stress vs. SCC Flaw Depth Relationship in Stress Corroded Alloy 7075-T651 Short Transverse Tensile Bars, 0.125-in. Diameter
Figure 5

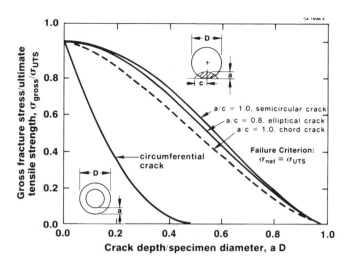

Normalized Limit Solutions for the Axial Loaded Cylindrical Tension Specimen with Various Crack Configurations and Depths
Figure 6

TABLE 4

FLAW SIZE MEASUREMENTS FROM FRACTURE SURFACES OF ALLOY 7075
SHORT-TRANSVERSE TENSILE BARS EXPOSED TO 3.5% NACL SOLUTION
BY ALTERNATE IMMERSION AND TENSION TESTED TO FAILURE

Temper	Spec. Diam. in.	Expos. Stress ksi	Expos. Time days	F/N (a)	Gross Fracture Stress(b) ksi	Max. Flaw Depth a_{max} in.	Flaw Area Fraction Ac/Ao
T651	0.125	0	2	0/5	76.3	0.0031	0
					71.7	0.0075	0.09
		0	9	0/5	69.0	0.0168	0.19
					67.4	0.0089	0.15
		20	2	0/5	72.2	0.0088	0.06
					62.1	0.0183	0.13
		20	4	0/5	61.3	0.0265	0.36
					35.0	0.0515	0.41
		20	6	4/5	58.9(c)	0.0199	0.32
		20	9	4/5	60.5(c)	0.0159	0.28
		30	2	0/5	68.0	0.0143	0.14
					53.9	0.0248	0.32
T7X1	0.125	40	2	0/5	68.6	0.0105	0.02
					59.7	0.0181	0.10
		40	4	0/5	58.1	0.0156	0.17
					49.9	0.0268	0.19
		40	6	1/5	61.4	0.0110	0.18
					45.2	0.0329	0.30
		40	9	1/5	59.2	0.0216	0.17
					47.5	0.0259	0.27
T7X1	0.225	40	2	0/5	70.9	0	0
					69.4	0.0137	0.01
		40	9	1/6	63.1	0.0236	0.10
					47.0	0.0380	0.16
		40	12	0/5	67.8	0.0128	0.07
					63.9	0.0135	0.07
T7X2	0.125	0	2	0/5	70.3	0.0046	0.01
					69.3	0.0048	<0.01
		0	9	0/5	67.6	0.0100	0.05
					65.8	0.0106	0.07
		40	2	0/5	71.3	0	0
					70.4	0	0
		40	9	0/5	68.9	0.0089	0.05
					66.5	0.0088	0.03

(a) Number of environmental failures out of number tested at the indicated exposure conditions.
(b) Highest and lowest breaking strength, respectively of specimens surviving the indicated exposure period, unless otherwise noted.
(c) Sole survivor, 4 out of 5 exposed specimens failed by the sixth day.

(e.g., Figure 4), and in a physical sense, models the dominant (largest) SCC flaw associated with final fracture.* Dimension a is considered to be the point of deepest SCC penetration, and the a/c = 0.8 aspect ratio approximates the equilibrium shape partial-thickness crack developed within a uniformly loaded tensile bar (2,17). As a first approximation, the critical flaw size at fracture can be estimated as the smaller value given by either a plastic limit-load or linear-elastic fracture mechanics (LEFM) failure criterion, as shown in Figure 5. The limit-load criterion predicts fracture when the stress on the uncracked ligament equals the material ultimate tensile strength (σ_{UTS}). Accordingly, the gross limit stress ($\sigma_L = P_L/Ao$) can be calculated as:

$$\sigma_L = \sigma_{UTS} \times [An/Ao] \qquad [4]$$

where P_L is the limit load, Ao the original cross-section area and An the uncracked ligament area. A graphical representation of the normalized limit load solution developed in this manner for the idealized geometry is shown in Figure 6 (2). In general, the limit load failure criterion best applies when the crack is so small

_____
* Though more than one SCC crack may be in evidence from the fractograph, these cracks, in reality, may occur on different planes. The possible case of crack link-up around the specimen circumference was evaluated as a 2-dimensional annular crack problem. Annular flaw depths predicted under this assumption tended to be smaller than actual sizes. Hence, the single elliptical-flaw approximates a worst case possibility for purposes of correlating flaw size with breaking strength.

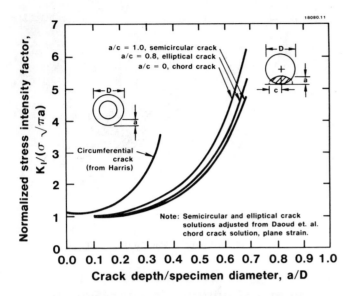

Stress Intensity Factor Solutions for Various Crack Configurations in an Axial Loaded Cylindrical Tension Specimen.

Figure 7

that the localized stress concentration is "washed out" by gross plastic deformation in the net section. If instead, crack dimensions are large relative to the scale of plasticity, then LEFM predicts the fracture stress and flaw size combination with reasonable accuracy. The LEFM failure condition is predicted when the crack-tip stress intensity factor, K, exceeds the material toughness (taken as Kic in this report). The elliptical-crack K-solution used in this study (2) was developed as a modification to the chord crack solution given by Daoud et al. (18). According to this procedure, K at the maximum crack depth, a, is estimated as follows:

$$Kellipse = Kchord \times (An,chord/An,ellipse) \quad [5]$$

where An,chord and An,ellipse, respectively, are the net (uncracked) areas of equal diameter tension bars, one containing a chord crack, and the other containing an elliptical crack of equivalent depth, a. Normalized K-values for the crack configurations of interest are given in Figure 7. The estimated elliptical-crack K-solutions shown in the Figure are in good agreement with finite element results subsequently obtained by Raju and Newman (17).

Elastic-plastic fracture mechanics (EPFM) was developed to extend the crack driving force concept to larger scale plasticity than allowed by LEFM (19 to 21). In principle, therefore, EPFM can be applied to better model the stress and flaw size relationship in the transition regime connecting the limit load and LEFM failure criteria as illustrated in Figure 5. The use of 3-D finite element methods to establish

elastic-plastic crack driving force solutions for surface flaws on the scale of interest has been demonstrated (22,23), and a solution applicable to the current specimen geometry is presently under development. In lieu of the availability of a rigorous solution, the following empirical equation was used to approximate the elastic-plastic failure criterion for purposes of this investigation.

$$\sigma_{EPFM} = \sigma_L [1 - (a/R)^q] + \sigma_{LEFM} (a/R)^s \quad [6]$$

where:

σ_{EPFM} = the EPFM fracture (or breaking) stress

σ_L = the fracture stress predicted by the limit load failure criterion, (Figure 6).

σ_{LEFM} = the fracture stress predicted by the LEFM failure criterion, (Figure 7).

a = flaw depth

R = the specimen radius

q, s = weighting exponents.

Values of q and s in the above expression are to be chosen to produce a smooth transition between and tangent to the limit load and LEFM failure criteria as illustrated in Figure 5. The values selected to these requirements for alloy 7075 and the two specimen diameters evaluated are given below for the assumed elliptical flaw having an aspect ratio a/c = 0.8:

Specimen Diameter, in.	q	s
0.125	2	4
0.225	1.5	4

Failure relationships developed in accordance with Equation 6, the above q and s values and mechanical properties (σ_{UTS} and Kic) of the three 7075 alloy temper variants (Table 1) are in good agreement with the experimental results. This is shown for 7075-T651 in Figure 5 and for the two specimen diameters of the intermediate T7X1 temper material in Figure 8. These findings demonstrate that maximum depth of SCC penetration can be reasonably and conservatively predicted from the measured breaking strengths. It follows, therefore, that SCC characterization can be extended to shallow and more natural flaws, while preserving many of the advantages of traditional fracture mechanics theory applied to specimens containing deep flaws. For example, the progression of early crack-like damage can be examined with time by removing samples for tension testing after selected intervals of exposure. Thus, from a single test method, it is possible to compare materials on the basis on their probabilities of initiating and propagating SCC flaws to an arbitrary depth or by their respective propagation rates, both being meaningful engineering descriptors of SCC damage.

Exposed 7075 tensile bar breaking strengths were converted to SCC flaw sizes in accordance

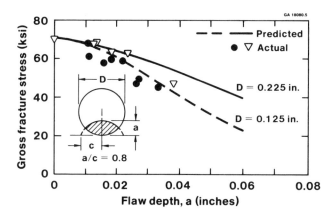

Fracture Stress vs. Maximum Flaw Depth in Stress Corroded Alloy 7075-T7X1 Tensile Bars of Different Diameters
Figure 8

with the above procedures. The results of these computations are reported in Reference (2) for all conditions evaluated. The selected illustrations that follow will serve to clarify advantages of the fracture mechanics interpretation. The range and mean of estimated flaw depths versus exposure time are shown for alloy 7075-T651 in Figure 9. These data

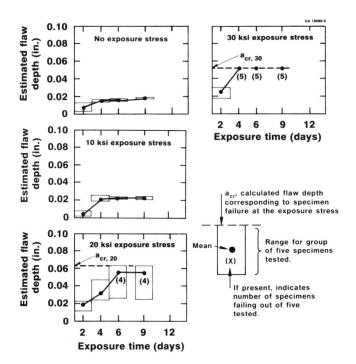

7075-T651 short transverse 0.125-in. diameter specimens exposed to 3.5% NaCl solution by alternate immersion.

Damage Progression During Exposure
Figure 9

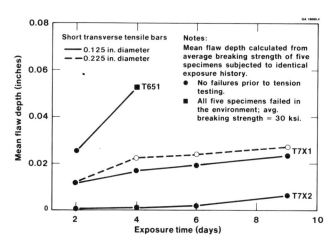

Effect of Temper on SCC Performance of Alloy 7075 Subjected to Alternate Immersion in 3.5% NaCl Solution at 30 ksi Exposure Stress
Figure 10

correspond to the individual groups of five replicate specimens subjected to identical exposure conditions. Whenever specimen failures occurred prior to completion of the designated exposure period, the breaking strength was equated to the exposure stress for purposes of estimating the maximum size SCC flaw, a_{cr}, the specimen could sustain for the given exposure interval. Mean trends in the 30 ksi exposure data for the three temper variants of alloy 7075 are shown in Figure 10. These results clearly illustrate that aging increase from the T651 to the T7X2 temper condition reduces the depth of SCC penetration with time, which is consistent with other performance rankings given in Table 3. SCC growth rates estimated from slopes of the curves in Figure 10 are shown in Table 3 to agree reasonably well with the plateau (K-independent) growth rate determined from conventional fracture mechanics tests using bolt-loaded DCB specimens. The 99 percent penetration limit noted in Table 3 is calculated by equating the breaking stress to the 99 percent survival stress, and it estimates the flaw depth that would not be exceeded in 99 percent of the exposed specimens.

When interpreted in conjunction with cracked body mechanics, the breaking load approach is able to normalize influence of specimen size on measured performance. This is demonstrated by equivalence of the 7075-T7X1 0.125 and 0.225-in. round tensile bar performance in Figure 10. In contrast, Figure 11 shows that specimen size biases are present in SCC ratings obtained by traditional pass-fail methods (24). The predicted curves of Figure 8 illustrate that small scale damage is more sensitively detected by the smaller specimen since its strength reduction is greater than that of the larger specimen containing an SCC flaw of equivalent depth.

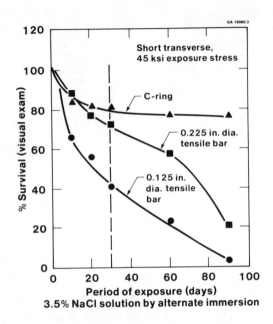

Short transverse, 45 ksi exposure stress

C-ring

0.225 in. dia. tensile bar

0.125 in. dia. tensile bar

Period of exposure (days)
3.5% NaCl solution by alternate immersion

Influence of Specimen Configuration on Pass-Fail Measure of Alloy 7075 SCC Performance
Figure 11

Evaluating alloys possessing different combinations of fracture toughness and strength is frequently encountered in trade-off studies to optimize structural material selection. Like the specimen size influence noted above, strength and toughness property differentials may bias alloy performance ratings made by traditional pass-fail methods (2). In accordance with Equation 6, the relative importance of strength and toughness to breaking strength depends on the specimen cross-sectional area and the size of flaw present. Alloy strength and toughness, however, may have little bearing on the size of the SCC flaw developed during the exposure portion of the test. A more meaningful alternative, therefore, would be to quantify SCC performances in terms of subcritical flaw dimensions, as in Figure 10, that are free of specimen size and the strength-toughness property biases.

SUMMARY

A promising new technique, the breaking load method, has been developed for accelerated stress corrosion testing of aluminum alloys. This new approach provides more discriminating information, with fewer specimens and shorter exposure times, than conventional pass-fail test procedures. The breaking load method utilizes data from tensile tests performed on replicate groups of smooth specimens after exposure to static stress and corrosive environments for various lengths of time (typically 4 to 10 days). The degree of degradation due to the stress corrosion attack is measured by comparing the specimen post-exposure strength against the material original tensile strength (no exposure).

Analysis of breaking load data by extreme value statistics enables calculation of survival probabilities and the estimation of a threshold stress, without depending on failures during the exposure.

An elastic-plastic fracture mechanics (EPFM) model is given that quantifies damage in the stress corroded specimen by an "effective flaw size" calculated from the measured breaking stress and the strength and fracture toughness properties of the test material. The effective flaw corresponds to the "weakest link" in the specimen test section at the time of the tension test, and it estimates the maximum penetration of stress corrosion attack. Breaking strength versus flaw depth combinations predicted by the EPFM model are shown to agree well with actual measurements. This feature enables progression of early crack-like damage to be examined statistically with time by removing samples for tension testing after selected exposure intervals. Thus, from a single test method, it is possible to acquire information to compare materials on the basis of their probabilities of initiating and propagating stress corrosion flaws to an arbitrary depth or by their respective stress corrosion cracking (SCC) rates, both being meaningful engineering parameters. An additional advantage to using flaw depth, or rate of growth, to examine SCC performance is that the effects of specimen size and alloy strength and toughness can be normalized. In contrast, the specimen lifetime and breaking strength are dependent on the above mechanical factors.

Comparisons were made of experimental results obtained by the breaking load method and by conventional procedures employing statically loaded smooth tensile bars and precracked specimens (DCB and WOL type). The material used in this comparative study was aluminum alloy 7075 aged to high, medium and low degrees of SCC susceptibility. The SCC ratings given by the breaking load method were found to be totally consistent with those given by the currently accepted test methods.

Though the breaking load method was developed to assess SCC, the approach has generic appeal for characterizing other damage processes that result in evolution and growth of "small" crack-like defects (e.g., fatigue and wear).

REFERENCES

1. D. O. Sprowls, "A Study of Environmental Characterization of Conventional and Advanced Aluminum Alloys for Selection and Design", Phase I - Literature Review, NASA CR-172387, August 31, 1984.

2. D. O. Sprowls, R. J. Bucci, B. M. Ponchel, R. L. Brazill and P. E. Bretz, "A Study of Environmental Characterization of Conventional and Advanced Aluminum Alloys for Selection and Design", Phase II - The Breaking Load Test Method, NASA CR-172387, August 31, 1984.

3. ASTM G47-79, "Standard Test Method for Determining Susceptibility to Stress Corrosion Cracking of High Strength Aluminum Alloy Products", 1985 Annual Book of ASTM Standards, Section 3, Vol. 03.02, Amer. Soc. Testing and Materials, p. 277.

4. ASTM G44-75, "Standard Practice for Alternate Immersion Stress Corrosion Testing in 3.5% Sodium Chloride Solution", 1985 Annual Book of ASTM Standards, Section 3, Vol. 03.02, Amer. Soc. Testing and Materials, p. 261.

5. ASTM B557-84, "Standard Methods of Tension Testing Wrought and Cast Aluminum and Magnesium Alloy Products", 1985 Annual Book of ASTM Standards, Section 3, Vol. 03.01, Amer. Soc. Testing and Materials, p. 64.

6. A. A. Griffith, "The Phenomena of Rupture and Flow in Solids", Phil. Trans. Roy. Soc., 221A, 1920, p. 163.

7. A. A. Griffith, "The Theory of Rupture", First International Congress of Applied Mechanics, Delft, 1924, p. 55.

8. E. J. Gumbel, "Les Valeurs Extremes des Distributions Statistiques", Ann. Inst. Henri Poincare, 4, 1935, p. 115.

9. A. M. Freudenthal, "The Statistical Aspects of Fatigue of Materials", Proc. Roy, Soc., A187, 1946, p. 416.

10. J. C. Fisher and J. H. Hollomon, "A Statistical Theory of Fracture", Tech. Pub. No. 2218, Metals Technology, AIME, 14, 1947, p. 5.

11. B. Epstein, "Statistical Aspects of Fracture Problems", J. Appl. Physics, 19, 1948, p. 140.

12. B. Epstein, "Application of the Theory of Extreme Values in Fracture Problems", J. Amer. Statistics Assoc., 43, 1948, p. 403.

13. E. J. Gumbel, Statistical Theory of Extreme Values and Some Practical Applications, U.S. Dept. of Commerce, Appl. Mathematics Series 33, 1954.

14. E. J. Gumbel, Statistics of Extremes, Columbia Univ. Press, New York, 1958.

15. E. J. Hahn and S. S. Shapiro, Statistical Models in Engineering, John Wiley & Sons, New York, 1968.

16. B. W. Lifka and D. O. Sprowls, "Significance of Intergranular Corrosion of High-Strength Aluminum Alloy Products", Localized Corrosion-Cause of Metal Failure, ASTM STP 516, Amer. Soc. Testing and Materials, 1972, p. 120.

17. I. S. Raju and J. C. Newman, Jr., "Stress Intensity Factors for Circumferential Surface Cracks in Pipes and Rods," Seventeenth National Symposium on Fracture Mechanics, Albany, NY, Aug. 7-9, 1984.

18. O. E. K. Daoud, D. J. Cartwright, and M. Carney, "Strain-Energy Release Rate for a Single-Edge Cracked Circular Bar in Tension", J. Strain Analysis, 13, 2, 1978, pp. 83.

19. J. W. Hutchinson, "Singular Behavior at the End of a Tensile Crack in a Hardening Material", J. Mechanics and Physics of Solids, 16, 1968, p. 1.

20. J. R. Rice and G. R. Rosengren, "Plane Strain Deformation Near a Crack Tip in Power Hardening Material", Mechanics and Physics of Solids, 16, 1968, p. 13.

21. P. C. Paris, "Fracture Mechanics in the Elastic-Plastic Regime", Flaw Growth and Fracture, ASTM STP 631, Amer. Soc. Testing and Materials, 1977, p. 3.

22. G. G. Trantina, H. G. DeLorenzi, W. W. Wilkening, "Three Dimensional Elastic-Plastic Finite Element Analysis of Small Surface Cracks", Engineering Fracture Mechanics, 18, 5, 1983, p. 925.

23. D. M. Parks and C. S. White, "Elastic-Plastic Line-Spring Finite Elements for Surface-Cracked Plates and Shells", J. Pressure Vessel Technology, 104, Nov. 1982, p. 287.

24. D. O. Sprowls, T. J. Summerson, G. M. Ugiansky, S. G. Epstein, and H. L. Craig, "Evaluation of a Proposed Standard Method of Testing for Susceptibility to SCC of High Strength 7XXX Series Aluminum Alloy Products", Stress Corrosion-New Approaches, ASTM STP 610, Amer. Soc. Testing and Materials, 1976, p. 3.

THE CORROSION-FATIGUE CRACK PROPAGATION BEHAVIOUR OF Mn-Ni-Al BRONZE PROPELLER ALLOYS

J. I. Dickson, L. Handfield, S. Lalonde,
M. Sahoo,* J. P. Bäilon
Départment de Génie métallurgique
École Polytechnique
Montréal, Canada
*PMRL-CANMET, Ottawa, Canada

Abstract

The fatigue propagation behaviour in air and in 3.5% NaCl aqueous solution of Mn–Ni–Al bronzes containing 12 and 14% Mn was studied for three material conditions: as–cast (AC), normalized (NSC) and quenched (HT). Accelerated cracking occurred in the salt solution at higher growth rates. At lower rates, corrosion–product induced crack closure caused decelerated propagation in this solution. The corrosion–fatigue mechanism was associated with facilitated initiation of fatigue cracking in α–grains and, in the NSC and AC materials, of decohesion at α–β interfaces proceeding ahead of the macroscopic crack front. The fatigue threshold increased with increasing β grain size as a result of greater roughness–induced crack closure effects. The effective value of ΔK_{Th} was constant within the experimental accuracy. The log da/dN vs log ΔK_{eff} data more clearly showed the accelerated cracking in the salt solution. The propagating crack experienced difficulty in crossing β–grains, which, in the alloys of larger grain size, resulted in oscillations in the lower portion of the crack propagation curves.

BECAUSE OF THE PRACTICAL IMPORTANCE OF CORROSION–FATIGUE, there is considerable interest in characterizing fatigue behaviour in aggressive environments and in improving our understanding of the mechanisms by which such environments shorten the fatigue life. A number of previous studies have been carried out on the fatigue behaviour in salt water of both Ni–Al [1–4] and Mn–Ni–Al [5–7] ship propeller bronzes. The present paper focusses on the influence of the environment (air compared to 3.5% NaCl aqueous solution), the load–ratio (R=0.1 and R=0.5), the cycling frequency (f=1Hz and f=20Hz), the material condition (as–cast material and two heat–treated conditions) on the fatigue propagation behaviour of two Mn–Ni–Al bronzes, differing essentially in their Mn content (12% and 14% Mn).

EXPERIMENTAL PROCEDURE

Plates 295 mm x 148 mm x 25 mm were cast in sand moulds employing a central riser. Typical compositions of the plates were 2.1% Ni, 7.9% Al, 3.3% Fe and either 12.0 or 14.3% Mn. Two compact tension specimens of thickness B=22.3 mm and width W=101 mm were machined from each plate. Some of the as–cast (AC) plates were normalized (NSC) at 900°C for two hours followed by slow–cooling (10°C/hr) to 230°C. Some plates were given a further heat treatment (HT condition) of a 2 hour anneal at 690°C followed by water quenching, developed to improve the impact properties of these alloys [8]. The fatigue tests were performed on either an MTS or an Instron servohydraulic machine, employing a sinusoidal waveform. A desk–top computer was employed for test control and data acquisition. The crack lengths were calculated from the sample compliance in the linear elastic portion of the load P vs crack mouth opening displacement (CMOD) curve and verified by optical measurements during the tests and/or on the fracture surfaces. The tests in the 3.5% NaCl solution were performed under freely corroding conditions by circulating this solution at a rate of 0.5 ℓ/min into an acrylic plastic cell, attached to the specimen employing electronic grade silicone glue. The salt solution in the 3.5 ℓ reservoir was changed every 48 h.

RESULTS AND DISCUSSION

MICROSTRUCTURES – These Mn–Ni–Al bronzes are multiphase alloys consisting primarily of grains of f.c.c. α–phase within a b.c.c. β–phase matrix. Iron–rich precipitates which appear black in optical micrographs are also present. In the AC and NSC materials, the β–layer between α–grains

Table 1: Microstructural Characterization

Alloy	α-grain size (mm)		β-grain size (mm)		% β-phase	
Condition	12% Mn	14% Mn	12% Mn	14% Mn	12% Mn	14% Mn
AC	0.05	0.03	0.5	0.3	22	36
N	0.10	0.08	3.5	2	19	24
HT	0.10	0.08	7	3	55	65

was generally thin but considerably thicker for the HT materials. In all three materials, some of the α-grains were elongated in Widmanstatten patterns, which aided measurement of the β-grain size in the materials (NSC alloys and 12% Mn AC alloy) for which the individual β-grains were difficult to distinguish. The α and β-grain sizes as well as the percentage of β-phase present for the different alloys and conditions are listed in Table 1.

CRACK GROWTH RATES - The log da/dN vs log ΔK curves for the 12% Mn AC alloy (Fig. 1) at R=0.1 show higher ΔK_{Th} values and slower crack growth at low ΔK in the 3.5% NaCl solution compared to air. At high growth rates, the f=20Hz curve for the salt solution rejoins the air curve while the f=1Hz curve for this solution crosses the air curve towards da/dN$\approx$2 x 10^{-5} and then shows a continuously larger difference in growth rates, suggesting corrosion-fatigue behaviour. The tendencies at R=0.5 in this 12% Mn AC material are similar. The crack growth curves in the salt solution indicates slower growth compared to air at the same ΔK value for the low growth rates and faster growth for the high rates. For the tests at f=1Hz in the NaCl solution, the curves for R=0.5 and R=0.1 rejoin each other near da/dN=2x10^{-4} mm/cycle. The 14% Mn AC material (Fig. 2) show similar tendencies as observed for the 12% Mn AC alloy. In the NaCl solution, the ΔK_{Th} values appear somewhat higher at f=1Hz than at f=20Hz and above da/dN$\approx$ 2x10^{-5} mm/cycle, the crack growth curve at 1Hz indicated higher growth rates than at 20 Hz and agreed well with the curve obtained, for the same test conditions, in the 12% Mn AC alloy.

The results indicate little influence of the Mn content (12 or 14%) on the crack growth rates in the two AC alloys. The lower rates in air were somewhat slower for the 14% Mn alloy, while for the tests in 3.5% NaCl solution, the high growth rates tended to be somewhat faster at the same ΔK value for the 14% Mn alloy.

The crack growth results for the normalized materials are presented in Figures 3 and 4. The two duplicate tests carried out in each environment show good agreement. The ΔK_{Th} value in the salt solution for the 12% Mn alloy (Fig. 3) tested at R=0.1 and f=20Hz is considerably higher than in air, while above da/dN$\approx$$10^{-4}$ mm/cycle somewhat faster crack growth is obtained in the 3.5% NaCl solution. At R=0.5, the growth curve above da/dN$\approx$$10^{-4}$ mm/cycle in the solution indicates faster cracking at f=1Hz than at f=20Hz. The difference in Mn content has little influence on the da/dN curves in the NSC

condition, the main difference being a somewhat faster crack rate for the 12% Mn alloy for intermediate ($\approx$$10^{-6}$ to 10^{-4} mm/cycle) rates.

Comparing the NSC and AC materials, the ΔK_{Th} value is higher and the low da/dN values indicate slower growth for the NSC condition, for both R-ratios and environments (Fig. 5). In air, essentially the same da/dN curves are obtained above da/dN$\approx$$10^{-4}$ mm/cycle. In the 3.5% NaCl solution, similar agreement at high growth rates is obtained for the tests at 20Hz and R=0.1 for the 14% Mn AC and both NSC materials, while the 12% Mn AC does not suggest accelerated growth compared to air. The tests at R=0.5 and 1Hz in NaCl solution also suggest that the NSC condition may be somewhat more susceptible to accelerated crack growth in the NaCl solution.

For the 12% Mn HT alloy (Fig. 6), three tests carried out in air at R=0.1 show similar ΔK_{Th} values but there was important dispersion at higher growth rates, which was associated with very crystallographic cracking and large fracture facets. At R=0.5, a similar ΔK_{Th} value is indicated to that at R=0.1, with faster crack growth compared to R=0.1 at somewhat higher da/dN values. Compared to air, the crack growth in the salt solution is decelerated at very low rates and, especially for the test at 1Hz and R=0.5, accelerated at high rates. Figure 7 shows similar relative da/dN results in the two environments for the 14% Mn HT alloy. The test in air for the 14% Mn HT alloy agrees for da/dN$\geq$$10^{-5}$ mm/cycle with two of the tests on the 12% Mn HT alloy. The HT alloys show better resistance to fatigue crack propagation both in air and in the 3.5% NaCl solution than either the AC or the NSC materials (Fig. 8).

FRACTOGRAPHIC OBSERVATIONS - The fracture surfaces produced in the salt solution and, except for the HT alloys, those produced at lower rates in air were covered with noticeable corrosion product. This product, except for that produced in the salt solution on the steep portion of the log da/dN vs log ΔK curve (da/dN$\leqslant$3x10^{-5} to 10^{-4} mm/cycle), was successfully removed by ultrasonic cleaning in a liquid soap-water solution [5]. The indication was that the corrosion product not removed corresponded to one produced by fretting under important crack closure effects.

For the AC and NSC materials, macroscopic facets corresponding approximately to the β-grain size could be seen on the fracture surfaces produced at lower rates, with these facets appearing more clearly for the tests in the NaCl solution. For the HT alloys, these

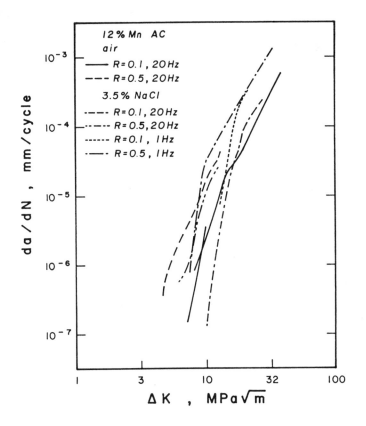

Fig.1 Crack growth results for the 12% Mn AC alloy.

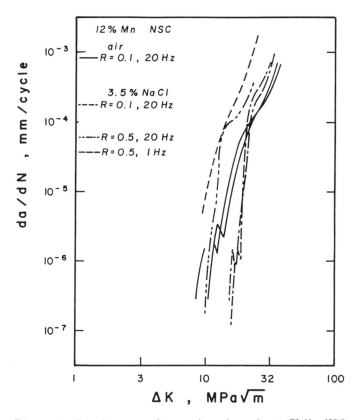

Fig. 3 Crack growth results for the 12% Mn NSC alloy.

Fig. 2 Crack growth results for the 14% Mn AC alloy.

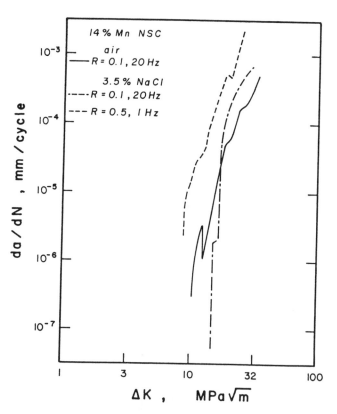

Fig. 4 Crack growth results for the 14% Mn NSC alloy.

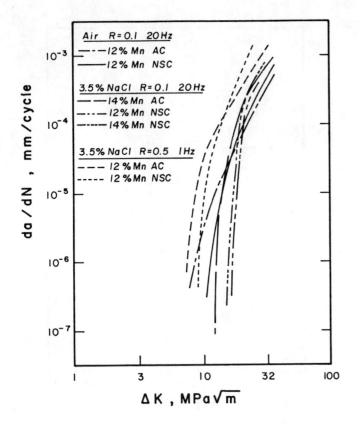

Fig. 5 Comparison of some crack growth curves for the NSC and AC alloys.

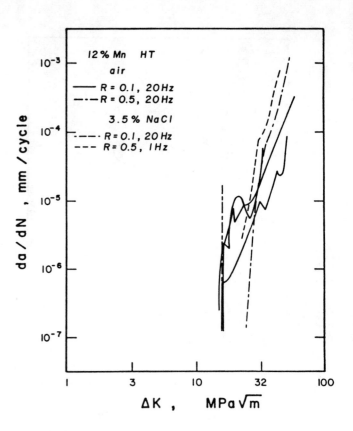

Fig. 6 Crack growth results for the 12% Mn HT alloy.

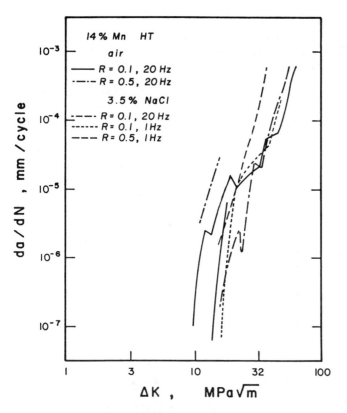

Fig. 7 Crack growth results for the 14% Mn HT alloy.

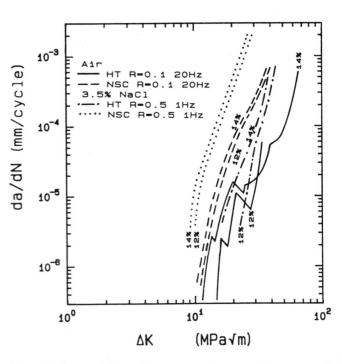

Fig. 8 Comparison of some crack growth curves for the HT and NSC alloys.

crystallographic facets were very clearly observed for the tests in air. In 12% Mn HT alloy tested in salt solution these facets appeared to be about twice as small as the β-grain size, which effect was apparently associated with important crack branching which occurred at low rates in this solution.

The microfractographic features observed by scanning electron microscopy for the AC and NSC materials have been described elsewhere [5]. The cracking occurs mainly in the α-phase. For tests in air, the propagation direction could be followed by striations and ridge lines and was seen to proceed in a straightforward manner from α-grain to α-grain, with the cracking changing from very crystallographic below and non-crystallographic above $da/dN \approx 10^{-4}$ mm/cycle. In contrast, when accelerated propagation was obtained in the salt solution, decohesion at α-β interfaces proceeded ahead of the macroscopic crack front, which facilitated initiation of fatigue cracking in the α-grains from different sites along these slots (Fig. 9). Some of the cracking at times even extended back towards the macroscopic crack front. On metallographic sections perpendicular to the fracture surface (previously coated with thick lacquer which was hardened to prevent rounding of the edges during polishing), this decohesion was observed not to extend more than approximately 20μm from the fracture surface probably because it is limited to the α-β interfaces and the α-phase is not continuous.

For the HT materials tested in air, the macroscopic facets, consisted of a large number of smaller facets, many of which appeared parallel. In the 14% Mn alloy, the α-grains appeared as islands presenting an irregular but crystallographic topography, within smoother β-facets covering the majority of the fracture surface. Cracking in the β-phase could be seen often to partially or completely bypass the α-grains (Fig. 10), with fatigue initiation in the α-grains then occurring from several sites along the α-β interfaces. For the 12% Mn HT alloy, almost 50% of the fracture surface passed through the α-phase. The β microfacets again were generally smoother than the crystallographic facets in the α-grains, which contained pronounced river lines (Fig. 11). Crack propagation appeared to lead in the β microfacets, where propagation was often parallel to the macropropagaton direction for that β-grain. From the α-β interfaces, cracking then spread laterally into the α-grains with initiation of cracking in an α-grain often occurring at several sites along the grain perimeter. At the higher rates, irregular fatigue striations were observed on the β-facets, with the interstriation distance finer near interfaces than in the centers of β-facets (Fig. 12). If these striations indicate crack front positions after successive crack advances, these observations suggest that, away from interfaces, the individual crack advances are longer but

occur less frequently possibly by being temporarily arrested on meeting new α-β interfaces. Striations were at times found on α-facets and indicated crack propagation rates 3-10 times lower than in the β-phase. For the HT alloys tested in NaCl solution, the fracture surfaces produced at the higher rates where corrosion-fatigue effects occurred could be cleaned sufficiently to observe the basic features. Cracking in the β-phase generally no longer tended to bypass α-grains. The fracture surface in the α-grains was generally smooth and often coincided quite well with that in the surrounding β-phase. The striations in the β-phase were generally more marked and appeared to have a more ductile aspect and a more regular spacing than those produced in air. The interstriation spacing in both phases was generally similar (Fig. 13) and generally indicated that the propagation proceeded smoothly across the interface. Some α-grains were surrounded by decohesion slots similar to those observed in the AC and NSC materials, and these occasionally caused fatigue initiation in such an α-grain to occur at several sites along the slot at the interface.

The metallographic observations of the fracture surface profiles confirmed the fractographic observations obtained on the HT alloys. For the tests in air, the crack path was rather devious especially in the α-grains. In the 12% Mn alloy, the path in the β-phase appeared to change frequently by small steps (Fig. 14), possibly as a result of the crack front frequently meeting α-grains, in which propagation was difficult. In the 14% HT alloy tested in air, the crack path was generally straighter in the β-phase, in comparison to the 12% Mn alloy, but often devious in the α-phase. In both alloys, the observations suggested that the cracking in α-grains quite often came from both sides of the grain perimeter and met in the middle after following a tortuous crack path (Fig. 15). Some secondary cracks could be seen to terminate at α-β interfaces on meeting α-grains. Some tendency was also noted for the crack path to be in the α-phase near or at times apparently at α-β interfaces. Segments of secondary cracks in the β-phase were often roughly parallel (Fig. 16) to one or two orientations of elongated α-grains arranged in Widmanstatten patterns or to straight, apparently crystallographic interfaces of non-elongated α-grains.

In both HT alloys tested in the NaCl solution, the crack path within both phases appeared straighter, with less crack deviation on entering α-grains. The orientations of different secondary cracks within a β-grain were more clearly parallel to the different orientations of elongated α-grains (Fig. 17). There was also less tendency for the crack path within α-grains to be situated near α-β interfaces.

Secondary cracks were observed to have considerable difficulty in propagating across

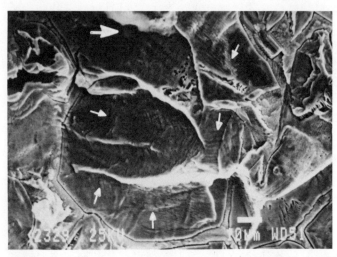

Fig. 9 Ductile striations produced at high da/dN in the 12% Mn NSC alloy tested in 3.5% NaCl solution (f=20Hz, R=0.1). The large arrow indicates the macroscopic propagation direction; the small arrows indicate local propagation directions in α-grains from initiation sites at decohesion slots at α-β interfaces.

Fig. 10 Bypassed α-grain by the faster crack propagation in the β-phase for the 14% Mn HT alloy tested in air.

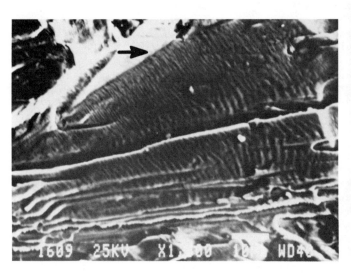

Fig. 11 Microfractographic aspect of the 12% Mn HT alloy tested in air. Some of the α-grains, which occur as round islands within the β-matrix can be easily recognized. The lateral propagation in the α-grains, often occurring from several initiation sites at the α-β interfaces, indicates that the microscopic crack front leads in the relatively flat β-microfacets.

Fig. 12 Variable interstriation spacings observed at high growth rates on a β-microfacet for the 14% Mn HT alloy tested in air. The interstriation spacing is largest towards the center of the microfacet.

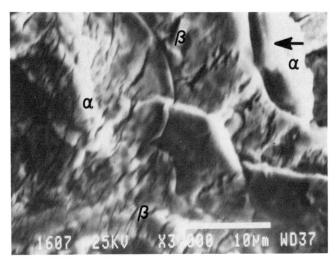

Fig. 13 Flattened ductile striations observed for the higher growth rates in the 14% Mn HT alloy tested in 3.5% NaCl solution. The microstructure has little influence on the local propagation rate.

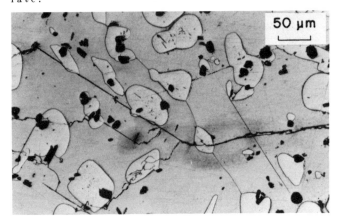

Fig. 15 Details of the path of a secondary crack in 12% Mn HT bronze tested in air. In this microregion, the microstructure is more typical of 14% Mn HT material. Note the α-grain presenting a very tortuous crack path.

Fig. 17 Relatively straight secondary cracks parallel to three orientations of elongated α-grains in 12% Mn HT alloy tested in 3.5% NaCl solution (f=20Hz, R=0.1).

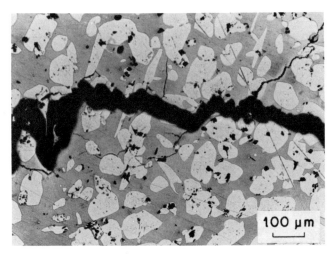

Fig. 14 Large and fine secondary cracks observed on the 12% Mn HT alloy tested in air. Some of the fine cracks stop on meeting α-grains. Some crack segments go along α-β interfaces and along interfaces with black Fe-rich precipitates.

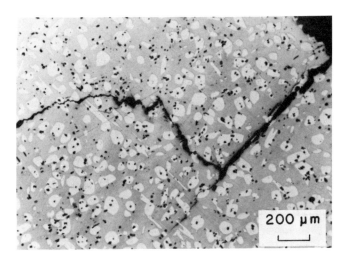

Fig. 16 Secondary crack in the 14% Mn HT bronze tested in air, with sections of the crack path in the β-phase tending to be parallel to two orientations of elongated α-grains.

β-grain boundaries in the HT alloys. For lower growth rates, the crack profile generally showed pronounced deviations in crossing β-boundaries, often accompanied by the formation of large secondary cracks near this boundary (Fig. 18). As ΔK and da/dN increased, the deviations at β-boundaries tended to be less severe and generally unaccompanied by important amounts of crack branching.

In the near-threshold region, very extensive crystallographic secondary cracking was observed in the region close to the fracture surface for the HT alloys tested in NaCl solution. In the α-phase the frequent 60° angles (Fig. 19) between these secondary crystallographic cracks indicated cracking on {111} planes, as verified by slip traces introduced by microhardness indentations. In the β-phase, these secondary cracks were parallel to the elongated Widmanstatten α-grains and not parallel to the secondary cracks in the α-phase. Slip traces identifiable by optical microscopy were not successfully produced in the considerably harder β-phase. The very important secondary cracking at low rates in the HT alloy tested in 3.5% NaCl solution, also appeared associated with crack deviations within β-grains, which effect can explain the smaller macroscopic facet size observed for tests in 3.5% NaCl solution compared to tests in air.

The blackish iron-rich precipitates present in the microstructure appear to have almost no influence on the straighter cracks obtained for the tests on the HT material in the NaCl solution. The more devious crack paths obtained in air showed a somewhat greater tendency (Fig. 14) to deviate slightly in order to go along the interface of a black particle situated very close to the crack path, but these particles still only appeared to have a very small influence on the crack paths, in keeping with their small size, with respect to the crack segments and to their degree of dispersion.

OSCILLATIONS IN da/dN – The crack growth curves generally obtained for the HT and NSC materials presented, towards da/dN values of 10^{-6} to 10^{-5} mm/cycle, oscillations (Fig. 20) in the log da/dN vs log ΔK curves. The growth rate, measured with an increasing ΔK procedure, would suddenly decrease by a factor of 2-5 and then after some time reincrease again and then show a similar sudden decrease. This effect was associated with the materials having a large β-grain size resulting in a large macroscopic fracture facet size. By the metallographic observations of the fracture surface profile, it was found that these sudden decreases in da/dN generally corresponded to crack lengths at which at mid-thickness the crack was observed to arrive at β-grain boundaries which it had difficulty to cross. For da/dN $\leq 10^{-5}$ mm/cycle, a pronounced crack deviation resulted at such a boundary, generally accompanied by a secondary crack of significant size (Fig. 18).

CRACK GROWTH AS A FUNCTION OF ΔK_{eff}. At low crack growth rates, crack closure effects have a particularly strong influence on the log da/dN vs log ΔK curves [9], since for a portion of the stress intensity range, the crack in the vicinity of the crack tip is "closed" and no cyclic damage in the region ahead of the crack tip is produced during this portion of ΔK. This ineffective portion can be subtracted from ΔK by measuring, K_{op}, the stress intensity factor at which the crack tip becomes opened. The partially closed crack has a compliance indicative of a crack length shorter than the true length. As the length portion over which the crack is closed decreases, the compliance increases accordingly. Once the crack tip is opened, linear elastic behaviour and a constant compliance are obtained until the plastification of the crack tip regions influence. The value of K_{op} and of $\Delta K_{eff} = K_{max} - K_{op}$ can thus be obtained by determining the value of K at which the compliance becomes constant on sufficiently precise recordings of the load vs crack mouth opening displacement hysteresis loops. Figures 21-24 summarize the log da/dN vs log ΔK_{eff} results obtained. The results obtained in air for the different tests on the AC and NSC materials are presented in Figure 21, which indicates that on a ΔK_{eff} basis there is even less difference in the crack growth curves for these two compositions and two material conditions, although the results suggest that the AC condition may have a very slightly higher resistance to fatigue propagation, which difference, however, cannot be considered to be significant. The differences in the log da/dN vs log ΔK curves for the AC and NSC materials are therefore primarily due to crack closure effects, despite the important differences in the size of the microstructural features. This aspect is also indicated by these curves being almost identical at high growth rates, where the crack closure effects become negligible.

The influence of the R-ratio on the log da/dN vs log ΔK curves obtained in air is also seen to only influence the ineffective portion, $K_{op} - K_{min}$, of ΔK and also to be related to crack closure effects.

Figures 22 and 23 report the log da/dN vs log ΔK_{eff} curves obtained in 3.5% NaCl for the two bronzes in the AC and NSC conditions respectively and compare these results with the curve typical of the results in air. For both conditions, all the results in the NaCl solution show on a ΔK_{eff} basis accelerated crack growth for da/dN$\geq 10^{-5}$ mm/cycle compared to the tests in air. For the high growth rates (da/dN$\geq 10^{-4}$ mm/cycle) in the NaCl solution, all the log da/dN-log ΔK_{eff} data indicate faster crack growth at f=1Hz than at f=20Hz. The crack growth data obtained at R=0.1 and 0.5 in the solution also either agree well on a ΔK_{eff} basis or show slower crack growth at R=0.5. The latter effect is the opposite of that which can be expected and suggests some difficulty in obtaining precise measurements of ΔK_{eff} for at least some of the testing conditions. Although the precision of

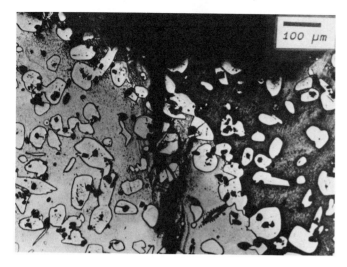

Fig. 18 Crack deviation and crack branching on meeting a β-grain boundary (12% Mn HT alloy, tested in 3.5% NaCl solution, R=0.1, f=20Hz).

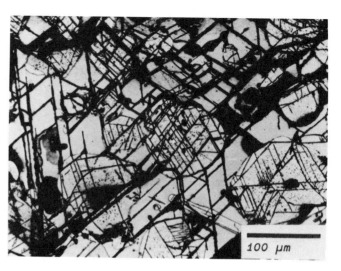

Fig. 19 Presence of pronounced crack branching in the near-threshold region (12% Mn HT alloy, tested in 3.5% NaCl solution, R=0.1, f=20Hz).

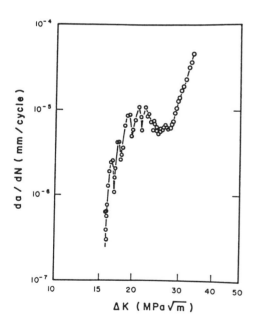

Fig. 20 Details of crack growth curve obtained for 12% Mn HT alloy tested in air, (R=0.1, f=20Hz).

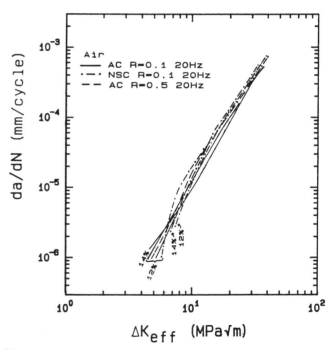

Fig. 21 Crack growth as a function of ΔK_{eff} for tests in air on AC and NSC materials.

the ΔK_{eff} results in the NaCl solution is far from perfectly satisfactory, these results do permit to see occurrence of corrosion-fatigue effects for the intermediate da/dN values (10^{-5} to 10^{-4} mm/cycle), for which the important amount of corrosion-product induced crack closure in the NaCl solution tends to hide the occurrence of this effect when the results are compared on a ΔK basis. The occurrence of this corrosion-product induced crack closure at low and intermediate rates in the salt solution was also demonstrated by not being able to remove the corrosion product from the fracture surfaces produced on the steep portion of the log da/dN vs log ΔK curve [5], which in this solution at times extended up to da/dN values of the order of 10^{-4} mm/cycle.

The log da/dN vs log ΔK_{eff} results for the HT material (Fig. 24) show greater dispersion in behaviour associated with the difficulty of measuring K_{op} with precision for these materials in which large macroscopic fracture facets were obtained. The results, nevertheless, serve to indicate accelerated crack growth occurring in the salt solution at lower da/dN values than can be observed from the log da/dN vs log ΔK curves.

The comparison on a ΔK_{eff} basis also indicates that the corrosion-fatigue behaviour obtained is also essentially a function of ΔK and of dK/dt as influenced by the cycling frequency and neither a function of K_{max} nor of the R-ratio.

THRESHOLD VALUES – The ΔK_{Th} values associated with the most reliable tests in both environments for a given manganese content and material condition and corresponding to da/dN values of 2×10^{-7} mm/cycle for R=0.1 and f=20Hz have been compared elsewhere [6] as a function of the β-grain size, which controls the macroscopic fracture facet size. These values generally show good agreement with a relationship (Fig. 25) in which ΔK_{Th} increases linearly with $d^{1/2}$, where d represents the β-grain size which controls the macroscopic facet size. The effective value of the threshold, $(\Delta K_{eff})_{Th}$, was also measured by subtracting the $K_{op}-K_{min}$ portion of ΔK, where K_{op} corresponds to the K value at which the crack tip starts opening. The $(\Delta K_{eff})_{Th}$ value for these conditions was found to be constant within the experimental accuracy at 4.0 MPa$\sqrt{m}$, although considerable dispersion in results were obtained particularly for the 14% Mn alloy and for the tests carried out in the salt solution. At R=0.5, the results on the AC and NSC material indicate considerably lower ΔK_{Th} values, resulting from the less important crack closure effect associated with this higher R-ratio. The measurements of ΔK_{eff} extrapolated to da/dN=2 x 10^{-7} mm/cycle again indicate a $(\Delta K_{eff})_{Th}$ value of approximately 4.0 MPa$\sqrt{m}$. The influence of the material condition and composition on the ΔK_{Th} values of these bronzes is thus seen to be largely determined by the roughness-induced crack-closure associated with the macroscopic fracture facet size, in spite of the important differences in the amount of β-phase present and

in the plasticity of the α and β phases. That the $(\Delta K_{eff})_{Th}$ values are within experimental accuracy the same in both environments further confims the strong influence of corrosion product-induced crack closure on the lower portion of the log da/dN vs log ΔK curve in the salt solution. This influence of the β-grain size, the R-ratio and the test environment on ΔK_{Th} is thus seen to be essentially an influence on the non-effective portion (i.e., $K_{op}-K_{min}$), with $(\Delta K_{eff})_{Th}$ being a constant within the experimental accuracy.

For the HT materials, the results on the 12% Mn alloy tested in air (Fig.6) suggest a similar ΔK_{Th} value at R=0.1 and at R=0.5, which result requires further verification. If it proves to be a reproducible effect, it would appear associated with the very large macroscopic fracture facets obtained in this material, for which the fretting corrosion product usually associated with threshold measurements [10] in air was not observed.

CORROSION-FATIGUE MECHANISM – As demonstrated by the fractographic and metallographic observations, the mechanism responsible for the accelerated crack growth rates in the NaCl solution for the AC and NSC material corresponds to fine narrow decohesion slots at the α-β interfaces proceeding ahead of the macroscopic crack front and helping to initiate ductile fatigue cracking in the neighbouring α-grains. While this mechanism was previously referred to as a dissolution mechanism [5], the crack growth rates involved are very high for such a mechanism and, at the highest rates measured, the corrosion-fatigue effect still appears to be increasing in importance as da/dN increases. This slotting therefore appears to be much more in agreement with a stress sorption mechanism. In these AC and NSC alloys, crack propagation in air proceeds in a straightforward fashion from α-grain to α-grain, while the occurrence of the slotting results in more irregular propagation.

For the HT materials, it is cracking in air which occurs in a less straightforward fashion, because more important amounts of cracking are occurring in the β-phase, and the crack experiences difficulty in crossing into the α-grains. In these HT materials, corrosion fatigue again facilitates crack initiation in the α-grains. Propagation in the α-grains is also facilitated and these lose their ability to hinder fatigue crack growth, resulting in more straigtforward cracking. This effect is at least partially responsible for the significant acceleration of the macroscopic crack propagation observed for the HT material at higher da/dN values in the NaCl solution. From the present study, it could not be determined whether this accelerated growth also resulted in part from a corrosion-fatigue effect in the β-phase.

INFLUENCE OF Mn AND β-PHASE CONTENT – As already noted, there is little difference in the fatigue propagation behaviour of the 12 and 14%

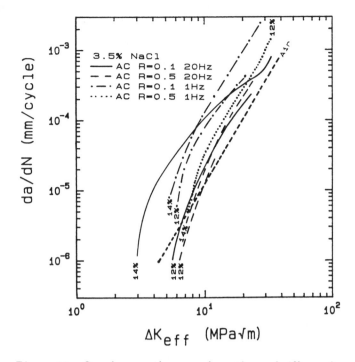

Fig. 22 Crack growth as a function of ΔK_{eff} for AC materials tested in 3.5% NaCl solution.

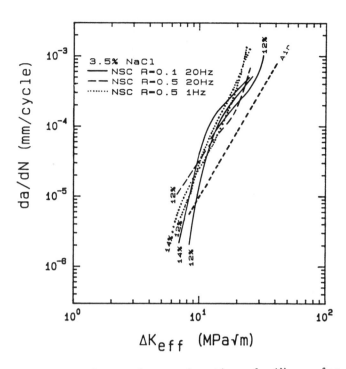

Fig. 23 Crack growth as a function of ΔK_{eff} for NSC materials tested in 3.5% NaCl solution.

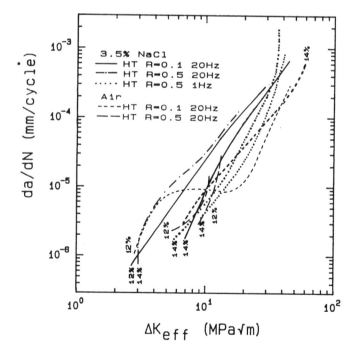

Fig. 24 Crack growth as a function of ΔK_{eff} for HT materials tested in air and in 3.5% NaCl solution.

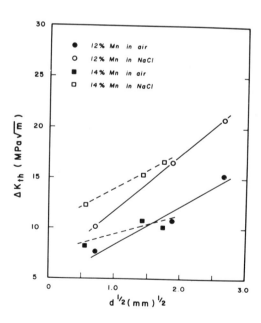

Fig. 25 Threshold values for the two alloys expressed as a function of the square root of the β-grain size, which depends on the material condition.

Mn alloys in the AC and NSC conditions. On a ΔK basis, the NSC alloys give slower crack growth at the low rates. This difference, which disappears when the comparison is done on a ΔK_{eff} basis, is the result of the difference in fatigue crack closure associated with the surface roughness of the two alloy conditions, this roughness at low rates being primarily a function of the β-grain size and of the size of the macrofacets.

Crack propagation for the HT condition is, at a given ΔK, considerably more difficult than for the AC and NSC conditions (Fig.8). The fractographic and metallographic observations indicate that this effect is associated largely with the more important amount of crack propagation that must then occur in the β-phase of the HT material, with the α-grains in the duplex microstructure hindering the fatigue propagation in air. This latter effect was much less important, but according to the metallographic observations, not completely absent for the tests in the NaCl solution, where the resistance to crack propagation became quite comparable on a ΔK_{eff} basis to that in air for the AC and NSC alloys. This result shows that a significant portion of the good fatigue propagation resistance of HT material also stems from the very crystallographic cracking producing large macroscopic fracture facets and giving rise to important crack closure effects.

The crack growth curves for the 12 and 14% Mn bronzes in the HT condition tested in air at f=20Hz and R=0.1 agree well with each other (Figs.6-8) for da/dN values $\geq 10^{-5}$ mm/cycle, although only one test was carried out on the 14% Mn HT alloy. The fractographic and metallographic observations showed that the α-grains more frequently encountered by the propagating crack in the 12% Mn HT alloy had a more important influence on the crack morphology and favoured a tortuous crack path in both phases (Figs. 10 and 14). The more tortuous crack path in the 12% Mn alloy did not appear to result in significantly slower crack growth rates compared to the 14% Mn alloy. The higher percentage of β-phase in the 14% Mn HT alloy may have compensated for the larger β-grain size of the 12% Mn alloy and for these other effects, since it appears clear that the presence of this harder β-phase is responsible for an important portion of the good resistance to fatigue crack propagation of these alloys in the HT condition.

CONCLUSIONS

The present study has shown that accelerated fatigue cracking is obtained at high crack growth rates when 12 and 14% Mn-Ni-Al bronzes are tested in 3.5% NaCl solution at 20Hz and especially at 1Hz, with this accelerated cracking resulting from facilitated crack initiation in α-grains. The differences in the ΔK_{Th} values and in the propagation results at low growth rates for the different (AC, NSC and HT) alloy conditions can be largely explained by differences in roughness-induced crack closure effects associated with the differences in the size of the β-grains and of the macroscopic fracture facets. The decelerated crack growth observed in the NaCl solution for the low da/dN values is associated with important corrosion-product induced crack closure effects. Because of the importance of the latter effect, the corrosion-fatigue accelerated crack growth can better be seen by comparing the crack growth data on a ΔK_{eff} basis. Above the near-threshold region, the increased percentage of β-phase in the HT materials results in a significant increase in the resistance to fatigue crack propagation. Oscillations in the lower portion of the crack propagation curves for the HT materials and the 12% Mn NSC alloy were shown to be the result of the crystallographic fatigue crack experiencing difficulty in crossing β-grain boundaries in these large grain size materials.

ACKNOWLEDGMENTS

The present study was supported by the Canada Center for Mineral and Energy Technology/Physical Metallurgy Research Laboratory through DSS contract 03 SQ.23440-2-9017 and the authors express their gratitude. The authors are indebted to Yves Blanchette for assistance in performing the ΔK_{eff} measurements and to Nicole Roy for the rapid and able typing of the manuscript.

REFERENCES

1. Collins, P,. and D.J. Duquette, Corrosion (NACE), 34, 119-124 (1978).
2. Mshana, J.S., O. Vosikovsky and M. Sahoo, Canadian Metallurgical Quarterly, 23 7-15 (1984).
3. Taylor, D. and J.F. Knott, Metals Technology, 9 221-228 (1982).
4. Parkins, R. N. and Y. Suzuki, Corrosion Science, 23, 577-599 (1983).
5. Dickson, J.I., L. Handfield, J.P. Baïlon and M. Sahoo, in "Fatigue 84", Vol. I, 191-197, EMAS Ltd Publishers, London (1984).
6. Dickson, J.I., L. Handfield, S. Lalonde and J.P. Baïlon, in "The Strength of Metals and Alloys", Vol. III, Pergamon Press, Oxford, in press.
7. Dickson, J.I., L. Handfield, J.P. Baïlon and M. Sahoo, in "Copper 1984" Symposium Proceedings, Vol. I, paper 22, CIM Conference of Metallurgists, Quebec, Canada, August 1984.
8. Couture, A. and M. Sahoo, in "Copper 1984" Symposium Proceedings, CIM Conference of Metallurgists, Quebec, Canada, August 1984.
9. Suresh, S. and Ritchie, R.O., in "Fatigue Crack Growth Threshold Concepts", 227-261, TMS-AIME (1984).
10. Dickson, J.I., J.P. Baïlon and J. Masounave, Canadian Metallurgical Quarterly, 20, 317-329 (1981).

PROTECTION OF Al-Zn-Mg WELDS AGAINST EXFOLIATION AND STRESS CORROSION CRACKING USING ALUMINIUM BASED METAL SPRAY COATINGS

N. J. H. Holroyd, W. Hepples, G. M. Scamans
Alcan International Limited
Banbury, Oxon, United Kingdom

PROTECTION OF AL-ZN-MG WELDS AGAINST EXFOLIATION
AND STRESS CORROSION CRACKING USING ALUMINIUM
BASED METAL SPRAY COATINGS

ABSTRACT

Localised corrosion problems have restricted the commercial exploitation of medium strength weldable Al-Zn-Mg alloys. This contribution therefore, presents the results of a detailed study on welded alloys to identify the electrochemical requirements for immunity to localised corrosion in service environments and records the progressive development of aluminium based spray coatings that generate the required protective conditions.

IT IS WELL documented that localised corrosion problems can occur in welded Al-Zn-Mg joints (1-5). Generally this is due to either exfoliation corrosion within a heat-affected region a few millimetres from the weld-bead (6) and/or a type of stress corrosion cracking which initiates at weld-toes and propagates into the interfacial region between the weld-bead and heat-affected zone (7-9). This region is commonly known as the 'white-zone' due to its etching behaviour in nitric acid (10).

Since potential-pH domains have been established for aluminium alloys in saline environments which provide immunity to both localised corrosion (e.g. exfoliation, pitting and stress corrosion) and general corrosion (11,12) it seems probable that similar domains will exist for Al-Zn-Mg alloy welded joints. These domains must be identified under conditions that closely simulate those experienced in service. Thus the approach adopted in this study to eliminate environment-sensitive fracture and corrosion problems in Al-Zn-Mg welded structures was to characterise the

local environmental conditions associated with weld-toe/white-zone cracking, to identify the potential domain for immunity in the characterised environment and then to develop an aluminium based spray coating to generate the required potential control.

EXPERIMENTAL

MATERIALS - Plates for testing were from welded plates prepared by laying a weld on 1 m lengths of 15 mm thick 7017-T651 plate using automated MIG welding with the following welding parameters: 342 Amps, 28 Volts, 0.62 m/min traverse rate, 10 m/min wire feed rate, 10 mm nozzle height and NG61 welding wire (2 mm dia). The measured compositions of the 7017-T651 plate, NG61 filler wire and the weld bead are quoted in Table 1.

Table 1 - Composition (wt%) of plate,
filler-wire and weld-bead

	Zn	Mg	Cu	Fe	Si
7017-T651 plate	4.75	2.0	0.14	0.17	0.08
NG61 filler-wire	0.2	5.2	0.10	0.18	0.09
Weld-bead	2.9	3.3	0.07	0.18	0.09

Spray wire for metallisation of the welded plate was produced by hot extruding d.c. cast 66 mm diameter billet (homogenised at 550°C for 16 hrs) down to 2 mm wire which was then coiled for use as feed-stock. Spray alloys were cast on two purity bases (99.5% and 99.9% aluminium).

Metal spraying of welds (weld-bead and 75 mm either side of weld) was accomplished by either arc or flame-spraying and was carried out by Metallisation Limited, Dudley, U.K. The conditions employed during spraying were: (a) arc-spraying - 200 amps, 28 volts, carrier gas compressed air and a coverage rate of about

10 m^2/hr, (b) flame-spraying: oxygen/propane fuel, compressed air, carrier gas and approximate coverage rate of 3.5 m^2/hr.

Prior to metal-spray-coating surfaces were grit blasted and the spray coatings thicknesses were typically 150 to 250 μm with good adhesion and low porosity.

SOLUTION CHEMISTRY STUDIES - The local solution chemistry conditions associated with various regions of welds were assessed using recently developed microchemical analysis techniques (13,14,15) using specimens as shown in Figure 1. The test environment (3% NaCl) was directly injected into each of the small drilled holes at the start of each experiment.

ELECTROCHEMICAL STUDIES -Test solutions were made-up using 'analar' grade reagents and distilled water and all experiments allowed natural aeration and were conducted at 20°C. All quoted electrochemical potentials refer to the saturated calomel electrode scale (SCE).

Potential-time data for spray-wires and spray-coated welds totally immersed in 500 cc of either 3% sodium chloride or 2% sodium chloride + 0.5% sodium chromate, pH 3 (acidified using HCl) was monitored using a hundred channel Hewlett-Packard data logger system with a high input impedance (> 10^{12} ohms) to prevent any polarisation effects during potential measurement. In a typical experiment the free corrosion potential (FCP) was measured for 500 hours and then a current density of + 50 μA/cm^2 was imposed and potentials were monitored for a further 100 hours. Corrosion characteristics were assessed visually throughout the experiments and polarisation characteristics were also measured using potentiokinetic techniques employing a sweep rate of 10 mV/min over a potential range of -1300 to -800 mV (SCE).

SLOW STRAIN RATE TESTING - Tensile specimens with gauge lengths containing undressed welds, Figure 1, were subjected to slow strain rate testing in a hard-beam tensile testing machine (16,17) using standard techniques (11,18) and in general a cross-head speed of 2.5 x 10^{-5} mm/s. Reference tests were conducted in dry air (complete immersion in magnesium perchlorate) as environmental effects can occur in laboratory air (19). All metal surfaces, other than metal sprayed regions or the corresponding underlying region in non-sprayed specimens, were isolated from the test environment using a rubber-based paint and precautions were taken to ensure that no galvanic effects could occur between the test specimen and the loading-frame.

Tests were conducted in the two test environments used in the electrochemical experiments. The acidified chromate/chloride electrolyte more closely simulates the service condition. Metal-sprayed specimens generally were tested at their free corrosion potential which was continuously monitored. Non-sprayed specimen testing was mainly under potentiostatic

control over a range of potentials -1300 to -900 mV (SCE).

A detailed fractrographic study was carried out on all the failed specimens.

RESULTS AND DISCUSSION

SOLUTION CHEMISTRY WITHIN WHITE-ZONE CRACKS - Typical environmental conditions within propagating cracks have been established as: pH 2.7, 2M Cl, 0.1M Al and below 0.001M for Zn, Mg, Cu and Cr. The local solution pH is controlled by dissolved aluminium via the hydrolysis process described by eq. (1):

$$Al^{3+} + H_2O = AlOH^{2+} + H^+ \quad \ldots\ldots eq. (1)$$

Solution chemistry development kinetics within small crevices for the various microstructural regions associated with a weld are given in Figure 2. The crevices are sufficiently small and constrained (1 mm dia, 15 mm deep) that the environmental conditions developed simulate conditions within cracks (14). The dissolved magnesium concentration developed in the 'white-zone' region is greater than that for the parent plate (Figure 2) e.g. after 24 hours concentrations are 3 x 10^{-3}M (35 ppm) compared with 4 x 10^{-5}M (0.5 ppm), which is consistent with the magnesium segregation reported to occur at 'white-zone' grain boundaries (8,9). This level of magnesium should not have any significant influence upon the 'cracking potency' of the environment and hence the local solution in the white-zone can be considered as equivalent to that formed during parent plate stress corrosion.

SLOW STRAIN RATE TESTING OF WELDS - Weld-toe and white-zone cracking is known to be promoted by slow strain rate testing when sufficiently low straining rates are employed (19) and as for SCC of parent plate material the upper limit of strain rate for crack initiation decreases with increasing cracking potency of the test environment (19,20). For example at a cross-head speed (CHS) of 2.5 x 10^{-5} mm/s weld specimens tested in laboratory air fail in a ductile manner away from the white-zone (the precise location depending upon specimen design), whereas a weld-toe/white-zone failure (WTC/WZC) generally results after testing in an aqueous chloride environment.

Mechanism of Weld-Toe/White-Zone Cracking (WTC-WZC) - Crack initiation of WTC/WZC is a complex phenomenon that involves at least a two-stage process; stage 1 involving the separation of the weld overlap region from the plate (the weld-toe region) and stage 2 involving the initiation of crack growth into the white-zone. The sequence of events thought to occur during weld-toe/white-zone cracking is as follows:

Stage 1 - in aggressive environments localised corrosion processes occur at the weld-toe and provide both a path to the white-zone and an acidified electrolyte (pH's of 3 were measured in 'activated' weld-toe regions. Alternatively since the weld-toe/plate interface is relatively weak stage 1 crack growth (weld-toe lifting) can be mechanically driven.

Stage 2 - once exposed the white-zone offers a highly SCC susceptible path for crack growth. However, a second initiation process is involved, as dressed welds (weld bead mechanically removed) are resistant to crack initiation even though the white-zones are exposed (19,21). This probably involves establishing local environmental conditions for white-zone crack growth which may be a slow process requiring the development of pits or crevices.

Slow crack growth in the white-zone can then progress until SCC in the parent plate, mechanical overload or crack arrest occurs.

The precise mechanism of slow crack growth in white-zones is, as for SCC in parent plate, not fully understood. However, a common hydrogen embrittlement mechanism is thought to occur and support for this proposition is given by the observation that both modes show reversible embrittlement produced by slow straining in a dry environment after pre-exposure to aqueous environments (11,19) and fractures produced in relatively mild environments, e.g. laboratory air, show little evidence of anodic dissolution.

A typical weld-toe/white-zone fracture promoted by slow strain rate testing a 7017-T651 weld specimen in an aqueous saline environment is shown in Figures 3 and 4. The optical micrograph of a cross-section through a fractured weld specimen (Figure 3) shows that both weld-toes have suffered corrosion and mechanical separation although white-zone cracking has only propagated from one toe and that SCC has initiated in the 7017 plate. The fractured weld-toe region (Figure 4) shows the corroded toe, the stage 1 to stage 2 transition and the intergranular fracture morphology typical of white-zone cracking which closely resembles that previously reported (22,23).

Influence of Potential - Slow strain rate test results (CHS 2.4 x 10^{-5} mm/s) over a wide range of potentials are shown in Figure 5. Significantly a potential range exists, -1200 to -1130 mV (SCE), where white zone cracking is extremely limited. Similar results were generated in 3% sodium chloride, in agreement with work reported by Reboul et al (23) for 7020 welds using constant load testing. The existence of the 'potential window' for WZC immunity is consistent with the experimental potential-pH diagram developed by Gimenez et al (12) for an Al-Mg alloy (5086) in saline environment in which a potential-pH domain for immunity to pitting was defined. In the present case we have added exfoliation and WTC/WZC susceptibility and so the immunity domain is further restricted (24).

Several important conclusions maybe drawn from these findings: (1) slow strain rate testing offers a rapid method of producing slow crack growth in Al-Zn-Mg weldments, (2) anodic polarisation accelerates initiation and growth of WTC/WZC, (3) weld-toe/white zone cracking initiates and propagates more readily in acidic than in neutral chloride environments and (4) a 'potential window' (-1200 to -1130 mV (SCE)) exists within which weld-toe or white-zone cracking does not readily occur even after a mechanical crack initiates at the weld toe.

ELECTROCHEMICAL PROTECTION OF AL-ZN-MG WELDS - The identified 'potential window' within which Al-Zn-Mg welds are virtually immune to environment-sensitive fracture establishes the electrochemical conditions required in service for slow crack growth problems to be prevented.

To generate complete protection the sprayed coating must provide polarisation characteristics such that the electrochemical potential is always within the immunity zone even when providing sufficient current to cathodically protect the white-zone, the exfoliation zone and the parent plate. In addition the resistance to 'parasitic corrosion' (localised corrosion of the spray itself) needs to be as high as possible to maximise the operative lifetime of the spray.

The approach adopted to develop the required aluminium spray coating to electrochemically protect Al-Zn-Mg welds involved three stages, namely; (1) selection of wire compositions and their electrochemical assessment in neutral and acidic saline solutions, i.e. FCP and polarisability measurements, (2) metal spraying of the most promising compositions onto welds, (3) electrochemical assessment and slow strain rate testing of sprayed welds.

Selection of Spray Wire Alloy Compositions - Aluminium anode alloys for sacrificial protection of steel structures have been successfully developed over the past twenty years (25-30). Zinc additions up to 5 wt% have been shown to progressively decrease the free corrosion potential of aluminium (25,31) as the zinc content is raised although the most negative potentials obtained are not in the required protection zone (25,26,32). The required potentials can however be reached by tin additions (25,26,33) but the potential obtained is highly dependent on the microstructural state of the tin (25,33) and on the purity level (34). In addition, the alloys polarise rapidly under load (25). Negative potentials are also obtained by additions of indium and mercury (26) and gallium has been added both as an activator and also to accelerate activation by tin (47).

Earlier work on sprayed aluminium coatings for the protection of both steel (35,40-43) and aluminium (33-39) was based on the use of aluminium-zinc alloys. Such sprayed coatings provided successful protection against pitting (35), exfoliation and stress corrosion cracking (37) in Al-Zn-Mg alloys and the benefit of a high aluminium purity level (99.5% minimum) was recognised.

This work was used as the starting point for developing improved aluminium spray coating alloys to protect against weld-toe cracking and all spray compositions were initially based on Al-4.5%Zn as a control composition to which additions of indium, tin and gallium were made. Alloys with the same activators were also cast on an Al-0.8%Mg base in view of the use of alloys of this type in battery anodes (44-47) and on a base of unalloyed aluminium. Two aluminium purity levels 99.5% and 99.9% were used to make the alloys.

ELECTROCHEMICAL ASSESSMENT OF SPRAY-WIRES - Free corrosion potential (FCP) time data for spray wires in 3% sodium chloride, Table 2, indicates that all of the selected wire compositions adopt and hold electrochemical potentials within the required potential limits of -1200 to -1130 mV (SCE) for times in excess of 100 hours.

In addition several wires also maintain the required potentials in the acidified saline solution. The number of compositions achieving this, however, is significantly reduced when an anodic current density of 50 $\mu A/cm^2$ is demanded.

METAL SPRAYING OF WELDS - The choice of alloy compositions for metal spraying was based upon the electrochemical and the parasitic corrosion performance of the spray wires. However, some alloys with the lower purity base were also included for comparison. Compositions selected were both arc- and flame-sprayed onto welded plate.

ELECTROCHEMICAL ASSESSMENT OF METAL SPRAYED COATINGS - Arc-Sprayed Coatings - Typical potential time data is given in Figure 6 and Table 3. All the arc-sprayed coatings listed in Table 3 maintained FCP's in 3% sodium chloride within the predicted 'potential window' for WTC/WZC protection, for times greater than 500 hours and most of the coatings also yielded 'safe' potentials when anodically polarised at +50 $\mu A/cm^2$. In the acidified saline environment, however, none of the arc-spray coatings generated 'safe' potentials when anodically polarised and hence none would be expected to provide complete protection against WTC/WZC in such environments.

Flame-Sprayed Coatings - Potential-time data (FCP and E at +50 $\mu A/cm^2$) for flame-sprayed coatings on Al-Zn-Mg welds are summarised in Table 5.

The results show a similar trend to those obtained with arc-spraying. However, as the deposited spray composition from a given spray wire composition can differ significantly depending upon whether arc or flame-spraying is involved any comparison between the sprays must take these differences into account.

Elemental losses occur by volatilization during arc-spraying which do not occur in flame-spraying, e.g. zinc levels are reduced from 4.5% in the spray-wire down to 0.7 to 1.0% zinc in an arc-spray coating and are maintained at around 4.45% in a flame-spray coating. This is not surprising since the maximum temperature involved in arc-spraying is in excess of 2800°C and is below 1000°C for flame-spraying. Typical losses during arc-spraying are indicated in Table 5 along with some vapour pressure data for the various elements in the sprayed coatings. In general the elemental losses promoted during spraying are in accordance with the vapour pressure data, i.e. during arc-spraying little or no tin, some gallium and significant levels of zinc, magnesium and indium volatilize and during flame-spraying only minimal losses occur in any element.

Certain trends are revealed when the data in Tables 3 and 4 are assessed in terms of the stable potentials adopted by spray-coatings when polarised at +50 $\mu A/cm^2$ and consideration is given to the elemental losses occurring during spraying; (1) the presence of 4.5% zinc as opposed to around 1% zinc, i.e. flame-spray versus arc-spray, is detrimental to polarisation characteristics in neutral 3% sodium chloride but beneficial in an acidified saline solution, see Figure 7, (2) the presence of zinc appears to override the influence of the other additions seen in the zinc-free alloys, (3) decrease in aluminium base purity level results in polarisation data moving to more positive values, (4) the polarisation characteristics of sprayed coatings containing zinc and other additions or just the other additions are generally more noble than the spray-wire.

Electrochemical Prediction of Influence of Sprayed Coatings Upon WTC and WZC - Applying the determined protection criterion to the polarisation data from arc and flame-sprayed coated welds in 3% NaCl and 2% NaCl + 0.5% Na_2CrO_4, pH 3 (Tables 4 and 5) in 3% sodium chloride all the coatings should provide complete protection (except perhaps the Al-In-Sn-Ga alloy which could be too cathodic and could lead to alkaline corrosion/hydrogen problems) whilst in the acidified saline environment only the flame-sprayed Al-In-Sn-Ga coating should provide complete protection.

However, as the spray coatings in general move potentials towards the required potential region some beneficial effects should result even though complete protection is not provided. In the acidified saline solution flame-sprays should offer significantly more benefit than arc-sprays with the possible exception of the Al-In-Ga flame-spray which is more readily polarised than the other flame-sprays and adopts potentials more characteristic of the arc-sprays.

SLOW STRAIN RATE TESTING OF SPRAYED WELDS - The slow strain rate (SSR) test data for sprayed welds is summarised in Figures 8 and 9. Qualitatively the results conform to the electrochemical predictions in that ranking is as expected, in the acidified saline environment complete protection is only generated by the Al-In-Sn-Ga flame-spray and in general the potentials adopted during SSR testing are similar to those promoted during galvanostatic polarisation at +50 μA/cm^2, Figure 9.

To increase the severity of the SSR test on flame-sprayed welds the test specimens were anodically galvanostatically polarised to +50 μA/cm^2. This demanded that the spray not only had to supply a current to exposed regions during the test but in addition had to provide a current to an external sink. As expected this moved the potentials adopted during testing in a positive direction and gave a more discriminative ranking of the protection generated by the flame-sprays, Figure 10.

In view of the above it is realistic to predict the protective capability of a spray on the basis of its electrochemical performance as long as it also has adequate mechanical and physical properties such as adhesion, ductility and wear resistance. Current work (24) indicates that modified spray-wire compositions will allow arc- as well as flame-sprayed coatings, to generate complete protection for Al-Zn-Mg welds.

CONCLUSIONS

- (1) Local environmental conditions associated with WTC/WZC in Al-Zn-Mg welds in service are acidic and similar to those of parent plate SCC cracks, (2) an electrochemical domain, i.e. 'safe' potential zone, exists within which Al-Zn-Mg welds are immune to all forms of corrosion including WTC/WZC and exfoliation, (3) Aluminium based sprayed coatings will maintain Al-Zn-Mg welds within the 'safe potential zone' and prevent service problems.

ACKNOWLEDGEMENT

The authors thank Mr. C.R. Wiseman for his invaluable assistance with the slow strain testing and Alcan Plate Limited, Birmingham for funding the research programme.

References

1. L.H. Chambers and D.C. Baxter, Engineer, 223, 518-520 (1967).

2. G. Bergholz, Conf. Proceed. 3rd Int. Congress on Metallic Corrosion, Vol.2, pp.367-376, Moscow (1969).

3. K.G. Kent, Metallurgical Reviews, No.147, 135-146 (1970).

4. J.M. Truscott, V.E. Carter and H.S. Cambell Journal of Metals, 99, 57-66 (1971).

5. G. Hollrigl, H. Birchel and H. Zoller, 'Proc. 3rd Int. Aluminium Extrusion Technology Seminar', Vol.1, pp.125-131, pub. Aluminium Association (1984).

6. R. Grauer and E. Wiedmer, Werkstaffe und Korrosion, 31, 45-48 (1980).

7. H. Cordier and I.J. Polmear, Int. Conf. EUROCOR 77, pp.607-613, London, 1977.

8. M.S. Rahman, H. Cordier and I.J. Polmear, Z. Metallkde, 73, 589-593 (1982).

9. H. Schimedel and W. Gruhl, Metall, 38, 32-37 (1984).

10. H. Cordier, M. Schippers and I.J. Polmear, Z. Metallkde, 68, 280-284 (1977).

11. N.J.H. Holroyd and D. Hardie, Corrosion Science, 21, 129-144 (1981).

12. Ph. Gimenez, J.J. Rameau and M.C. Reboul, Corrosion, 37, 673-682 (1981).

13. N.J.H. Holroyd, G.M. Scamans and R. Hermann "Embrittlement by the Localised Crack Environment", Ed. R.P. Gangloff, AIME, pp.327-347, Warrendale (1984).

14. N.J.H. Holroyd, G.M. Scamans and R. Hermann Int. Conf. Proceed "Corrosion within Pits, Cracks and Crevices", H.M.S.O. - to be published in 1986.

15. N.J.H. Holroyd and R. Hermann, to be published.

16. R.N. Parkins, "Stress Corrosion Research", pp.1-27, (ed. M. Arup and R.N. Parkins), Netherlands, Pub. Sijthoff and Noordhoff, (1979).

17. R.N. Parkins, "Stress Corrosion Cracking - The Slow Strain Rate 'technique', ASTM-STP 665, pp.5-25, Philadelphia (1979).

18. N.J.H. Holroyd and G.M. Scamans, "Environment-Sensitive Fracture: Evaluation and Comparison of Test Methods", ASTM-STP 821, pp.202-241, Philadelphia (1984).

19. N.J.H. Holroyd and D. Hardie, Unpublished Results.

20. N.J.H. Holroyd and D. Hardie, 'Hydrogen Effects in Metals', Ed. I.M. Bernstein and A.W. Thompson, AIME, pp.449-457 (1981).

21. M. Pirner, Aluminium, 53, 674 (1977).

22. Y.S. Kim and S. Pyun, Brit. Corros. J., 18, 71-75 (1983).

23. M.C. Reboul, B. Dubost and M. Lasherms, Corrosion Science, 25, 999-1018 (1985).

24. N.J.H. Holroyd, W. Hepples and G.M. Scamans - to be presented to Corrosion Science (1986).

25. R.A. Hine and M.W. Wei, Materials Protection, 3, 49-55 (1964).

26. J.T. Reding and J.J. Newport, Materials Protection, 5, 15-18 (1966).

27. T.J. Lennox, R.E. Groover and M.H. Peterson Materials Protection, 10, 39-44 (1971).

28. W.B. MacKay, "CORROSION", Vol.2, Ed. L.L. Shrier, Section 11, pp.21, pub. Newnes-Butterworths, London (1976).

29. J.B. Bessone, R.A.S. Baldo and S.M. de Micheli, CORROSION, 37, 533-540 (1981).

30. M.C. Reboul, P.H. Gimenez and J.J. Rameau, Corrosion, 40, 366-371 (1984).

31. V. Sivan, J. Electrochem. Soc. (INDIA), 27, 181-185, (1978).

32. I.L. Muller and J.R. Galvele, Corrosion Science, 17, 995-1007 (1977).

33. K. Tohma, H. Kudo and Y. Takeuchi, J. of Japanese Light Metals, 34, 157-164 (1984).

34. K.B. Pai, M.D. Gandhi, K.M. Pai and D.L. Roy, J. Electrochem. Soc. (INDIA), 29, 27-30 (1980).

35. D.J. Scott, Trans. Inst. Metal. Finishing, 49, 111-112 (1971).

36. V.E. Carter and H.S. Cambell, J. Inst. Metals, 89, 472-475 (1960-61).

37. V.E. Carter and H.S. Cambell, Brit. Corros. J., 4, 15-20 (1969).

38. W.E. Ballard, "Corrosion", Vol.2, Ed. L.L. Shreir, 2nd Ed., Sec. 13, pp. 78-91 (1976).

39. W.E. Ballard, "Metal Spraying", pub. Griffin, London, 4th Ed., pp.437 (1963).

40. F.C. Porter, The Engineer, 211, 906-907 (1961).

41. T.P. Hoar and O. Radovici, Trans. Inst. Metal. Finishing, 42, 211-222 (1964).

42. R. Kreienbuehl, P. Kunzmann and W. Widmer, Proceed. '8th Thermal Spraying Conference', pub. American Welding Society, pp.436-442, Miami Beach, Florida (1976).

43. V. Vesely and J. Horky, as ref. 42, pp.430-435.

44. T. Valand and G. Nilsson, Corrosion Science, 17, 931-938 (1977).

45. A.R. Despic, D.M. Drazic, M.M. Purenovic and N. Cikovic, J. of Applied Electrochemistry 6, 527-542 (1976).

46. Y. Hori, J. Takao and H. Shamon, Electrochemica. Acta, 30, 1121-1124 (1985).

47. M.J. Pryor, D.S. Deir and P.R. Sperry, U.S. Patent 32406088 (1966).

48. 'Comprehensive Inorganic Chemistry', Ed. J.C. Bailar, H.J. Emeleus, Sir. R. Nyholm and A.F. Trotman-Dickinson, Pub. Pergamon Press (1975).

49. Handbook of Chemistry and Physics, 66th Edition, Ed. R.C. Weast, Pub. CRC Press (1985).

Table 2 - Polarisation data for spray-wires in aqueous saline environments. Potential data in brackets is for 99.5% purity base alloys and all other data is for 99.9% purity base.

	SP Al	Al-4.5Zn	Al-Zn-In	Al-Zn-In-Ga	Al-In-Ga	Al-In-Sn-Ga
3% NaCl						
FCP (100 hr)	-1110	-1250 (-1154)	-1294 (-1164)	-1268	-1539	-1565
+50 μA/cm^2 (70 hr)	-715	-983		-1145	-1553	-1567
2% NaCl + 0.5% Na$_2$CrO$_4$ pH 3						
FCP (100 hr)	-724	-1097	-1166	-1175	-1185	-1200
+50 μA/cm^2 (70 hr)	-715	-1110	-1080	-1080	-1196	-1266

Table 3 - Polarisation data for arc-sprayed coatings on Al-Zn-Mg welds immersed in aqueous chloride environments. Figures in brackets refer to 99.5% aluminium purity based alloys whilst all other data is for 99.9% purity base alloys.

Composition of Wire	3% NaCl		2% NaCl + 0.5% Na$_2$CrO$_4$ (pH 3)	
	FCP (mV) (100 hr)	+50 μA/cm^2 (mV) (70 hr)	FCP (mV) (100 hr)	+50 μA/cm^2 (mV) (70 hr)
Al-4.5Zn	-1230 (-1156)	-1206 (-1110)	-1131 (-1060)	-959 (-894)
Al-4.5Zn-In	-1240 (-1180)	-1188 (-1097)	-1163 (-836)	-911 -
Al-4.5Zn-In-Ga	-1232	-1198	-947	-935
Al-0.8Mg-Sn-Ga	-1299	-1200	-947	-915
Al-In-Ga	-1436	-1136	-1087	-917
Al-In-Sn-Ga	-1303	-1184	-918	-913

Table 4 – Free-corrosion potential and potential at +50 μA/cm² for flame-sprayed Al-Zn-Mg welds immersed in aqueous saline environments.

	3% NaCl		2% NaCl + Na₂CrO₄ pH 3	
	FCP (mV) (100 hr)	+50 μA/cm² (mV) (70 hr)	FCP (mV) (100 hr)	+50 μA/cm² (mV) (70 hr)
Al-4.5Zn	-1206	-1119	-1072	-1060
Al-4.5Zn-In	-1160	-1139	-1048	-1046
Al-4.5Zn-In-Ga	-1223	-1132	-1119	-1097
Al-In-Ga	-1240	-1215	-987	-957
Al-In-Sn-Ga	-1374	-1379	-1220	-1143

Table 5 – Vapour pressure data (48,49) for the various elements in the aluminium based sprays together with approximate values of elemental losses during arc-spraying.

Element	Boiling Point (°C)	Temp °C for V.P. = 10^{-3} atm/°C	V.P. at 100°C (mm/Hg)	% Loss in Arc-spray
Al	2467	1510	4.70×10^{-4}	–
Zn	907	479	–	80
Mg	1090	584	–	90
In	2080	–	1.22×10^{-2} (at 960°C)	75
Sn	2270	1584	8.3×10^{-3}	0
Ga	2403	–	5.8×10^{-3}	20

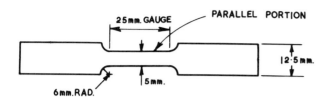

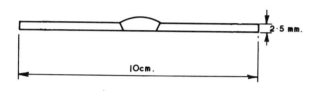

(a) slow strain rate testing

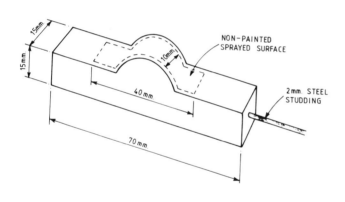

(b) electrochemical experiments

Figure 1 - Specimen designs.

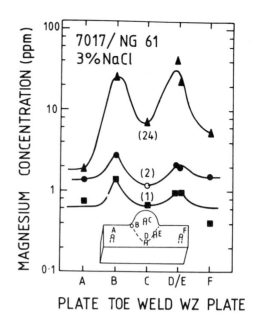

(a) magnesium

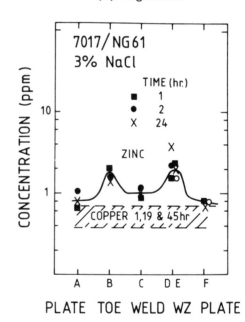

(b) zinc and copper

Figure 2 - Solution chemistry development inside small crevices for various microstructural regions associated with a weld.

299

Figure 3 - Optical micrograph of a cross-section of a bead-on-plate weld specimen after slow strain rate testing in 3% NaCl + 0.5% Na$_2$CrO$_4$, pH 3 at the free corrosion potential.

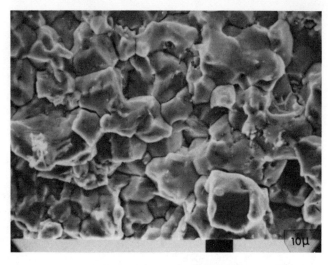

(c) white-zone cracking.

Figure 4 - Slow strain rate test failure showing (a), (b) and (c).

(a) cracking at the weld-toe showing evidence of corrosion.

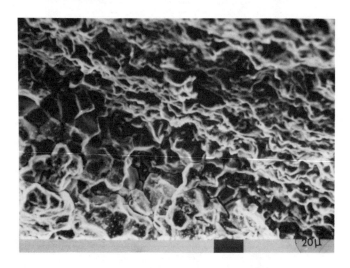

(b) weld-toe to white-zone transition

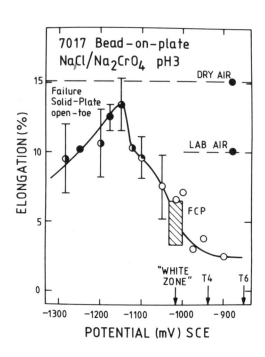

Figure 5 - Influence of potential upon elongation and fracture mode for welded specimens slow strain rate tested (CHS ~ 2.5 x 10^{-5} mm/s) in 2% NaCl + 0.5% Na$_2$CrO$_4$, pH 3.

300

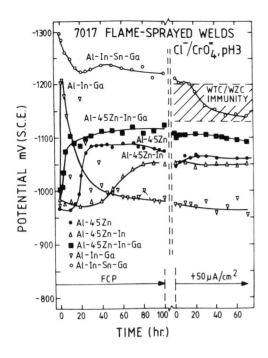

(a) Flame-Sprays

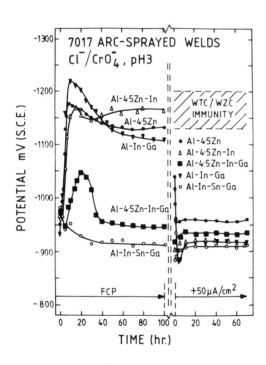

(b) Arc-Sprays

Figure 6 — Potential-time behaviour for arc and flame-sprayed welds at the free-corrosion potential and $+50\ \mu A/cm^2$ in 2% NaCl + 0.5% Na_2CrO_4, pH 3.

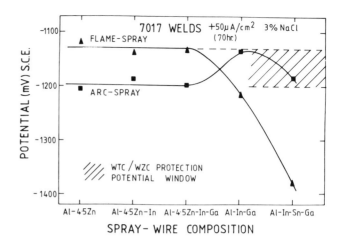

(a) 3% NaCl

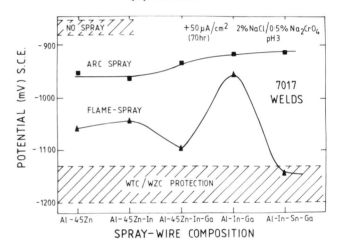

(b) 2% NaCl + 0.5% Na_2CrO_4, pH3

Figure 7 — Potentials obtained by arc and flame-spray coated welds after 70 hours immersion at a current density of $+50\ \mu A/cm^2$.

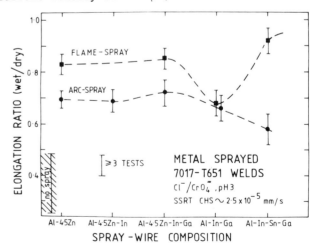

Figure 8 — Influence of arc and flame-spray coatings on slow strain rate testing in 2% NaCl + 0.5% Na_2CrO_4, pH3.

301

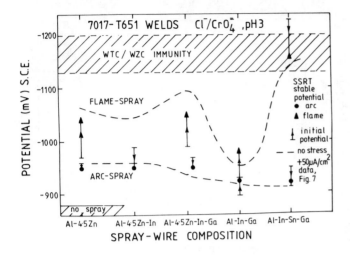

Figure 9 – Comparison of the electrochemical potentials adopted by arc and flame-spray coated welds at +50 $\mu A/cm^2$ with those measured during slow strain rate testing in 2% NaCl + 0.5% Na_2CrO_4, pH 3.

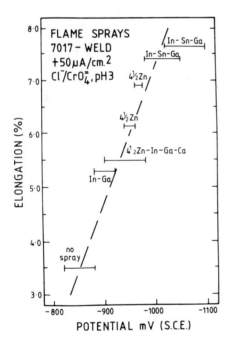

Figure 10 – Elongations and electrochemical potentials for flame-spray coated welds slowly strained to failure (CHS $\sim$ 2.5 x 10^{-5} mm/s) in 2% NaCl + 0.5% Na_2CrO_4, pH 3 whilst anodically polarising at +50 $\mu A/cm^2$.

TO PUNCH OR TO DRILL—
THAT IS THE QUESTION

J. S. Snodgrass, R. E. Zinkham
Reynolds Metals Company
Richmond, Virginia, USA

ABSTRACT

This project was conducted to demonstrate, both qualitatively and quantitatively, the reason for using drilled holes instead of punched holes in high strength aluminum alloys. Holes were drilled and punched in alloys 2024, 6061, and 7075. Stress corrosion cracks developed in 2024-T3, 2024-T351, 7075-T6, and 7075-T651 punched hole walls from the residual stresses introduced by the punching operation. Residual stress measurements in the 7075-T651 material by X-ray diffraction showed short transverse tensile stresses greater than the threshold for SCC in the punched hole walls. The drilled hole had compressive residual short transverse stresses. Punch clearance was found to be a significant factor with higher residual stress with larger punch clearance. Material thickness (two gauges tested, 2.9 mm and 6.4 mm or 0.115 in and 0.250 in respectively) was not a significant factor. Alloy 6061 suffered no stress corrosion cracking.

Alloy and temper combinations with recognized susceptibilities to short transverse stress corrosion cracking must be drilled rather than punched.

IN PRODUCTION FABRICATION of parts from aluminum sheet and light gauge plate, round holes must often be made to accommodate fasteners. The two primary processes for making these holes are punching and drilling. From an economic, high production standpoint, punching is the preferred method. However, from metallurgical and corrosion standpoints, punching is often replaced by more costly drilling.

Drilling is used because it is thought to introduce smaller residual stresses in the hole wall than punching. Because no clear demonstration of the presence and effect of residual stresses from punching has been found, the present project was undertaken. Such a demonstration is expected to be useful in the education of fabricators as to the proper application of hole making methods.

The purpose of this work is to demonstrate why drilled holes are preferred over punched holes in high strength aluminum alloys in light plate and heavy sheet gauges.

The selection-of-materials-process for the demonstration considered several points. A variety of alloys was desired to show, in a broad fashion, the response of several alloy families to the hole forming methods. Thus, the alloys selected were 2024, 6061, and 7075. At least two of the alloys, 2024 and 7075, were expected to demonstrate stress corrosion cracking susceptibility in the selected tempers. The third alloy, 6061, was not expected to show stress corrosion cracking.

The remainder of this paper will describe the materials used; the procedures used to make the holes, to perform stress corrosion tests, and to perform residual stress analyses; the results; and the conclusions.

PROCEDURES

MATERIAL SAMPLES - The materials used were nominal 6.4 mm (.25 in) plate and 2.9 mm (.115 in) sheet of alloys 2024, 6061, and 7075 (see Table 1). Specimens were cut from plant produced materials. Figure 1 schematically shows a typical specimen that was 38 mm (1.5 in) X 190 mm (7.5 in) with five symmetrically arranged hole locations designated A - E.

Table 1 - Basic Materials

Number	Alloy/Temper*	Original Gauge mm	(in)	Actual Gauge mm	(in)	Surface
06	7075-T651	6.73	(0.265)	6.73	(0.265)	Mill Finish
07	6061-T651	6.22	(0.245)	6.22	(0.245)	Mill Finish
08	2024-T351	6.43	(0.253)	6.43	(0.253)	Mill Finish
09	7075-T6	3.18	(0.125)	2.92	(0.115)	Machined**
10	6061-T6	3.18	(0.125)	2.92	(0.115)	Machined**
11	2024-T3	3.18	(0.125)	2.92	(0.115)	Machined***

*Confirmed from production plant records, nominal chemistries and
 properties available (1)
**Machined to have comparable gauge and surface to number 11
***Machined to remove alclad layers.

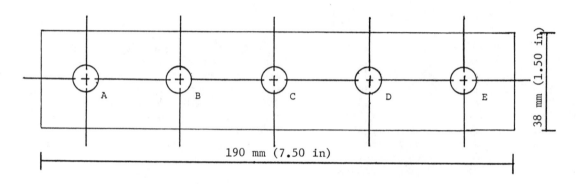

Fig. 1 - Schematic Sample Layout

All holes designated A were drilled with a standard 12.7 mm (1/2 in) twist drill and no coolant. Holes designated B, C, D, and E were punched with a 12.7 mm (0.500 in) diameter flat punch. The die clearances used were calculated from the following equation:

$$\% \text{ Die Clearance} = \frac{\text{Die Dia.} - \text{Punch Dia.}}{2 \times \text{Gauge}} \times 100$$

Die clearance details are shown in Table 2.

Punching was performed on each piece, one hole at a time, with no coolant or lubricant being applied. The punch was driven by a modified power shear with the following description:
Make: Pacific Press & Shear Company
Model: J 90-10
Rating: 10 Gauge Steel by 3 m (10 ft 0 in) long
Control Setting: Mode Selector--Single Stroke
    Manual Advance Speed Selector--
    Rapid Advance/Normal Press
STRESS CORROSION CRACKING (SCC) TESTING - The effect of punching and drilling on SCC resistance was demonstrated by exposing each material with the five holes (one drilled and four punched) to 30 and 60 day alternate immersion exposure per ASTM Standard G44. (2) Prior to exposure the specimens were degreased and waxed along the specimen edges to protect the identification markings and to avoid extraneous corrosion of the edges. The standard exposure conditions were 10 minutes immersion in 3.5% sodium chloride-deionized water solution each hour and the remaining 50 minutes of each hour drying in air at 27 ± 1°C (80 ± 2°F) and 45% (± 6%) relative humidity.

Specimens were examined for stress corrosion cracks each working day for the first two weeks of exposure. Special emphasis was given to the areas around the holes and the walls of the holes. After about two weeks of exposure sufficient corrosion products from pitting of the 2024 and 7075 had accumulated on the specimen surfaces to prevent observation of cracking. Thus, after 30 days exposure one complete set of specimens was removed from exposure and was cleaned by immersion in concentrated nitric acid. Similarly, after 60 days exposure a second complete set of specimens was removed and cleaned. All exposed surfaces were examined with a low power stereoscopic microscope at magnifications up to 50 X. Photomicrographs were made of significant areas. To facilitate examination of the hole walls the specimens were split along the longitudinal center line. Metallographic sections through cracks in the hole walls of the 2024 and 7075 were mounted, polished, and photographed to help confirm that the mode of cracking was stress corrosion.

Table 2 - Punch Die Clearance Details

Hole	Nominal Clearance (%)	Die Dia. mm	(in)	Actual Clearance (%) 7075	6061	2024
	Heavy Gauge (approximately 6.4 mm or 0.250 in)					
B	2.5	13.08	(0.515)	2.8	3.1	3.0
C	5.0	13.21	(0.520)	3.8	4.1	4.0
D	8.0	13.72	(0.540)	7.5	8.2	7.9
E	12.0	14.22	(0.560)	11.3	12.2	11.9
	Light Gauge (approximately 2.9 mm or 0.115 in)					
B	2.5	12.85	(0.506)	2.6	2.6	2.6
C	5.0	12.95	(0.510)	4.3	4.3	4.3
D	8.0	13.21	(0.520)	8.7	8.7	8.7
E	12.0	13.46	(0.530)	13.0	13.0	13.0

STRESS ANALYSIS – To further demonstrate that sufficient residual axial stresses (short transverse stresses) can be present around punched holes to cause the observed cracking, X-ray diffraction was used to determine the magnitude and distribution of the stresses. The approach used was to make four X-ray diffraction measurements at the hole wall surface, .13, .25, and .38 mm (.005, .010, and .015 in) into the specimen from the wall. The subsurfaces were exposed by electropolishing the hole wall in a nitric acid-methanol electrolyte. Details of the experimental procedures used and corrections applied to the results are contained in a private report prepared by Lambda Research Incorporated for Reynolds Metals Company. Residual stress determinations were made on three holes in an unexposed 7075-T651 specimen. The holes measured were the drilled hole (A), the minimum punch clearance hole (B), and the maximum punch clearance hole (E).

RESULTS AND DISCUSSION

STRESS CORROSION CRACKING (SCC) TESTING – The results of the SCC test exposures demonstrate that sufficient residual stresses exist in the walls of punched holes to cause SCC in susceptible materials. Transverse or laminar cracking was observed in the walls of the punched holes in alloys 2024-T3, 2024-T351, 7075-T6, and 7075-T651. Tables 3-5 summarize qualitatively the corrosion observed on the hole walls. While random pitting corrosion was found on the exposed flat surfaces of all specimens, no radial cracking was found around the holes.

Table 3 contains information on alloy 2024. In the heavier gauge material both pitting and cracking became more severe with increased exposure time. All of the punched holes developed cracking while the drilled hole did not appear to crack. Figure 2A shows the wall of drilled Hole A after 60 days and Figure 2B shows a cross section through corroded areas. While the predominant mode of corrosion is pitting, there are a few small cracks or corroded grain boundaries visible in the photomicrograph of Figure 2B. Figure 3A shows the wall of punched Hole E after 60 days and Figure 3B shows a cross section through the corroded areas. Two modes of corrosion are evident, pitting and corrosion cracking. The deformation of the hole wall by the punching operation can be seen in Figure 3B. This information demonstrates the fact that 2024 in T3 type tempers may experience SCC in the walls of punched holes but not in the walls of drilled holes.

Table 3 - Alloy 2024 Hole Wall Corrosion Information

Material	Gauge mm	(in)	Exposure (days)	Hole(s)	Extent of Corrosion* Pitting	Cracking
2024-T351	6.43	(0.253)	30	A-C	M	N
				D-E	M	SL
			60	A	SV	N
				B-E	SV	SV
2024-T3	2.92	(0.115)	30	A-E	M	N
			60	A	SV	N
				B-E	SV	SL

* N = Not found, SL = Slight or Shallow,
  M = Moderate or significant depth and size,
  SV = Severe or deep and large.

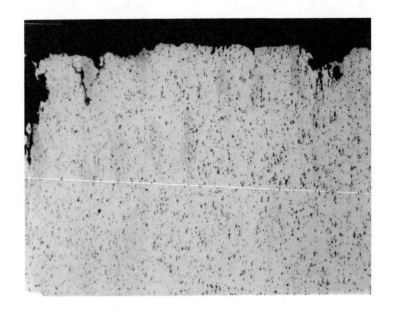

A. Surface, 25 X

B. As-Polished Cross Section, 50 X

Fig. 2 – 2024-T351 Drilled Hole Wall After 60 Days Alternate Immersion Exposure

306

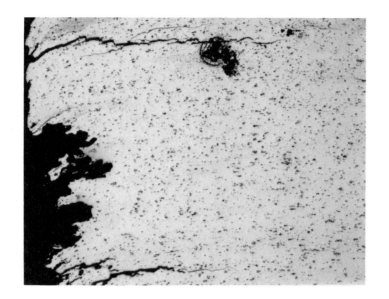

A. Cracked Surface, 25 X

B. As-Polished Cross Section, 50 X

Fig. 3 - 2024-T351 Punched Hole Wall After 60 Days Alternate Immersion Exposure

307

Table 4 contains information on alloy 6061. In both gauges and for both exposure times, no significant corrosion was experienced. Figure 4 shows the condition of drilled Hole A after 60 days. The markings left by the twist drill are clearly visible. Figure 5 shows the roughened but uncorroded surface of punched Hole E after 60 days. Thus, this relatively corrosion resistant alloy experienced neither SCC nor significant pitting.

Table 4 - Alloy 6061 Hole Wall Corrosion Information

Material	Gauge mm	(in)	Exposure (days)	Hole(s)	Extent of Corrosion* Pitting	Cracking
6061-T651	6.22	(0.245)	30	A-E	SL	N
			60	A-E	SL	N
6061-T6	2.92	(0.115)	30	A-E	SL	N
			60	A-E	SL	N

* N = Not found
  SL = Slight or Shallow

Fig. 4 - 6061-T651 Drilled Hole Wall After
         60 Days Alternate Immersion Exposure

Magnification: 25 X

Fig. 5 - 6061-T651 Punched Hole Wall After
60 Days Alternate Immersion Exposure

Magnification: 25 X

Table 5 contains information on alloy 7075. Cracking became evident within the first two weeks of exposure in the punched holes. The pitting was elongated and deep. Figure 6A shows the wall of drilled Hole A in the heavier material after 60 days exposure. The pits and drill marks are visible. Figure 6B shows a cross section from the same hole with only pits present. Figure 7A shows the wall of punched Hole E in the heavier material after 60 days exposure. The severe pitting and cracking are evident. Figure 7B is a photomicrograph of a section through the wall of Hole E showing the extensive, branching stress corrosion cracks and the deformation caused by the punching.

Table 5 - Alloy 7075 Hole Wall Corrosion Information

Material	Gauge mm	(in)	Exposure (days)	Hole(s)	Extent of Corrosion* Pitting	Cracking
7075-T651	6.73	(0.265)	30	A	M	N
				B-E	M	SV
			60	A	M	N
				B-D	M	SV
				E	M	M
7075-T6	2.92	(0.115)	30	A	M	N
				B-D	M	SL
				E	M	M
			60	A	SV	SL
				B-E	SV	SV

* N = Not observed, SL = Slight or Shallow,
  M = Moderate or significant depth and size,
  SV = Severe or deep and large.

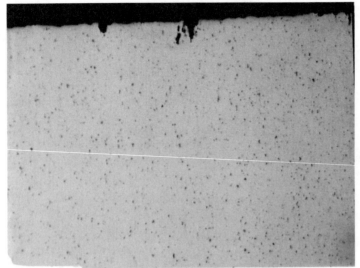

B. As-Polished Cross Section, 50 X

A. Surface, 25 X

Fig. 6 — 7075-T651 Drilled Hole Wall After 60 Days Alternate Immersion Exposure

310

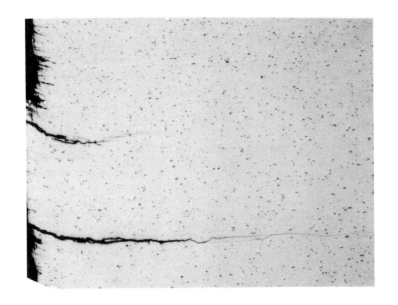

B. As-Polished Cross Section, 50 X

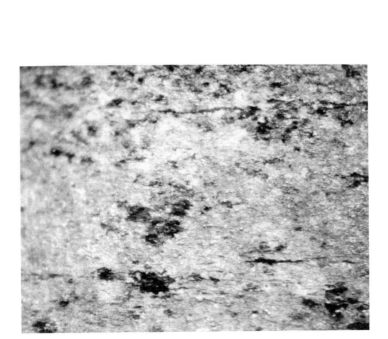

A. Surface, 25 X

Fig. 7 – 7075-T651 Punched Hole Wall After 60 Days Alternate Immersion Exposure

Figures 2-7 have illustrated typical examples of the corrosion responses of three alloys to drilled and punched holes. The observed corrosion responses are consistent with published literature such as ASTM G64-85. (3) Both 2024-T3 type materials and 7075-T6 type materials have poor short transverse SCC resistance. This behavior was not expected on the thinner 2024-T3 because of the rapid quench usually achieved at gauges 3.18 mm (0.125 in) and thinner. On the other hand, 6061-T6 type materials have excellent short transverse SCC resistance.

Thus, punching appears to introduce sufficient residual short transverse tensile stress to cause SCC in susceptible materials. Drilling appears to introduce either no short transverse tensile stress or short transverse compressive stress.

STRESS ANALYSIS - In order to more clearly understand the observed cracking, it was necessary to determine the magnitude and distribution of axial or short transverse stresses in typical hole walls. Holes A, B, and E in an unexposed 7075-T651 specimen were examined. While examinations of all alloys, gauges, and holes was desirable, it was decided that for the stated purpose of demonstrating the effects of punching and drilling that the limited examinations indicated above were a cost effective compromise.

The results of the X-ray diffraction residual stress determinations at four points into the walls of the three holes are summarized in Figure 8. The approximate range of errors for the stress values is ±14 MPa (±2 ksi).

Drilled Hole A was found to have a residual compressive short transverse stress gradient from the wall surface (-286 MPa or -41.5 ksi) to a radial distance of .37 mm (0.0147 in) from the wall surface (-19 MPa or -2.8 ksi). This demonstrates the reason for the absence of observed SCC even in alloy/temper combinations that are normally susceptible to SCC.

Hole B, punched with minimum clearance, was found to have a short transverse stress gradient from compression (-148 MPa or -21.5 ksi) at the wall surface to tension (75 MPa or 10.9 ksi) at .41 mm (0.0162 in) radially from the wall surface. It is suggested that once pitting corrosion reached a depth where the tensile stress exceeded the SCC threshold (approximately 48 MPa or approximately 7 ksi), (4) SCC began.

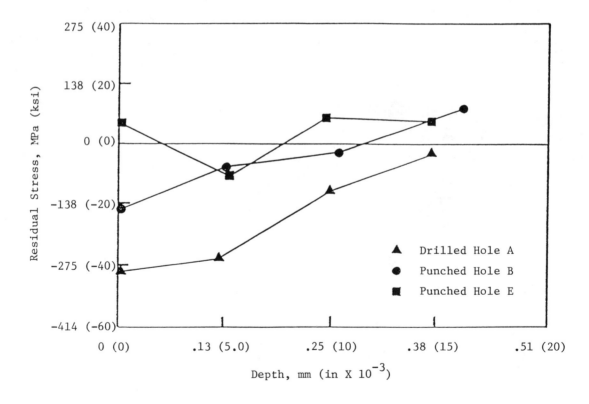

Fig. 8 - Axial Residual Stress vs Depth
AA 7075-T651
Mid-Thickness Location

312

Hole E, punched with maximum clearance, was found to have predominantly residual short transverse tensile stresses from the wall surface (50 MPa or 7.2 ksi) to .37 mm (0.0147 in) (52 MPa or 7.5 ksi) radially into the specimen from the wall surface. There was a compressive stress of -70 MPa (-10.2 ksi) found at .13 mm (0.0052 in) into the specimen from the wall surface. In spite of this apparently anomalous compressive short transverse stress, SCC proceeded rapidly and extensively (Figure 7B) in the wall of this hole.

A comparison of the residual stresses and approximate times to cracking has shown that Hole E with the greatest residual tensile stress cracked first. The other punched holes cracked after a somewhat longer time. This behavior appears to fit the stress distributions shown in Figure 8 in that Hole B with tensile stress some .25-.38 mm (0.010-0.015 in) away from the hole wall cracked only after pitting corrosion proceeded to the depth of threshold tensile stress (approximately 48 MPa or 7 ksi) for SCC.

CONCLUSIONS

The work described has clearly demonstrated the following points regarding the effects of punching and drilling on the SCC resistance of selected aluminum alloys:

1. Punching can introduce sufficient residual short transverse tensile stresses in 2024-T3 and 7075-T6 type aluminum materials to cause stress corrosion cracking.

2. Drilling can introduce residual short transverse compressive stresses which can provide resistance to stress corrosion cracking.

3. Based on the two gauges tested, metal thickness did not appear to be a significant factor.

4. Punch to die clearance was significant to the extent that increased clearance caused increased residual short transverse tensile stress and lower SCC resistance.

5. Alloys such as 6061-T6 with high resistance to SCC can be punched or drilled without observable SCC.

ACKNOWLEDGEMENTS

The authors wish to acknowledge the able assistance of Mr. R. O. Conrad of Reynolds Metals Company Mill Products Division for punching technology, Mr. Paul Prevey and Mr. Robert Bolin of Lambda Research, Inc. for X-ray diffraction technology, Messrs W. L. Davis and R. M. Howard for corrosion technology. Our sincere thanks goes to the management of Reynolds Metals Company Corporate Quality Assurance and Technology Operation Metallurgy Laboratory for their support in conducting the work and presenting this paper.

REFERENCES

1. Aluminum Standards and Data, The Aluminum Association, Inc., Washington, D. C., 1985.
2. Annual Book of Standards, Volume 03.02, ASTM, Philadelphia, PA, 1985, pp. 261-266.
3. Annual Book of Standards, Volume 03.02, ASTM, Philadelphia, PA, 1985, pp. 349-353.
4. Aluminum--Properties and Physical Metallurgy, Edited by John Hatch, American Society for Metals, Metals Park, Ohio, 1984, p. 271.

STRESS CORROSION CRACKING PROPERTIES
OF 2090 Al-Li ALLOY

E. I. Meletis
University of California
Division of Materials Science and Engineering
Davis, California, USA

ABSTRACT

The stress corrosion cracking (SCC) properties
of 2090 Al-Li alloy (Al-2.9Cu-2.2Li-0.12Zr) were
investigated in 3.5% NaCl solution. The alloy
was tested in the peak-strength condition and
in two overaged tempers. The SCC velocity as
a function of the stress intensity factor (K_I)
was obtained by testing precracked double cantilever beam (DCB) specimens in alternate immersion.
Slow-strain rate experiments were also conducted
in air and total immersion by using smooth cylindrical specimens.

The results from both testing methods show that
the degree of susceptibility depends on the aging
condition, the peak-aged temper being the most
resistant. Overaging induces susceptibility due
to the development of strain at the grain boundaries. This strain is a result of phase interactions and transformations and is produced by
the shearing of δ' by growing θ' and the conse-
quent transformation of θ' to T_1. Possible
embrittlement mechanisms are discussed.

THERE IS RENEWED INTEREST in Al-Li alloys in recent years, mainly because these alloys offer
certain advantages over conventional high strength
7XXX and 2XXX series aluminum alloys. The Al-
Li alloys exhibit high specific strength and high
specific modulus which is translated into fuel
savings in aerospace applications. Although the
stress-corrosion resistance of aerospace materials
is a critical property, it has attracted little
attention in the case of Al-Li alloys.

Early work by Rinker and co-worker [1,2]
on alloy 2020 (Al-4.5Cu-1.1Li-0.5Mn-0.2Cd) indicated that aging the alloy from the underaged
to the peak-aged condition results in significant
reductions in the plateau velocities. This
behavior was attributed to the electrochemical
potential difference between the grain boundary
T_1 precipitate and the matrix, that provides
a driving force for preferential precipitate

dissolution. Christodoulou, et al. [3] investigated the SCC behavior of the binary Al-2.8Li
alloy. They determined that the peak-aged temper
was more susceptible than the underaged temper,
whereas the overaged alloy was almost immune
to SCC. It was concluded that hydrogen embrittlement may play a role in the cracking mechanism.

Pizzo et al. [4] studied the SCC susceptibility of P/M Al-Li alloys containing Cu and
Mg. It was suggested that the stringer oxide
particles present have a determining effect on
the stress-corrosion susceptibility. Recently,
Vasudevan et al. [5,6] conducted an extensive
comparison of the stress-corrosion resistance of
several Al-Li-Cu-Zr alloys. Their results on
alloys tested in the peak-strength condition
showed a decrease in K_{ISCC} values by increasing
the Li content or the Li/Cu ratio. On the other
hand, aging had a smaller effect on both the
apparent K_{ISCC} and the plateau velocity of
Al-2.9Cu-2.1Li alloy. Finally, Holroyd et al.
[7] studied the SCC behavior of an Al-Li-Cu-Mg
alloy. It was concluded that the peak-aged temper
was more resistant than the underaged temper,
where overaging resulted in a highly stress-corrosion resistant alloy.

The present report is part of an in-depth
study with the overall objective to determine the
effects of composition and microstructure on the
SCC behavior of Al-Li alloys. It is hoped that
this investigative effort will contribute towards
more closely defining the operative stress corrosion cracking mechanism in Al-Li alloys. Four
ingot alloys are under investigation: Al-2.2Li-
0.12Zr, Al-2.9Cu-2.2Li-0.12Zr, Al-3Li-0.12Zr and
Al-3Li-1Cu-0.12Zr. In this paper, the results of
the commercial 2090 Al-2.9Cu-2.2Li-0.12Zr alloy
are reported.

EXPERIMENTAL

All specimens for this work were made from
commercially produced 2090 plate, 2.54 cm thick.
The alloy had been cast, hot rolled, solutionized,
quenched in water, stretched and aged. This

material was supplied by ALCOA Technical Center and was received in the peak-strength (T651) condition. The microstructure of the as-received plate consisting of flattened, parallel grain boundaries, is presented in Figure 1. The alloy was tested in three different tempers. These were the as-received peak-aged (PA) and two over-aged tempers designated in this report as OA1 and OA2. Overaging treatments consisted of reheating the as-received alloy in constant temperature oil baths for periods of 14h (OA1) and 90h (OA2) at 190°C. The hardening response of the alloy to isothermal aging at 190°C and the position of the three tempers are shown in Figure 2. The two overaged tempers were selected on the basis that OA1 has identical hardness (R_B 88) and comparable tensile strength to PA temper, whereas OA2 has a considerably lower hardness (R_B 80) and strength.

Stress corrosion crack propagation measurements were made using DCB specimens cut from the plate in the S-L orientation. The DCB specimens were 2.54 cm wide (plate thickness), 1.3 cm high and 12.5 cm long. The load was provided by opposing bolts inserted across the notch with a steel ball between the bolts to give uniform loading. The specimens were given pop-in pre-cracks of 0.5 cm to 0.7 cm by gradually increasing the initial deflection. The following equation was used to determine K_I values as a function of crack length (a):

$$K_I = \frac{ED\sqrt{3}}{4\sqrt{H}\,(a/H + 0.673)^2}$$

where E = modulus of elasticity, 79 GPa for 2090 [5],

H = half-width of the specimen (1.3 cm), and

D = deflection of specimen arms at the load line.

Valid plane strain samples were obtained in this study since the Brown and Strawley [8] criterion $b > 2.5\ (K_{IC}/\sigma_{YS})^2$ was satisfied for all three tempers tested. The fracture toughness (K_{IC}) in the short transverse direction is approximately 16.5 Mpa$\sqrt{m}$ and does not change with overaging [5]. The bolt-end of the DCB specimens was coated with a protective lacquer to prevent contact with the test solution. The environmental exposure consisted of alternate immersion in 3.5% NaCl solution. The samples were held in a plastic net and totally immersed in the solution for 10 minutes each hour. Crack lengths were measured on both surfaces for each sample using a traveling optical microscope. Three samples were tested for each temper with the total testing period exceeding 1200h.

Stress corrosion cracking susceptibility of the three tempers was also assessed by using the slow-strain rate technique (SSRT). Miniature, smooth cylindrical specimens 3 mm in diameter were machined with their tensile axis parallel to the short transverse direction of the plate. These tests were conducted by totally immersing the gauge length of the specimens in 3.5% NaCl

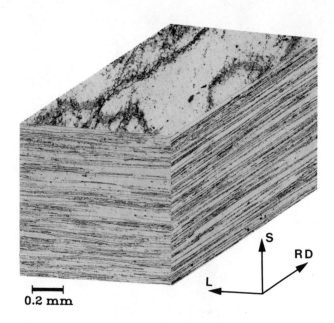

Fig. 1 - Grain structure of the as-received 2090 plate.

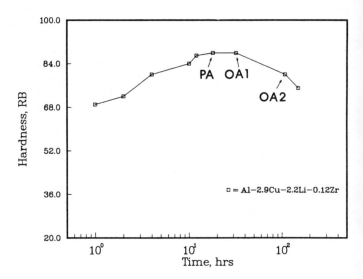

Fig. 2 - Hardening curve for 2090 Al-Li alloy aged at 190°C. The PA, OA1, and OA2 tempers tested are indicated on the curve.

solution. Two specimens from each temper were tested without solution in the laboratory environment to determine the tensile properties. The strain rate used for both the SCC and tensile specimens was 2×10^{-7} S^{-1}.

Transmission electron microscopy was utilized to characterize the microstructure and types of precipitates present in the as-received material prior to testing. Thin foils were also made from areas just below the tensile and stress-corrosion fracture surfaces for TEM studies. Electropolishing of the foils was carried out in a 1:3 mixture of nitric acid in methanol at -20°C and at a potential difference of 16V. Scanning electron microscopy (SEM) was used to examine the tensile and stress-corrosion fracture surfaces.

RESULTS AND DISCUSSION

MICROSTRUCTURAL CHARACTERIZATION - The microstructure of the 2090 plate, as-received, was found to be unrecrystallized, with a very large apparent grain size and a flattened grain morphology, typical of rolled high strength aluminum alloys. Large constituent particles were distributed in a network that was flattened and stratified by rolling. These intermetallics were in the approximate size range 1 μm to 8 μm in their longest dimension and were identified by energy dispersive x-ray (EDX) analysis to be mainly Al_7Cu_2Fe.

The fine scale microstructural details, including dispersoid particles, subgrains, and strengthening precipitates were studied by TEM and selected area electron diffraction (SAD). The precipitation sequence in ternary Al-Li-Cu alloys has been shown to be as follows [9,10]

$$\alpha \begin{array}{l} \longrightarrow \quad \delta'(Al_3Li) \quad \rightarrow \quad \delta(AlLi) \\ \\ \longrightarrow \quad G.P.\ Zones \quad \rightarrow \quad \theta'' \rightarrow \theta' \rightarrow \theta(Al_2Cu) \\ \\ \longrightarrow \quad T_1(Al_2CuLi) \end{array}$$

Analysis of the matrix and precipitate SAD patterns (Figure 3a) confirmed the superlattice structure of the alloy. This super structure is due to δ' phase precipitation which is an ordered $L1_2$ type phase, possessing a cube/cube orientation with respect to the matrix. The δ' phase has a spherical shape (Figure 3b) and is coherent with the matrix since its lattice parameter is similar to aluminum and its formation results in small misfit strain [9]. The streaking parallel to the two [100] orientations shown in the SAD of Figure 3a is due to the plate-like θ' precipitates. The precipitates were present in the alloy microstructure with a rather uniform distribution, Figure 3c. This phase has a tetragonal structure with (001) θ' ‖ (001)Al, [001] θ' ‖ [001]Al and a large misfit strain along one direction. The ternary T_1 was the third type of hardening precipitates observed in the microstructure. This phase forms as hexagonal platelets with a (111) habit plate, Figure 3d.

Observations indicated that when T_1 precipitates were found on subgrain boundaries, they were of larger platelet diameter than the T_1 found in the interior of the subgrain. In Figure 4a, a subgrain boundary which is decorated with large T_1 precipitates can be seen. This is in agreement with previous reports that in the early stages of aging, the T_1 phase precipitates preferentially at grain or subgrain boundaries [1]. Another feature shown in Figure 4a are the two δ precipitates in the upper section of the micrograph. A central dark field (CDF) at higher magnification from these precipitates is shown in Figure 4b. Such δ precipitates were observed in a considerable number of subgrain boundaries. This phase has a cubic structure and is surrounded by a high density of misfit dislocations suggesting that δ is semicoherent with the matrix [11]. The unrecrystallized structure of the alloy is a result of the Zr additions. This element has limited solubility in Al and forms Al_3Zr precipitates that are insoluble at the homogenization temperature. The major function of these small coherent precipitates is to retard subgrain boundary migration and coalescence as shown in Figure 5a and thus to inhibit recrystallization [12]. Finally, another precipitate type was observed in the microstructure with an inconsiderable volume fraction that more than likely was T_2 (Al_6CuLi_3) or T_B ($Al_{7.5}Cu_4Li$) ternary and it is shown in Figure 5b.

STRESS CORROSION CRACKING - The results of the stress corrosion crack propagation rate experiments are presented in Figure 6. None of the three tempers tested exhibited a well defined plateau crack velocity (region II) mainly due to the low fracture toughness of the alloy. The SCC threshold stress intensity for the PA, OA1 and OA2 tempers were determined to be 7.3, 5.74 and 2.08 Mpa√m, respectively. A comparison in terms of SCC growth rates and K_{ISCC} values clearly shows that overaging has a detrimental effect on the stress-corrosion resistance of the alloy. This conclusion is strengthened by the fact that the results also indicate some type of proportional relationship between SCC resistance and overaging time. A 14h overaging treatment results in a small difference in the stress-corrosion behavior, i.e., apparent plateau velocity and K_{ISCC}, between the PA and OA1 tempers. On the other hand, severely overaging the PA temper by 90h results in a large decrease in the SCC resistance of the OA2 temper.

The stress corrosion velocity data for 7075-T651 aluminum alloy is also included in Figure 6 for comparison. It can be seen that the 2090-T651 alloy is more resistant or at least as resistant as the 7075-T651 alloy to stress corrosion cracking. Alloy 2090 exhibits slightly lower apparent plateau velocity and higher K_{ISCC} than 7075 alloy. It should be noted however, that the plateau region for alloy 7075 extends from 13-19 MPa√m whereas alloy 2090 exhibits limited plateau region mainly due to its lower fracture toughness compared to 7075 alloy. It is also interesting to point out that the stress-corrosion resistance of 7XXX alloys is generally improved by overaging where the opposite was observed for 2090 Al-Li alloys.

Optical microscopy examinations of the crack at the surface of the DCB specimens showed that cracking was progressing exclusively along grain or subgrain boundaries, as illustrated in Figure 7. Intergranular SCC is a common characteristic of rolled high strength aluminum alloys when tested in the short transverse direction. Transgranular SCC in these alloys can be promoted by testing along the rolling direction [13]. All DCB specimens were mechanically fractured after stress-corrosion testing; SEM of the fracture surfaces verified the intergranular mode of propagation of the stress corrosion

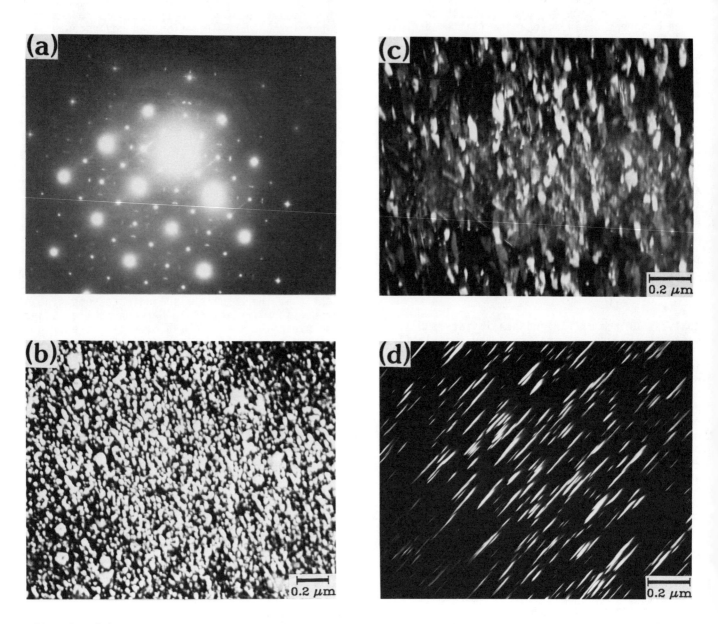

Fig. 3 - (a) SAD pattern showing the superlattice structure of 2090 and central dark field image of (b) δ', (c) θ' and (d) T_1 precipitates.

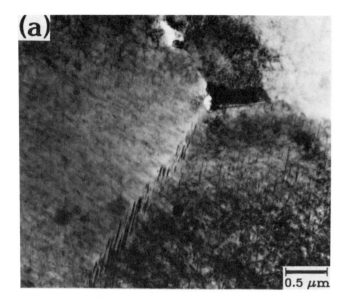

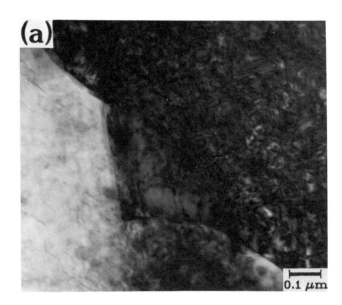

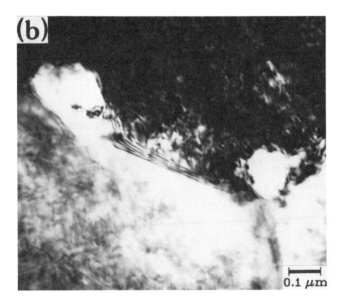

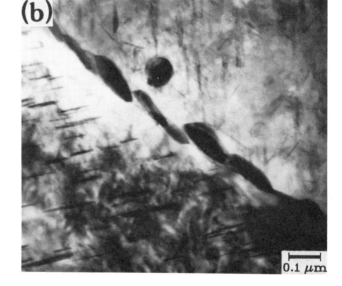

Fig. 4 - TEM micrographs showing (a) T_1 and δ precipitates (b) CDF of the two δ precipitates shown in (a).

Fig. 5 - TEM micrographs of (a) an Al_3Zr precipitate pinning a grain boundary and (b) T_2 or T_B precipitates at a grain boundary.

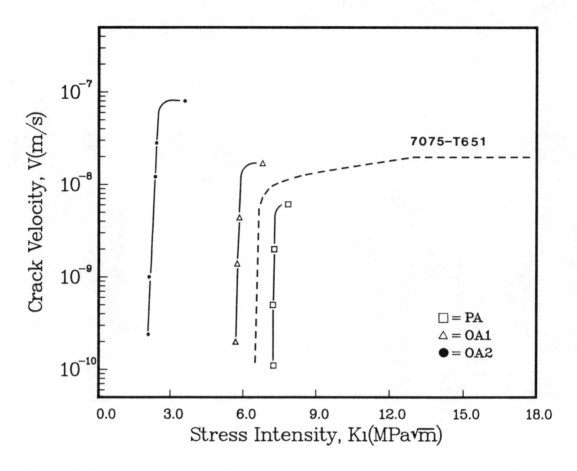

Fig. 6 - Stress Corrosion crack growth rate versus stress intensity for the three tempers of 2090 and 7075-T651 alloy.

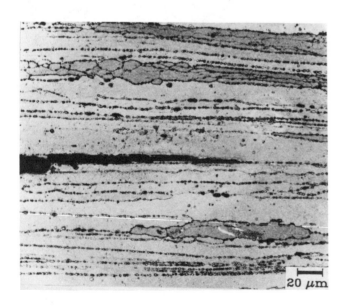

Fig. 7 - Optical micrograph of a typical stress corrosion crack propagating intergranularly in a 2090 DCB specimen.

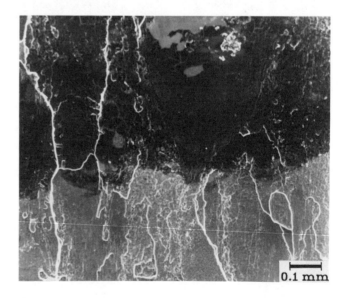

Fig. 8 - Fracture surface of a DCB specimen showing the SCC/Tensile Fracture interface.

cracks. Observations made at the SCC/overload fracture interface showed that intergranular cracks in the stress-corrosion region propagated along the same grain and subgrain boundaries during the overload failure (Figure 8). Examinations at higher magnifications revealed a difference in the morphology of these two fracture modes. Stress-corrosion fracture surfaces were smooth with subgrain boundaries visible while tensile fracture surfaces exhibited microroughness, as illustrated in Figure 9. It should be noted that the region in front of the stress-corrosion crack tip (Figure 9) was not embrittled since it exhibited microductility identical to that observed in regions away from the SCC/tensile fracture interface.

Testing by using the SSRT provides several parameters which can be used to characterize stress-corrosion resistance. These are maximum stress, reduction in area, total elongation, and fracture energy. In the present study, fracture of all tensile and stress-corrosion specimens occurred in the elastic region, thus stress at failure was the only criterion remaining available for SCC evaluation. The stress-corrosion susceptibility was determined from the ratio of fracture strength in the NaCl solution versus tensile strength in laboratory air. The results presented in Table 1 show the tensile strength, stress-corrosion strength and their ratio for the three tempers tested. Increased susceptibility to SCC is indicated by a reduced strength ratio. It is evident from the slow-strain results that overaging induces stress-corrosion cracking susceptibility in the 2090 Al-Li alloy. Since a ratio near one indicates immunity to SCC, the results suggest that PA and OA1 tempers are relatively stress-corrosion resistant, where the OA2 temper is mostly susceptible.

It is important to note at this point that the same general trend in the SCC susceptibility with aging has been shown by both testing methods. Moreover, besides this qualitative agreement, the two methods also indicate a type of quantitative agreement between SCC susceptibility and length of overaging time. It is concluded therefore, that the two testing methods were found in excellent agreement in evaluating the stress-corrosion properties of the 2090 alloy.

Slow strain rate fracture surfaces of the three tempers tested were examined with SEM. All tensile and stress-corrosion fracture surfaces exhibited an intergranular mode of fracture similar to that observed in the DCB specimens. Figures 10 and 11 present fracture surfaces of PA temper tested with the SSRT in laboratory air and 3.5% NaCl solution, respectively. The similar appearance of the micrographs indicates that no change in fracture mode occurred due to the electrolyte. This can also be seen by comparing Figures 12 and 13 which are fractographs of the OA2 temper tested in laboratory air and 3.5% NaCl solution, respectively,

In all SCC specimen surfaces some corrosion attack was evident. The extent of this corrosion attack was increased with aging time; the OA2

Fig. 9 - Scanning electron micrograph of the stress-corrosion and tensile fracture interface.

Table 1 - Slow Strain Rate Testing Data

	PA	OA1	OA2
Tensile Strength, MPa	307	294	189
S-C Fracture Stress, MPa	273	250	55
S-C FS/TS, Ratio	.89	.85	.29

temper was the most attacked, despite the fact that OA2 specimens were exposed in the electrolyte for shorter periods of time due to their higher SCC susceptibility. This higher electrochemical reactivity accompanying overaging may have important implications on the SCC mechanism operating in these alloys. For example, a dissolution model can account for the obtained results since overaging may produce a microstructure which is more prone to corrosion and therefore be more susceptible to SCC. The presence of finely distributed AlCl$_3$ corrosion products on the fracture surfaces as determined by EDX analysis also supports this mechanism. However, there is an alternative explanation. A closer examination of the SCC fracture surfaces shows that dissolution has a rather localized nature being present mostly as pits in PA and OA1 tempers, Figure 12b, and as peripheral attack around the cylindrical OA2 specimens, Figure 13a. Such type of attack more than likely serves as a crack initiation or reinitiation stage, rather than a crack propagation stage [14,15]. This is shown in Figure 13b where stress-corrosion fracture initiated from a dissolution site and progressed through the specimen with limited signs of dissolution. The difference in the appearance

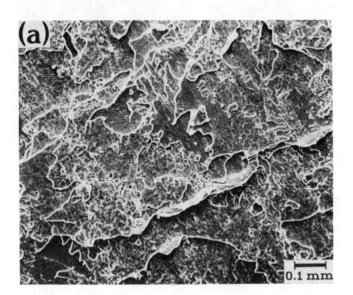

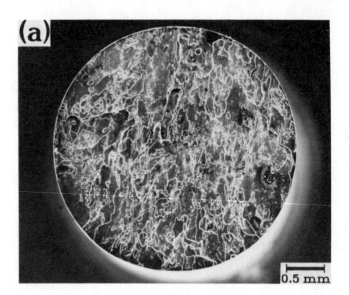

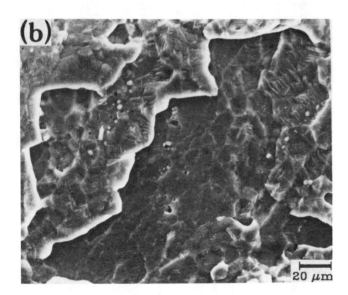

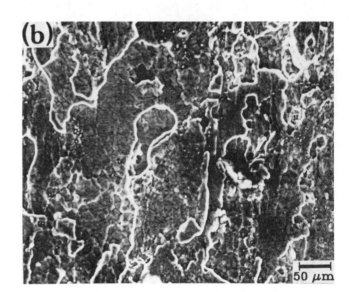

Fig. 10 - Scanning electron micrograph of a PA tensile fracture surface showing (a) intergranular fracture and (b) microductility.

Fig. 11 - Scanning electron micrograph of a PA SCC specimen showing (a) the entire fracture surface and (b) high magnification of a dissolution site.

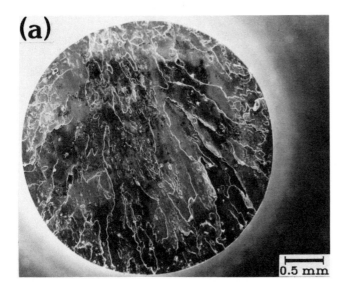

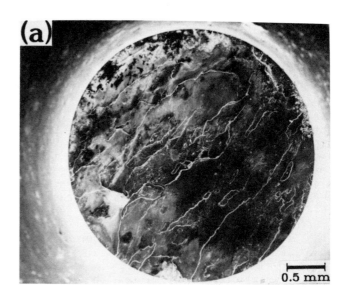

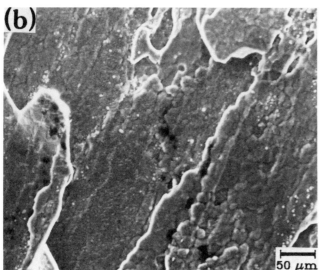

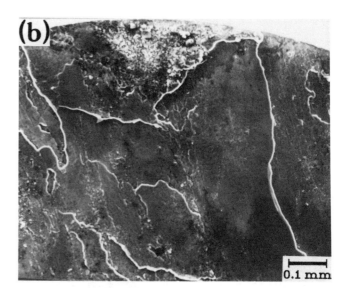

Fig. 12 - SEM of a tensile OA2 specimen showing (a) the entire fracture surface and (b)
detailed fracture surface morphology.

Fig. 13 - SEM of a SCC OA2 specimen showing
(a) the fracture surface and (b) crack initiation at a corrosion·attacked area.

between these two areas is quite obvious in the fractograph. In this case, stress-corrosion cracks may advance with either hydrogen embrittlement or highly localized chemical dissolution. It should be emphasized though, that the greater SCC susceptibility of the overaged tempers can be explained by a shorter crack initiation period due to higher corrosion attack resulting from microstructural changes occurring during aging.

Transmission electron microscopy studies from areas just below the fracture surfaces of slow strain rate specimens revealed the occurrence of several microstructural changes. First, it was confirmed that by increasing aging time, besides the coarsening of the δ' precipitates, more δ precipitates nucleate and grow. Although quantitative analysis was not performed, the TEM observations showed that the number of δ precipitates increased substantially in the OA2 temper compared to PA temper. Figure 14 illustrates an area in an OA2 specimen where several δ precipitates were observed. The presence of more δ in the microstructure is significant since this phase incorporates more Li than δ' and is anodic with respect to the matrix [16]. Therefore, exposure to a corrosive environment would be expected to result in preferential δ dissolution along grain boundaries. In spite of that, several dissolution slots along grain boundaries, resulting from electropolishing the foil, were observed with the TEM which were surpassing δ precipitates leaving them unaffected. This observation really questions the compatibility of a mechanism based on the preferential dissolution of precipitates along grain boundaries.

Another microstructural change occurring with aging was the gradual transformation of the θ' precipitates to T_1 precipitates. In fact, no θ' were present in the OA2 temper indicating that

Fig. 15 - TEM of the OA2 temper showing T_1 precipitates impinging a grain boundary. Coarse δ' precipitates can also be seen.

the transformation had been completed as shown in Figure 15. The above transformation is believed to consist of two steps. First, the θ' precipitates grow with aging and cut the δ' precipitates encountered. This phase interaction is possible since the δ' phase wets the θ' phase. This transformation requires certain orientational rearrangements, since θ' forms on (001) and T_1 on (111), and may be accompanied with δ' dissolution and reprecipitation.

The result of the above microstructural changes is the development of strain at the grain boundaries of the overaged tempers. More specifically, this strain is due to the grain boundary impingement by the transformed T_1 precipitates, Figure 15, and the shearing of the δ' precipitates, Figures 16a and b. Once the δ' precipitates have been sheared as a result of the previously mentioned transformation, the effective barrier to slip is reduced and further deformation on that plane is favored, thus slip tends to be confined to narrow bands. These intense slip bands impinge upon grain boundaries creating a stress concentration. In Al-Li alloys, the ordered structure of the δ' precipitates further encourages this planar localization of slip. Also, some minor contributions to this grain boundary strain may be offered by the strain surrounding the δ precipitates since substantially more δ was found in the overaged tempers.

It is important to note that this grain boundary strain can explain the greater SCC susceptibility of the overaged tempers regardless of the embrittlement mechanism operating. It is expected that the presence of a stress at grain boundaries will enhance a hydrogen embrittlement or dissolution process along grain boundaries.

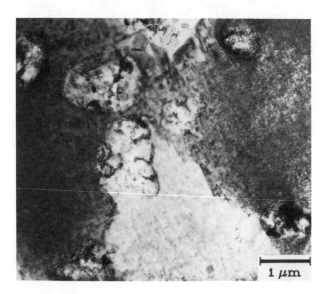

Fig. 14 - TEM of an OA2 specimen showing a large number of δ precipitates at subgrain boundaries.

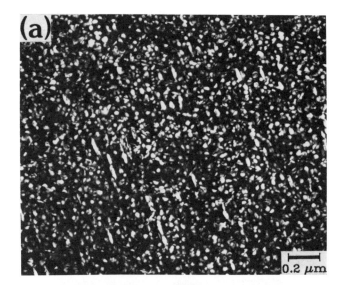

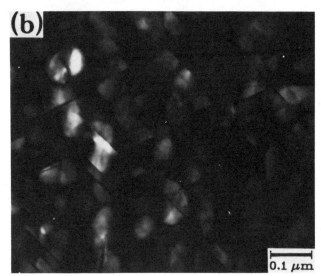

Fig. 16 - CDF of sheared δ' precipitates (a) overall view (b) higher magnification of an area shown in (a).

Finally, the T_1 phase is less corrosion resistant than the Cu-rich θ' phase and therefore the overaged tempers are also expected to exhibit a shorter crack initiation stage due to faster dissolution compared to high strength temper.

CONCLUSIONS

1. This study has shown excellent agreement between the two testing methods used to determine the SCC properties of 2090 Al-Li alloys.
2. The degree of SCC susceptibility is dependent upon the aging condition with overaging inducing susceptibility.
3. The 2090-T651 alloy exhibited slightly better SCC resistance than the 7075-T651 alloy.
4. The greater SCC susceptibility of the overaged tempers is due to microstructural changes resulting from aging. These microstructural changes involve θ' and δ' phase

interactions, the θ' → T_1 transformation, the shearing of δ', and the formation of more δ precipitates. The presence of more active phases at grain boundaries and the development of grain boundary strain is suggested to reduce the SCC resistance of the overaged tempers.

ACKNOWLEDGEMENTS

The author wishes to thank IIT Research Institute and Georgia Institute of Technology for supporting this research. The author would also like to thank the ALCOA Technical Center for providing the testing material. Comments on the microstructural characterization by Professor T.H. Sanders, Jr., are greatly appreciated.

REFERENCES

1. Rinker, J.G., "Effect of Precipitation Heat Treatment on the Microstructure, Toughness and Stress Corrosion Crack Propagation Resistance of Aluminum Alloy 2020," Ph.D. Thesis, Georgia Institute of Technology, Sept. 1982.
2. Rinker, J.G., M. Marek and T.H. Sanders, Jr. "Aluminum-Lithium Alloys II," p.597, Eds. T.H. Sanders, Jr. and E.A. Starke, Jr., TMS-AIME, Warrendale, PA (1984).
3. Christodoulou, L., L. Struble and J.R. Picken ibid p.561.
4. Pizzo, P.P., R.P. Galvin and H.G. Nelson, ibid p.627
5. Vasudevan, A.K., R.C. Malcolm, W.G. Fricke and R.J. Rioja, "Resistance to Fracture, Fatigue and Stress-corrosion of Al-Cu-Li-Zr Alloys," final report NADC Contract N00019-80-C-0569, June 1985.
6. Vasudevan, A.K., P.R. Ziman, S.C. Jha and T.H. Sanders, Jr., "Stress Corrosion Resistance of Al-Cu-Li-Zr Alloys," paper presented at the 3rd International Aluminum-Lithium Conference, University of Oxford, England, 8-11 July, 1985.
7. Holroyd, N.J.H., A. Gray, G.M. Scamans and R. Hermann, "Environment-Sensitive Fracture of Al-Li-Cu-Mg Alloys," ibid.
8. Brown, W.F., Jr. and J.E. Strawley, ASTM STP 410 (1966).
9. Silcock, J.M., J. Inst. Metals 88, 357-364 (1959-60).
10. Noble, B. and G.E. Thompson, Metal Sci. J. 6, 167-174 (1972).
11. Williams, D.B. and J.W. Edington, Met. Sci. 9, 529-532 (1975).
12. Sanders, T.H., Jr. and E.A. Starke, Jr., "Aluminum-Lithium Alloys II," p.1, Eds. T.H. Sanders, Jr. and E.A. Starke, Jr., TMS-AIME, Warrendale, PA (1984).
13. Meletis, E.I., "Fracture: Measurement of Localized Deformation by Novel Techniques," p.87, Eds. W.W. Gerberich and D.L. Davidson, TMS-AIME, Warrendale, PA (1985).
14. Meletis, E.I., "Crystallographic Characterization of Transgranular Stress Corrosion Cracking of Face-Centered Cubic Metals and Alloys," Ph.D. Thesis, Georgia Institute of

Technology, Atlanta, GA, Dec. 1981.

15. Meletis, E.I. and R.F. Hochman, Corros. Sci.
    24, 843-862 (1984).
16. Niskanen, P., T.H. Sanders, Jr., M. Marek
    and J.G. Rinker, "Aluminum-Lithium Alloys,"
    p.345, Eds. T.H. Sanders, Jr. and E.A.
    Starke, Jr., TMS-AIME, Warrendale, PA(1981).

REPAIR WELDING OF CRACKED FREE MACHINING INVAR 36

Maxwell Pevar
General Electric Company
Space Systems Division
Materials Engineering Unit
Valley Forge, Pennsylvania, USA

Abstract

Invar 36 has a low coefficient of thermal expansion. The alloy is nominaly 35% Ni, 1% Co, 0.12 C, 0.90 Mn, 0.35 Si Bal Fe. The alloy is difficult to machine due to its work hardening tendencies. An easier machining grade of the alloy (known as Free-Machining Invar) is not recommended for welded fabrication for the Selenium makes the alloy prone to hot-short, i.e., the weldment undergoes stress cracking during the cool down from the melt. The paper deals with Laser Welding of the Free-Machining grade and in particular with a repair welding procedure. The repair method utilizes spacial filler wire repairs without the need to grind out the cracks present in the weldment. The specific joint investigated was an off-set butt joint. (see Figure 6) with a 1.016 mm (0.040 inches) deep off-set within a 1.78 mm (0.070 inch) thick wall.

AT GENERAL ELECTRIC SPACECRAFT OPERATION, King of Prussia, PA Carpenter Free-Cut Invar "36" has been utilized in components requiring its low temperature coefficient of expansion properties. Laser welding is the selected joining process. Filler material was not used. Laser welding provides a concentrated penetrating heat source resulting in a narrow, heat affected zone and therefore less distortion and residual stress than achievable by conventional arc welding processes. Laser welding was used successfully on many components. However, a rash of fine cracks were detected on more recently fabricated parts. They appeared to run down the center of individual pulsed laser spot welds. An investigation of several cracked weldments disclosed the cracks progressed from the surface in an inter-granular mode. This condition was similar to "hot short" found in ingot casting.

A review of Carpenter Technology's literature as follows, gave no hint as to the cause of these cracks.

Carpenter in their 1966 brochure on this alloy states "welding: Carpenter Free-Cut Invar "36" can be welded by the conventional methods. Caution must be taken so as not to over heat the molten metal. This will avoid spattering of the molten metal and pits in the welded area. When the filler rod is required, Invarod is recommended."

A metallurgical investigation of the weld cracks and a review of Hot Shortness established the cause of cracking and set the stage for the repair welding procedure described herein.

METALLURGICAL INVESTIGATION OF WELD CRACKS

The surface cracks observed on the pulsed laser weldments are shown in Figure 1. The cracks running along the center of the weldment are shown in Figure 1a (20X magnification) and the radial cracks within the pulsed spot weld are shown in Fibure 1b at 50X magnification. The surface of the weld was ground down slightly polished and etched to better disclose the weld crack pattern as shown in Figure 2. A cross section through the weld showed these cracks to progress intergranularly along columnar grain within the weld metal. A concentration of selenium was present along these grain boundaries.

HOT SHORTNESS

These intergranular cracks are hot short cracks which occur in the grain boundaries during solidification of a casting when the last to solidify portion of the metal is ruptured due to the presence of a "weakening" material (such as excess sulfur or selenium added for free machining) on the grain boundaries during the cool down from the melt. The free machining ingredient, selenium, is present in stringer form in the base metal. See Figure 3. The selenium migrates to the grain boundaries during solidification since it is the lower melting point material. The crack patterns radiate from the shrink cavity in the center of each pulse weld on a microscopic scale and along the center line of the weld on a macro scale. The base metal along the weld acts as a heat sink during solidification and applies the stresses which cause the macro cracks parallel to the weld at its center line.

A brief literature survey disclosed the following: The 1948 Metals Handbook (ASM) define hot shortness as: "Brittleness in Hot Metal". The 8th Edition Vo. 10, Page 71 of the Metals Handbook in discussing welding of Beryllium describes "that the precipitation of an aluminum rich phase had melted during welding. This introduced a hot shortness condition that combined with thermal along the large columns grains of the fusion zone." The Fifth Edition, Section One, Page 4.95 of Welding Handbook in discussion of nickel alloys states: "In nickel and high nickel alloys sulfur and lead must be controlled to prevent hot short."

An inventory of Carpenter Free-Cut Invar "36" was made and a chemical analysis was made of each heat and compiled in attached Table I. The heat number appears to the left and the bar size appears to the right. Heat 44264 which cracked during laser welding had the highest Selenium and Sulfur content.

REPAIR WELD CONSIDERATION

The writer came up with the possibility of repair welding the joints with the use of a filler rod weld pass using filler rod which would lower the Selenium content of the weldment. The main risk would be that the cracks might progress into the base metal prior to the weldment melting down. Normally a repair welding procedure would require the cracked material be ground out completely before attempting a repair weld, since in many metal systems the cracks progress with the re-application of heat. It was decided to try to repair weld directly without grinding out the cracks. A meeting was held with Carpenter Technology personnel. They mentioned that welding was not a recommended method for Free-Cut Invar since they often were advised of problems in joining this grade. However, they had some 0.889 mm (0.035 inch) diameter Invarod available.

It's composition is listed in Table I under weld wire. It had almost zero selenium but more sulfur than desired. It was decided to try the material on test pieces of the questionable alloy heat having cracks present in the weldment.

REPAIR WELD PROCESS DEVELOPMENT

The laser system used is a Raytheon Solid State Nd:Yag laser capable of 400 Watts average power. This laser is incorporated with a Ratheon Laser Machining Center and Anorod controls. There are two CRT's; one for video and one for positioning display. See Figure 4. The one piece of equipment missing to try out the weld repair was a wire feed system. Such a system for this equipment was not available. However, Mr. W. Gelches, the laser welder operator, designed and built his own unit which interfaced very well with the Ratheon equipment. Initially test pieces from bad heat 44264 and good heat (lower Selenium) were welded and compared. The bad heat developed surface cracks; the good heat was crack free. A repair weld pass was made on both bad and good welded specimens (without grinding away the cracks) using the Invarod filler wire. The bad specimens was markedly improved and the good specimen remained crack free. Optimization of weld parameters was achieved with several additional specimens of bad material. A finalized weld schedule was prepared and included in a laser weld repair procedure. A cross-section through a repair weld is shown in Figure 5. It is to be noted that in this joint configuration the function of the weldment is two fold, structural and sealing. The structural portion was achieved with deeper penetration of short length weldment (see Figure 5a) whereas the sealing was achieved by all around welding. Figure 5b is a sealing weld; full penetration to encompass the joint off-set was not a requirement.

SUMMARY

A practical example of repair method in cracked Free Machining Invar weldments has been described herein. Each joint configuration, i.e., metal thickness and geometry will require its own weld schedule. There will be limitations with this process where the thickness of the joint members will be beyond the penetration of the filler metal remelt capability. Prior grinding out of the cracked metal at the weldment might be required in this situation. It would be ideal if the repair method were not needed by selection of low selenium heats of the

Free Machining Invar. However, with small
quantity purchases, the supplier finds it
difficult to formulate his alloy chemistry to
accomodate a weldable grade. It is estimated
that Selenium content in the alloy under
0.20% will render the alloy weldable by the
laser welding method.

ACKNOWLEDGEMENTS

The writer is grateful to contributions
of GE's spacecraft advance manufacturing
group support in particular W. Gelches,
Process Specialist, and M. F. Haley, Process
Engineer, and to M. Porter, Metallographer
of the Materials and Process Engineering
Laboratory.

This manuscript was based on work done
under a contract number DSCS (F)470-81-C-004.
The U.S. Government retains a nonexclusive,
royalty-free license to publish or reproduce
the published form of this contribution, or
allow others to do so, for U.S. Government
purposes.

TABLE 1

G.E. MATERIAL

Heat No.	C	Mn	Si	P	S	Cr	Ni	Mo	Cu	Co	Se	Ti	Size Dia.
36313	.055	.77	.25	.006	.003	.21	35.90	.06	.06	.11	.21		(2")
36306	.058	.76	.27	.008	.003	.16	35.97	.03	.05	.13	.20		(1 5/8")
36691	.051	.73	.24	.004	.003	.16	35.91	.03	.06	.13	.22		(1")
29352	.035	.76	.21	.009	.003	.12	35.99	.02	.09	.33	.21	.002	(3")
44264	.046	.65	.22	.010	.004	.21	35.94	.02	.08	.12	.25	<.01	(1 5/8")
43482	.052	.74	.28	.008	.003	.12	36.00	.04	.04	.05	.19		
52130	.053	.74	.26	.008	.002	.10	35.82	.01	.02	.03	.22		(2 3/4")
33750	.052	.75	.23	.007	.002	.14	35.99	.01	.02	.12	.20		(1")

WELD WIRE

Heat No.	C	Mn	Si	P	S	Cr	Ni	Mo	Cu	Co	Se	Ti	Size Dia.
D18413	.068	2.69	.14	.006	.012	.04	35.82	.01	.04	.06	<.001		(.035")

(Has .053 Al, .71 Ti, 0.04 Sn, <0.001 Pb)

Bad Welding Heat Has Highest Sulfur and Selenium

(Has 0.02 Al and <0.002 Ca and <0.001 Mg)

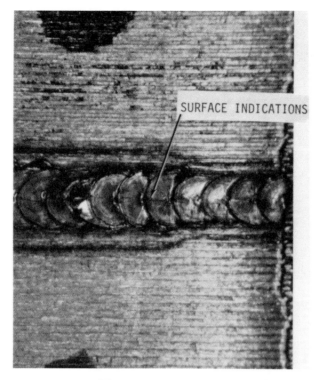

FIGURE 1a   MAGN: 20X

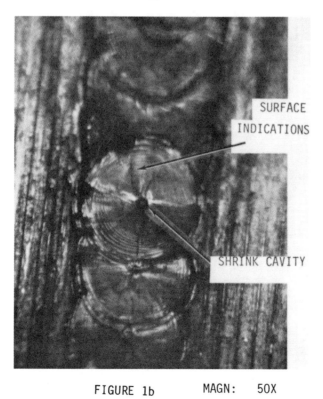

FIGURE 1b   MAGN: 50X

FIGURE 1

331

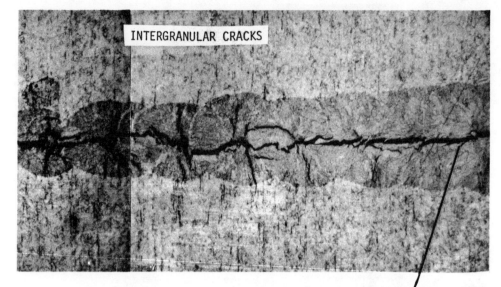

INTERGRANULAR CRACKS

MAGN: 50X

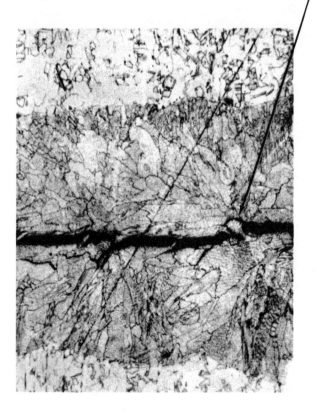

MAGN: 100X

FIGURE 2

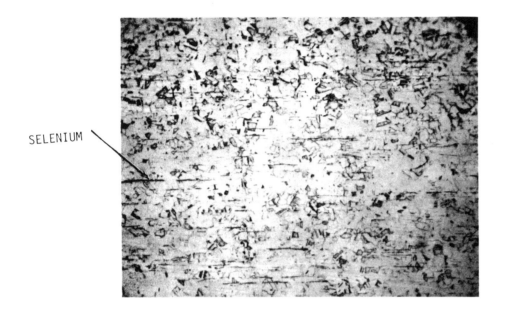

SELENIUM

MAGN: 500X

GENERAL MICROSTRUCTURE

FIGURE 3

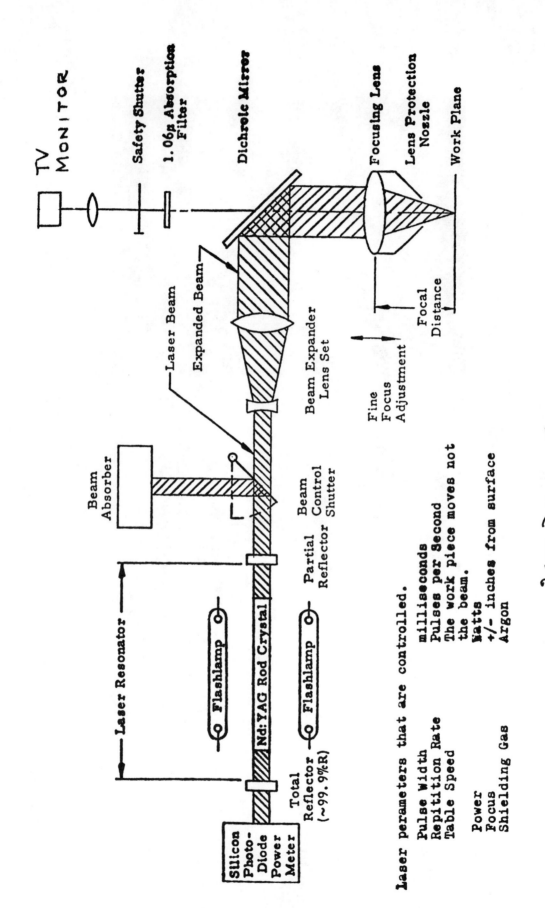

BEAM DELIVERY SYSTEM
RAYTHEON Nd:YAG LASER

FIGURE 4

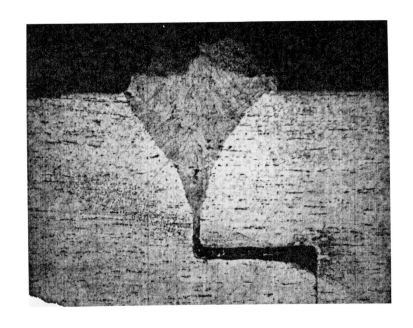

TACK WELD   MAGN: 50X
FIGURE 5a

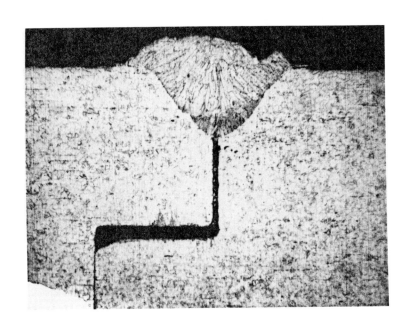

SEAM WELD   MAGN: 50X
FIGURE 5b

FIGURE 5

FILLER WIRE REPAIR OF INVAR 36 FILTER

PERFORMANCE OF A NEW CORROSION RESISTANT PRESTRESSING STRAND

Augusto S. Sason
Florida Wire & Cable Co.
Jacksonville, Florida, USA

ABSTRACT

Corrosion of reinforcing materials in certain bridge deck panels has led to Federal Highway Administration directives precluding the use of prestressed concrete strand in certain applications. Corrosion problems in other pre-tensioned and post-tensioned prestressed concrete applications have also surfaced.

To meet this problem, Florida Wire and Cable Company has developed epoxy coated strands referred to as Flo-Bond and Flo-Gard. Data will be presented from tests performed regarding combined characteristics of corrosion resistance, controllable bond transfer characteristics, anchoring, handling performance and practical applications in both pre-tensioned and post-tensioned prestressed concrete structures.

THE ORIGIN OF COATED STRAND

A decree by the Federal Highway Administration early in 1981 prohibited the use of prestressed concrete strand in certain bridge deck panels and prestressed members because of the lack of effective corrosion resistance. A program for the development of corrosion resistant strand was accelerated at Florida Wire and Cable and, at the PCI Convention, November 1982, a system of epoxy coated corrosion resistant strands was introduced under the trade names of Flo-Bond and Flo-Gard.

One of the products was a strand designed to offer corrosion resistance in combination with bond transfer characteristics that are equal to or exceeding bare strand capability. It is sold under the trade name Flo-Bond. This strand is intended for use where prestressed concrete has been excluded until now because of corrosion.

Also introduced was Flo-Gard, a strand designed exclusively for corrosion resistance, primarily in unbonded post-tensioning applications.

REQUIREMENTS FOR COATED STRAND

The requirements for coated prestressed strand greatly exceed the requirements for coatings widely used on epoxy coated rebar. Prestressed concrete strand is utilized under high tension with significant elongation. Prestressed concrete strand is a seven-wire construction with interstices between the wires. Furthermore, the individual wires show some relative movement in the bending of the strand.

While the coating for prestressed strand must have chemical and corrosion resistance equal to or exceeding the coatings widely used for epoxy coated rebar, as covered by ASTM and FHWA specifications, it must also display the ductility and toughness required in the handling and application of prestressed concrete strand.

The epoxy coating on rebar requires a coating thickness of only 6 to 9 mils and allows holiday frequency of up to 2 per foot.

The thickness of the Flo-Bond coating, on the other hand, is approximately 30 mils. This thickness is required in order to completely bridge the interstices of the seven wires in the prestressed strand and to obtain the goal of near zero holidays. (See Fig. 1)

We recognized that, in developing a coated prestressed strand, we were interjecting two new interfaces: the polymer-steel interface and the polymer-concrete interface. We needed a coating that would bond tightly to the steel and allow the transfer of stress from the strand to the concrete, with no yielding at either interface.

To assure a strong bond between the steel and the polymer, we utilized a proprietary cleaning process which provides an extraordinarily clean surface. The application of the coating also imposed a modest thermal history to the strand.

To assure a strong bond between the concrete and the polymer, we impregnated the coating with a grit which would obtain those results. The method of polymer and grit application made extraordinary demands on the flow time and gel time of the polymer.

For all of these reasons, the coating on prestressed concrete strand required a brand new formulation of cross-linked polymers.

TESTS ON COATED STRAND

The chemical corrosion resistance of this brand new formulation has been tested by certified laboratories and indicates full resistance to cathodic disbonding, chloride ion permeability, resistance to applied voltage and chemical resistance to calcium chloride and sodium hydroxide. The coating has also resisted any deterioration after being subjected to full(Fig2) tension in a salt spray cabinet for 3,000 hours.

Additionally, the mechanical properties of tensile strength, bond strength, bond transfer length, stress relaxation, fatigue endurance, and gripping characteristics were evaluated.

Although the application of the coating imposes a modest thermal history to the strand, there is no change in the mechanical properties of the strand before and after the coating treatment. See Figure 3. The temperatures the strand is exposed to during this treatment are not high enough nor long enough to affect the metallurgical or mechanical properties of the strand, including stress relaxation.

Utilizing the laboratories of Wiss, Janney & Elstner in Northbrook, Illinois, long-term bond transfer tests were conducted with 4 inch square by 8 foot long prestressed concrete beams, each containing one 1/2" 270 KSI strand, stressed to 75% of nominal ultimate tensile strength.

Using Whittemore gage techniques, the before and after prestress microstrain was measured and analyzed to provide the computer generated bond transfer curve shown in Figures 4 and 5.

In this example, the bond transfer length for bare strand was 30 inches live end and 29 inches dead end. Similarly, the bond transfer length for a typical grit impregnated coated strand is 30 inches live and 31 inches dead end.

Subsequently, these same bond transfer test specimens were retested after 14 months to determine the effect of time on these properties. The coated strand showed significantly less increase in transfer length during this time period than did the control specimen with bare strand.

The integrity of the steel to polymer bond is further substantiated by the fact that, in tensile tests of this product utilizing standard gripping chucks, we get higher chuck efficiencies than with bare strand.

The dynamic fatigue properties of the coating were tested using a 200,000# capacity electrohydraulic testing machine. The specimens repeatedly survived cyclic loading from 60 to 66% of ultimate strength for up to a half million cycles and from 40 to 80% of ultimate strength for 50 cycles. Throughout the demanding test, the specimens passed the holiday test with no evidence of cracks or breakdown in coating.

To examine the tenacity of the steel-fusion bonded polymer interface, we tested bond strength in concrete, fatigue endurance, and gripping characteristics. Extensive comparative pull-out tests were run. In these tests we used a standard 6" diameter by 12" long cylinder into which samples of various strands were imbedded the entire depth. The pull-out tests consisted of pulling in tension against the concrete cylinder until initial slip was observed. Figure 6 illustrates the equivalence of bond strength in bare and coated strand.

Sectioning of the cylinders after tests were completed revealed that an entirely new mechanism for bond transfer is occurring.

In bare strand slipping occurs by a spiral action or unscrewing of the strand in the helical cavity originally formed. This is shown by the glossy surface (Figure 7) caused by the movement of the strand against the concrete and the spiral hills mirroring the spiral interstices of the strand.

In the case of the polymer coated product, there is no such twisting of the strand. (Figure 8) The chalk markings of the blue coating against the smoothed down hills of the spiral grooves suggests that the mechanism for failure is, in fact, unidirectional sliding as opposed to untwisting. Pullout data for these tests shown in Figure 7 illustrate the equivalence of bond strength.

APPLICATIONS FOR COATED STRAND

The final test is, as always, putting the product to work in practical applications. The first opportunity was in the manufacture of tapered utility poles by Hughes Company in Orlando, Florida.

A tapered utility pole can take advantage of both the improved bond transfer characteristics and the corrosion characteristics, by placing the strand closer to the surface with less concrete cover than would be required with a non-corrosion resistant strand. We set up our experiment, utilizing all coated strand, some fully bonded prestressed, others with unbonded ends, post-tensioned.

We were informed that this particular pole with normal strand would be expected to withstand a deflecting force, at the end normal to the axis of the pole, of no greater than 6,500 lbs., and a deflection no greater than 9 ½'.

The actual deflection at breaking (Figure 9) was 13 ½', and the tension required to break was 9,500 lbs., both figures clearly exceeding the performance of normal strand.

Failure occurred in concrete compression and the integrity of the coated strand was beyond expectation.

We have since made a test of bridge deck panels, in cooperation with the Kentucky Department of Transportation, in which the strand was demonstrated to have easily met the strength requirements of prestressing.

The epoxy coated strand, Flo-Gard, was used to join segmental concrete dock cells in a salt water environment by post-tensioning methods. The anchor assemblies are capped and filled with elastomer sealant. The first floating dock post-tensioned with Flo-Gard is used by the ferry boat system between Norfolk and Portsmouth, Virginia.

The specifications for replacement of the Lancaster Street Bridge, over Harford Run in the City of Baltimore, called for epoxy coated strand which meets the properties of Flo-Bond. The coated strand was used in both precast prestressed concrete piles and hollow box beams. This is a joint project of the city, the state, and the FHWA.

Flo-Bond strands have been used as the pretensioning tendons in 24 inch double tees that span 49 feet over an area of a chemical plant where corrosion of uncoated strands would have created a hazardous condition.

In Oregon, Flo-Bond strand is being used in a bridge (Figure 10) that replaces one that had corroded and thrown off chunks of concrete. The Oregon DOT has been using epoxy coated reinforcing steel, but in this bridge, the Hubbard Creek Bridge, the Oregon DOT specified for the first time that the prestressed strand must also be epoxy coated. The Flo-Bond strand is used in bulb tee girders that are six foot high and 140 foot long. There are 26 strands in both the web and the bulb. Since Oregon has many more bridges that need to be replaced, the DOT will be closely monitoring the Flo-Bond strand in the Hubbard Creek Bridge.

Flo-Bond strand has frequently been used in prestressed fender pilings. In Florida, such pilings are being used in the bridge at Eau Gallie, on the east coast, and in the bridge at Apalachicola.

A special experimental fender pile design, using Flo-Bond strand, was cast for the Navy, which will test the special fender piles in Norfolk, Virginia. Because the Navy wanted "soft" piles, piles that would bend, not break, when accidentally hit with great force by a berthing ship, the final stress on the strand was less than half of the normal stress. That was done to provide "reserve" strength. In order to seat the wedges on the coating, the strands were tensioned and locked off on shims at 50% of ultimate tensile strength and then lowered to a force of 25% of the ultimate tensile strength by lifting the anchor and removing the shim. The wedges which were seated at the higher force were thus not disturbed.

Flo-Bond strand is also being used in cable stays for the Quincy, Illinois, bridge (Figure 11) over the Mississippi River. This is the first application of coated strand to cable stays. The Flo-Bond is contained in a polyethylene tube which, after the strand has been placed, will be pumped full of grout. This use of Flo-Bond coated strand provides additional protection against corrosion. Prior to being used, the Flo-Bond strand was tested for fatigue resistance by up to 4 million cycles of stress.

In California, Flo-Bond strand is being used in soil anchors that keep the earth beneath a private home on the Pacific Palisades from sliding into the sea. The Flo-Bond strands, 100 feet in length, are fed through casings to a predetermined area. Then grout is pumped into the casing, which is withdrawn, and, after a wait, more grout is pumped in. This grout expands into the soil and forms a 30 foot anchor. At the outer end, the Flo-Bond strand is anchored into a concrete retaining wall. The 70-feet of free strand between is then stressed so that the soil beneath the home is compressed to keep it from shifting.

PRODUCT AVAILABILITY

Epoxy coated strand is available in a full range of sizes, from 5/16" to .600 diameter, stress-relieved and low-relaxation grades. A 250 K.s.i. one quarter inch epoxy coated wire is also available. The epoxy coated wire is so ductile that it can be wrapped around its own diameter with no damage to the coating.

FUTURE DEVELOPMENTS

Meantime, testing continues. At the Florida DOT Corrosion Research Center, Flo-Bond strand and bare strand, embedded in concrete cylinders, is undergoing electrically accelerated corrosion testing in salt water. The bare strand failed on the seventh day of testing due to corrosion. After almost three years, the cylinders with Flo-Bond strand show no discernable deterioration. One of the Flo-Bond cylinders was broken open for inspection. No corrosion was taking place at any point along the strand.

Also at the Florida DOT, test panels with Flo-Bond strand and bare strand are being subjected to alternating wet and dry salt air and salt water spray. The voltage potential of each strand is read five days a week. After 17 months, there was no sign of corrosion.

CONCLUSIONS

The preceding tests and applications show that the epoxy coated prestressed concrete strand performs as well or better than bare strand. The coating has rendered the strand virtually free of corrosion.

But, of course, Flo-Bond and Flo-Gard strand have already been accepted by the Federal Highway Administration for experimental use in the same marine and salt-prone environments where the Federal Highway Administration is still prohibiting bare strand. So that Flo-Bond and Flo-Gard prestressed strand can make available the efficiencies of prestressed concrete . . . where it is otherwise forbidden. Flo-Bond and Flo-Gard may be used in any prestressing application where you want to virtually eliminate the danger of corrosion—and the danger of premature failure.

Figure 1
One-half inch diameter epoxy coated strand.
Coating thickness over crowns of wire is 30± mils.

Figure 2
Uncoated and epoxy coated strand specimens after
3000 hours in salt spray. There was no corrosion
on the coated specimen.

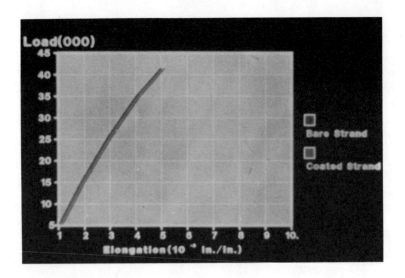

Figure 3
Comparison of tensile test data of strand as
normally produced and after coating application,
showing no effect on properties due to coating
process.

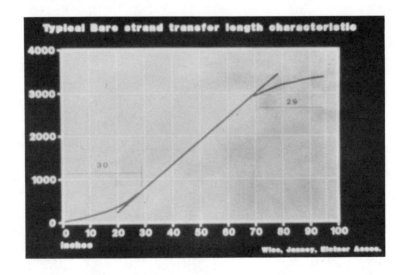

Figure 4
Bond transfer test data for bare strand using
Whittemore Gauge Technique on 10 foot beam.

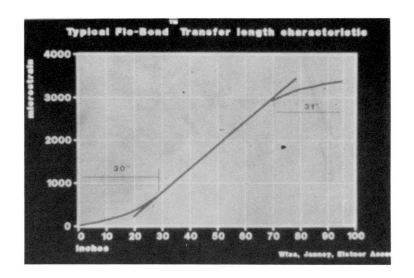

Figure 5
Bond transfer test data for coated strand using
Whittemore Gauge Technique on 10 foot beam.

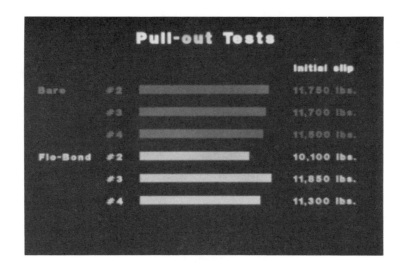

Figure 6
Pull Out Test using 6" diameter by 12" long cylinders
comparing coated strand to bare strand.

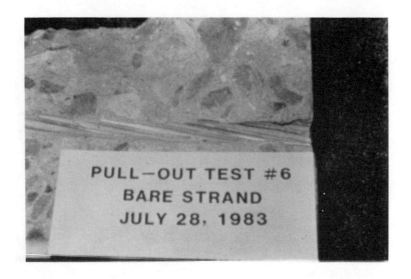

Figure 7
Section of cylinder with bare strand shows slippage
occurs by a spiral action or unscrewing of the strand
in the helical cavity originally formed.

Figure 8
Section of cylinder with coated strand shows no such twisting.
The markings suggest a unidirectional sliding action as
opposed to untwisting.

Figure 9
Tapered utility pole with coated strand undergoes
destructive testing. Deflection at breaking was
13½' (versus 9½' for bare strand) and tension at
breaking was 9,500 pounds (versus 6,500 pounds for
bare strand.)

Figure 10
Hubbard Creek Bridge, Orford, Oregon. This coastal
bridge replaces a bridge that had corroded. The
Hubbard Creek Bridge is the first bridge on which
the Oregon DOT specified coated prestressed strand.

Figure 11
The Quincy (Illinois) Bridge over the Mississippi.
The bridge is the first cable stay bridge in which
coated strand is being used in the cable stays.

THE PRESSURIZED THERMAL SHOCK ISSUE
AT H. B. ROBINSON UNIT 2 PLANT

J. H. Phillips
Carolina Power & Light
USA

ABSTRACT

A program was developed for the H. B. Robinson, Unit 2 Nuclear Plant to ensure that Pressurized Thermal Shock (PTS) type transients would not pose safety concerns related to crack propagation in reactor vessel welds and that further, PTS considerations would not restrict plant operational life. The program included insertion of a low leakage reactor core, use of partial shielded fuel assemblies on core flats to reduce fast neutron fluence at critical vessel welds, an extensive transient analysis program, additional data on copper impurity levels in welds, heating of reactor storage tank water, and improved training emphasizing PTS events. A program result is that the critical lower girth weld will not reach the NRC circumferential weld screening criterion of 300°F over the life of the plant.

INTRODUCTION

The Carolina Power & Light Company H. B. Robinson Plant is a three-loop Westinghouse PWR which began operation in 1971 and which has had a relatively high fast neutron exposure to the reactor vessel wall as a consequence of the length of operation and high wall neutron fluxes. This paper summarizes the actions and evaluations that were carried out by Carolina Power & Light Company to establish the potential for Pressurized Thermal Shock (PTS) events at this plant and to evaluate safety margins for future operation.

BACKGROUND

The issue now referred to as Pressurized Thermal Shock (PTS) was made more visible by transients which occurred at Rancho Seco and Three Mile Island in 1978 and 1979.

Figure 1 is a graphical depiction of this concern. The figure shows schematically how fast neutron irradiation increases the reference nil-ductility transition temperature (RT_{NDT}) such that the vessel can be in a brittle fracture temperature region during potential reactor transients. If a transient quickly lowers the temperature of an irradiation reactor vessel into a temperature range of degraded toughness while simultaneously maintaining high pressures, a pre-existing crack or flaw could propagate in the vessel. Figure 2 illustrates the type of transients that can cause rapid cooldown of the reactor vessel while still maintaining high pressures.

The NRC initiated a staff technical evaluation of the PTS issue which resulted in a report completed in November 1982 (Reference 1). This report provided the basis for a rule published in the Federal Register (Reference 2). This report is a result of work by the utility owner's groups, the NRC evaluation group and NRC consultants working together to understand this multidisciplined and complicated issue.

The NRC initiated research as part of the Unresolved Safety Issue A-49 Pressurized Thermal Shock Project to further evaluate this issue. This effort has produced a plant specific PTS risk analysis of a Babcock & Wilcox (B&W) plant (Oconee), a Combustion Engineering (CE) plant (Calvert Cliffs), and a Westinghouse plant (H. B. Robinson Unit 2). Loss Alamos National Laboratories, Idaho National Laboratory, Oak Ridge National Laboratory, and Brookhaven National Laboratory have participated in the analysis of these plants. Pacific Northwest Laboratories is evaluating the probable effects of a reactor vessel with a through-wall crack and performing a cost benefit study of various

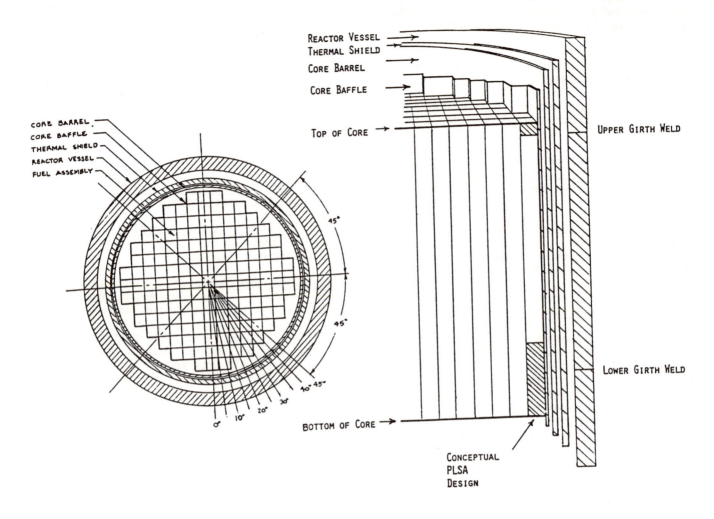

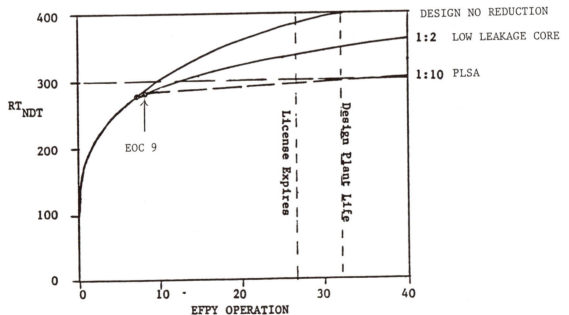

FIGURE 3 FLUX REDUCTION EFFECT ON RT_{NDT}

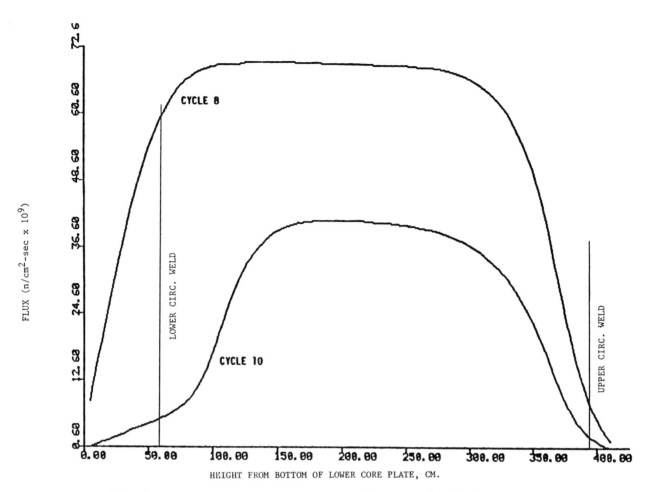

FIGURE 4   AXIAL FLUX DISTRIBUTIONS FOR CYCLE 8 AND CYCLE 10 AT THE
CENTER OF THE CORE FLAT.   COMBINED EFFECT OF LOW LEAKAGE
CORE AND PLSA.

2. Plant Specific Analyses

A plant specific analysis was initiated in early 1982 with the Nuclear Safety Analysis Center at EPRI. This consisted of a joint NSAC-CP&L effort to determine the effects of one of the most severe PTS transients on HBR Unit 2. The PTS transient was the small break loss of coolant accident (SBLOCA). The result of the analysis was that this SBLOCA would not initiate a flaw through the design lifetime of the plant.

CP&L also contracted with Westinghouse to perform a complete probabilistic PTS analysis of HBR Unit 2. The analysis was based on work accomplished by the Westinghouse Owners Group. The purpose of this analysis was to assess the PTS risk and to identify mitigating changes to the HBR Unit 2 plant. The result of this analysis is summarized in Figure 5. The curve shows that for the projected end-of-life RT_{NDT} of $300^{O}F$ for the critical circumferential weld, the risk associated with crack propagation is 7×10^{-7} per reactor year. Also shown on this figure is a curve developed by the NRC from a longitudinal weld which led to

screening criteria. At the screening criterion of $270^{O}F$ for a longitudinal weld the risk of flaw extension is 6×10^{-6} occurrences per reactor year. It can be seen that a HBR Unit 2 circumferential weld at the screening criteria is less risk than the longitudinal weld at the screening criteria. Also, it can be seen on Figure 5 that the best estimate surface RT_{NDT} which is much less than $260^{O}F$ for HBR Unit 2 has a very low risk as a result of PTS.

The benefit of risk reduction by reducing the RT_{NDT} of the reactor vessel can be seen. Actions or modifications that alter the transient severity can alter the curve. For instance, heating the refueling water storage tank would lower the curve. The heated water would reduce the thermal shock thereby reducing the severity of the transient.

The NRC requested that CP&L participate in the USI A-49 Pressurized Thermal Shock Analysis of HBR Unit 2. CP&L provided plant specific drawings, procedures, setpoints, plant tours, access to plant operators, etc. As a result of CP&L's participation, we were given project status updates,

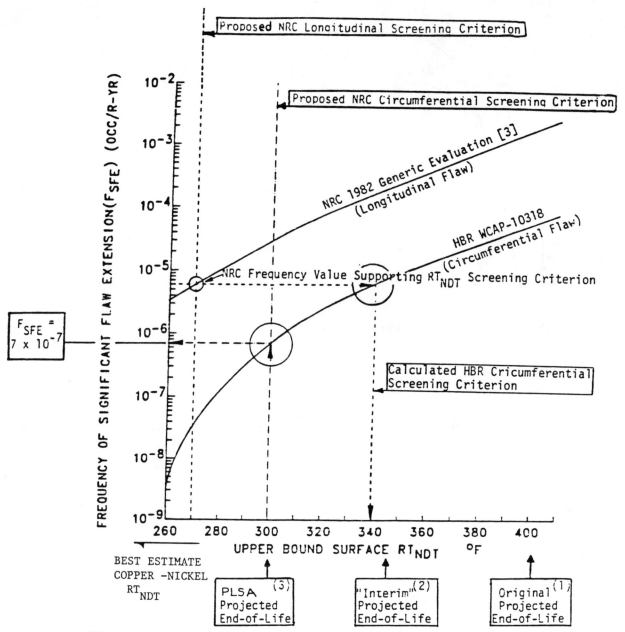

(1) Based on original core configuration with "upper bound" Copper and Nickel.

(2) Based on currently installed "Low Leakage Core" with "upper bound" Copper and Nickel, not including additional planned reduction.

(3) Based on currently installed "Low Leakage Core" and planned flux reduction using "Part Length Shielding" with assumed "upper bound" Copper and Nickel.

FIGURE 5   DETERMINATION OF TOTAL RISK OF FLAW EXTENSION FROM PRESSURIZED
THERMAL SHOCK OF THE H.B. ROBINSON UNIT 2 REACTOR VESSEL

review/comment, and access to the plant specific computer models of the HBR Unit 2 plant.

The USI A-49 is a comprehensive analysis of the HBR Unit 2 plant. It consists of:

o An extensive thermal and hydraulic RELAP5 computer model.

o An event tree analysis to identify PTS type transients.

o Probabilistic analysis to calculate transient event probabilities including human factor probabilities.

o Common mode analysis.

o Probabilistic fracture mechanics analysis.

o Mitigating measures review.

o Sensitivity and uncertainty analysis.

The analysis is completed at this date, results are in agreement with the Westinghouse analysis--the risk of vessel failure at HBR Unit 2 is very low.

3. Material Property Investigations

Material property investigations were initiated in the early phases of the CP&L PTS program to confirm vessel copper concentrations. An extensive search of historical data for copper content in the core region materials was completed. Chemical analysis for copper was not performed by the reactor vessel manufacturer on the core region welds, but it was discovered that a weld in the head of the reactor vessel was made of the same weld wire and flux as the inaccessible critical core region weld. A "boat" sample of this weld was removed from the reactor vessel head weld in early 1984. The result of this analysis revealed that the best estimate of copper content is approximately 0.20 percent instead of 0.34 percent copper content that had been assumed. This lower copper content has been supported by a statistical study of similar welds and a chemical evaluation of a section of the Millstone Surveillance weld as documented by EPRI in the report EPRI NP-3573-SR (Reference 4). The EPRI statistical study evaluated welds from other plants made in the same time period and with the same weld wire and flux as the Robinson Unit 2 core region welds. This report also documents the chemical analysis of a cross section of the Millstone Surveillance weld. The Millstone Surveillance weld was also made of the same weld wire and flux combination as

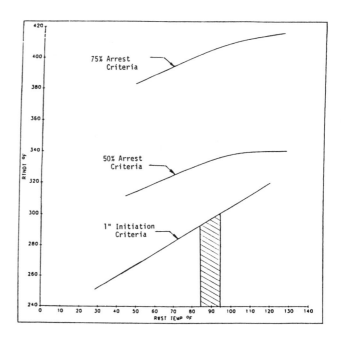

FIGURE 6 RT_{NDT} VERSUS RWST WATER TEMPERATURE FOR H.B. ROBINSON UNIT 2
REACTOR VESSEL

one of the Robinson Unit 2 core region welds.

4. Safety Injection Water Heating

CP&L committed to heat the refueling water storage tank (RWST). The RWST is the source of safety injection water so heating this water will reduce the amount of thermal shock to the vessel. Figure 6 shows the effect of the heated injection water on a postulated small break loss of coolant accident. Heating the RWST above 90°F would preclude the initiation of a 1 inch flaw as a result of the postulated severe transient.

5. Training

The training program for PTS at HBR Unit 2 was upgraded. In addition, many of the procedures were upgraded to better deal with PTS-type transients. Much of this work was accomplished through the Westinghouse Owners Group. Training and procedure revisions are important since postulated PTS transients assume operator action to mitigate the events.

Figure 7 is a curve created as part of the emergency response guideline by the Westinghouse Owners Group. This curve illustrates an important finding of the PTS research, i.e., the final transient temperature is the most important parameter that will determine crack propagation for rapid cooldown transients. This curve

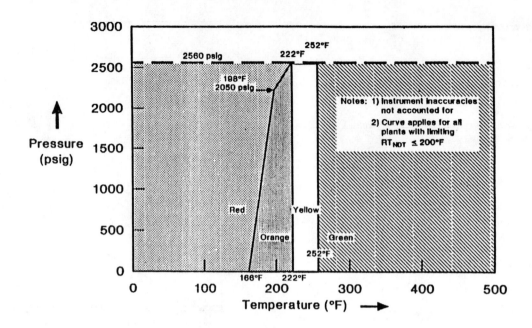

FIGURE 7   CATEGORY 1 PLANT OPERATIONAL LIMITS CURVE

indicates that for a PTS transient, a flaw will not propagate as long as the final temperature remains above 222°F at a pressure of 2560 psig. This curve was drawn for a plant with an RT_{NDT} of 200°F.

The importance of the final transient temperature has been confirmed by several investigators. Note that Figure 7 was based on a step cooldown-type transient.

Therefore, even though the behavior of a given plant may be very unique during a PTS transient; the final minimum temperature appears to be a common denominator in preventing flaw propagation.

CONCLUSIONS

The use of a low leakage reactor core combined with the use of specially shielded elements at core flats has significantly reduced the projected end-of-life RT_{NDT} temperature for the critical lower girth weld at H. B. Robinson.

Establish copper impurity levels for the lower girth weld has resulted in lower projected end-of-life RT_{NDT} values.

Plant transient analysis results indicate a very low PTS risk from the studied transients.

The combined results from the Robinson analyses indicate that the critical reactor vessel welds will not reach the NRC - PTS screening criteria over the life of the plant.

REFERENCES

1. "Pressurized Thermal Shock," NRC Policy Issue SECY-82-465, November 23, 1982.

2. "Analysis of Potential Pressurized Thermal Shock Events," Nuclear Regulatory Commission, 10CFR Part 50, Final Rule, Federal Register, Vol. 50, No. 141, July 23, 1985, pp. 29937 - 29945

3. "Effects of Residual Elements on Predicted Radiation Damage to Reactor Vessel Material," Regulatory Guide 1.99 – Revision 1, U.S. Nuclear Regulatory Commission, Washington, DC, April 1977.

4. "Robinson Unit 2 Reactor Vessel: Pressurized Thermal Shock Analysis for a Small-Break LOCA," NP-3573-SR, Electric Power Research Institute. Palo Alto, CA, August 1984.

RESIDUAL STRESS ANALYSIS OF TYPE 304 AUSTENITIC STAINLESS STEEL PIPE WELDMENTS

P. Mark Simpson
J. A. Jones Applied Research Co.
Charlotte, North Carolina, USA

The strength and ductility of 304 stainless steel piping make it very durable and forgiving. However, when the material is sensitized and subjected to tensile residual stress and aggressive high-temperature aqueous environments, failure may result from intergranualr cracking (IGSCC). Research has determined that by reducing any one of these three contributing factors, the cracking can be minimized or remedied (Reference 1). New field welding techniques have been developed to minimize sensitization and/or stress. In order to judge the relative merits of different welding schedules, a knowledge of inside diameter (ID) surface and through-thickness residual welding stresses is required. In addition, this quantitative information is needed expeditiously to keep pace with weld schedule development. Two methods used to determine the residual welding stresses that meet the time, accuracy, and cost requirements are:
° blind-hole drilling
° layer-removal

Blind-hole drilling measures surface stresses, while the layer-removal technique measures through-thickness residual stresses. Both methods are supported by automatic data acquisition to reduce time and labor costs.

BLIND-HOLE DRILLING - When initial field weld schedules are being developed, surface residual stress information is useful to verify pipe weldment stresses. The blind-hole stress relief technique described by Rendler and Vigness (Reference 2), and detailed in ASTM E-837-81, has been used in the following weld schedule studies related to IGSCC. This process involves:
° Applying strain gage rosettes
° Aligning a drilling fixture
° Drilling the hole (using a high-speed drill or an air-abrasive jet)
° Reading strain changes
° Reducing data (Reference 3)

This technique produces timely information and has been shown to be fairly accurate for this application based on unpublished results of a Society for Experimental Mechanics (SEM) round-robin study of blind-hole drilling on 304 stainless steel. Rosette layout for this technique depends on the evaluation needs. Blind-hole drilling can be used on the outside and inside surfaces of both 12-inch and 28-inch pipe weldments. Data reduction is fairly straightforward and can be computed on a pocket calculator; however, a large number of gages are most efficiently analyzed and reported through automated data acquisition and computer-based reduction. Figure 1 shows the results of three 28-inch weldment studies demonstrating the ability of blind-hole drilling to rank weld schedules based upon ID surface stress.

The special problems of measuring stain by blind-hole drilling with an air-abrasive jet have been well documented (References 4, 5, 6, and 7). Two particular points requiring caution in analyzing blind-hole drilling data of 304 stainless steel weldments include:
° Non-uniform strain fields (especially in the heat-affected zone (HAZ) area)
° Possible plastic deformation (due to strains greater than 50 percent of yield strength)

In addition, calibration constants must be determined carefully (Reference 8). The results shown in Figures 2 and 3 respectively were for a zero stress round-robin coupon and an analysis performed on a spool section of pipe as-fabricated with no field weld. The purpose of these measurements was to determine the repeatability of stress measurements at low stress and the "starting" state of stress of the base material prior to field weld parameter development. The results show a precision bandwidth of $\pm$ 1000 psi and low fabrication stresses.

LAYER REMOVAL - Fundamentals of Welding (Reference 9) lists several methods for quanitative measurement of residual stresses. The method used and described here is one pioneered by Rosenthal and Norton (Reference 10). In this technique, and area of weldment is analyzed through the following steps:

° Fitting with strain gages rosettes
° Removing a beam section from the weldment
° Splitting the beam section
° Layering from the inner surface to the outer instrumented surface

Through-thickness stresses are calculated from the stains measured at each step. This layer removal technique is depicted in Figure 4. Due to practical limitations of analyzing pipe residual stresses, a plate coupon is used rather than a beam. This modified technique, as described by the Society of Automotive Engineers (SAE) (Reference 11) and as used by Argonne National Laboratories (Reference 12), results in reasonable success and correlates well with finite element models. By using a three-element delta rosette on the plate coupon, the bi-axial state of stress transverse to the pipe surface can be determined through the thickness of the weldment. In order to obtain reasonably accurate results, several principles and assumptions must be used. First, a linear redistribution of stresses through the thickness is assumed with each removal process; this does not present problems except during the splitting step (see Figure 7, Reference 10). Second, the "jump" in stress measured during the split can be refined by using Dr. E. F. Rybicki's "Consistent Splitting Model" (Reference 13) which compensates for normally unaccounted stresses removed in the saw kerf. Third, St. Vinant's Principle must be observed when determining the physical plate dimensions. Lastly, to insure proper interpretation of final residual stresses, machine-induced stresses must be accounted for and minimized.

Butt welds produce tri-axial residual stresses, but major interest is in the axial and circumferential, or hoop, pipe directions. In addition to varying in magnitude through the thickness, the bi-axial stress magnitude vary in relation to the distance from the weld fusion line. The area of most concern is approximately the first 1/4-inch of base metal adjacent to the weld fusion line, the HAZ. Most IGSCC has occurred in the HAZ indicating, among other things, the area of severest tensile, residual weld stresses. Since strain gage rosettes average strain over an area and the area of most concern, the HAZ, is so small, a very small rosette such as a Measurements Group EA-09-030YB-120 must be used. The rosettes are typically laid out in the location and number as shown in Figure 6. This particular gage orientation allows the stress state to be determined across the weldment as well as through the thickness. In order to increase the information on actual residual weld stresses and to minimize loss of information due to gage failure, two coupons located 180 degrees apart are used.

The volume of strain gage data to be reduced to usable information is enormous. Each of the 28 rosettes (or 84 gages) must be analyzed for:

° Temperature compensation
° Quarter bridge non-linearity
° Transverse sensitivity
° Reduction to stress
° Reduction to stress at each layer for through-thickness residual stresses

The numerous strain gage readings per test alone (about 2200) require the use of automated data acquisition and reduction for timely information retrieval. In order to gather and reduce strain gage data, a Measurements Group System 4000 is used to monitor as many as 100 strain gages at a time. Because the System 4000 can monitor and record data almost instantaneously, the machining sequence shown in Figure 4 is performed almost continually. In addition, the System 4000 monitors in real time for excess strains related to fixturing and machining of the coupon. Continuous operation requires the use of a dedicated milling machine for the maximum productivity of four weeks, start to finish. The data reduction for the through-thickness residual stresses is a specialized analysis not incorporated in the System 4000 software. Presently, this part of the data reduction is performed using a Digital VAX 11/750 and plots produced with DISPLA software. See Figures 7-10 for representative experimental results of a through-thickness stress analyses.

CONCLUSIONS - The results of blind-hole drilling indicated the presence of small amounts of residual stress in the as-fabricated 12-inch spool piece. In addition, Figure 5 reveals the existence of stresses apparently induced during the through-wall machining steps; however, the final values of the through-wall residual stresses, shown in Figure 7-10, indicate that a major percentage of the residual stresses were weld-related. The net effect was that favorable compressive residual stresses were achieved by the chosen welding technique. These experimental results are in agreement with analytical predictions shown in Figure 11.

Fundamentals of Welding (Reference 9) describes the Rosenthal and Norton layering technique as a "fairly accurate... troublesome, time consuming, and completely destructive method." This technique is indeed tedious and time consuming, and as compared to such methods as the blind-hole technique; however, the sectioning method has proven to be extremely useful for providing quantitative information on through-thickness residual stresses in heavy-section pipe weldments. By employing an automated data acquisition system, the sectioning technique becomes more feasible since the time required to complete the experiment is reduced. In fact, nine such through-wall analyses have been performed in the exceptional time frame of nine months, and the blind-hole technique has been used to analyze ten weldments in less than one year. This indicates that the two experimental methods, blind-hole drilling with air-abrasive jet and the parting-out and layer removal technique, though somewhat difficult, can be used to obtain useful information in a timely fashion.

REFERENCES

1. Proceedings: Seminar on Countermeasures for Pipe Cracking in BWR's EPRI WS-79-174, Volume 1, Special Study Project WS-79-174, Workshop Report. May 1980.

2. N. J. Rendler and I. Vigness, "Hole-drilling Strain-gage Method of Measuring Residual Stress". Paper Presented at 1966 SESA Spring Meeting held in Detroit, Michigan, on May 4-6.

3. ASTM E837-81, "Standard Method for Determining Residual Stresses by the Hole-drilling Strain-gage Method," 1983 Annual Book of ASTM Standards, Vols. 03.01 and 12.02.

4. S. Redner and C. C. Perry, "Factors Affecting the Accuracy of Residual Stress Measurements Using the Blind-hole Drilling Method", Measurements Group, Inc. Post Office Box 27777, Raliegh, North Carolina, 27611, USA.

5. J. P. Sandifer and G. E. Bowie, "Residual Stress by Blind-hole Method with Off-center Hole". Paper was presented at 1976 SESA Spring Meeting held in Silver Spring, Maryland, on May 9-14.

6. R. A. Kelsey, "Measuring Non-uniform Residual Stresses by the Hole Drilling Method". Presented at the Annual Meeting of the Society for Experimental Stress Analysis in Chicago, Illinois. November, 1985.

7. F. Witt, F. Lee. W. Rider, "A Comparison of Residual-stress Measurements Using Blind-hole, Abrasive-jet and

7. Trepan-ring Methods," Experimental Techniques, February 1983.

8. G. S. Schajer, "Application of Finete Element Calculations to Residual Stress Measurements," Journal of Engineering Material and Technology, April 1981, Vol. 103, pp. 157-163.

9. K. Masubuchi, Residual Stresses and Distortion", Fundamentals of Welding, Welding Handbook, 7th Ed., Vol. 1, Chapter 6, 1981.

10. D. Rosenthal and J. T. Norton, "A Method of Measuring Tri-axial Residual Stresses in Plates", presented at the Twenty-fifth Annual Meeting, AWS, Cleveland, Ohio, Oct. 16-19, 1944.

11. Report of Iron and Steel Technical Committee, "Methods of Residual Stress Measurement - SAE J936", Society of Automotive Engineers, Inc., Dec. 1965.

12. W. J. Shack, W. A. Ellingson, L. E. Pabis, "Measurement of Residual Stresses in Type-304 Stainless Steel Piping Butt Weldments", EPRI Research Project 449-1, NP-1413, Phase Report, June 1980.

13. E. F. Rybicki, J. R. Shadley, and W. S. Shealy. "A Consistent-splitting Model for Experimental Residual-stress Analysis". Paper presented at 1981 SESA Fall Meeting in Keystone, Colorado, on Oct. 11-14.

14. F. W. Brust and R. B. Stonesifer, "Effect of Weld Parameters on Residual Stresses in BWR Piping Systems", EPRI Research Project 1174-1, NP-1743, pp. 6-7, Figure 6-8, Final Report, March 1981.

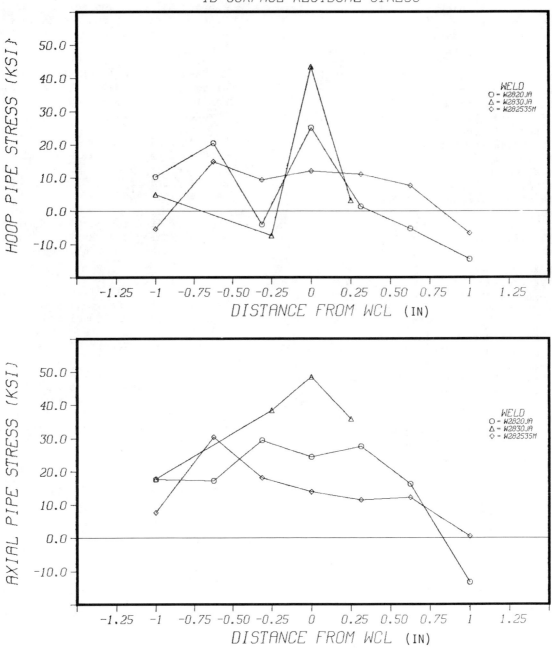

Figure 1. 28-inch weldment study

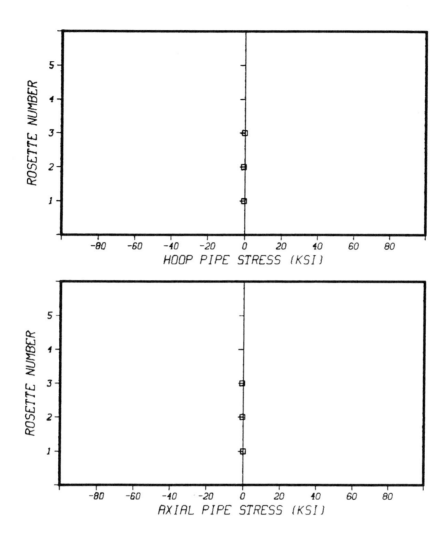

Figure 2. Summary Plot of Blind-hole
Analysis on SEM Round-robin Study.
Result of Three Holes Drilled on a
Zero Stress Sample of 304 Stainless
Steel.

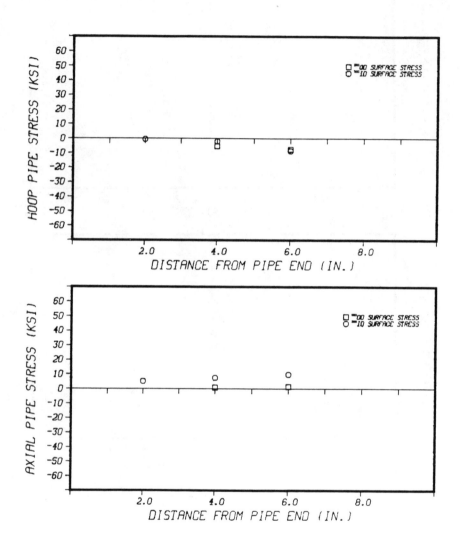

Figure 3. Blind-hole Results of an
Analysis of an as-fabricated Pipe
with No Weld, Indicating Low Levels
of Residual Fabrication Stresses.

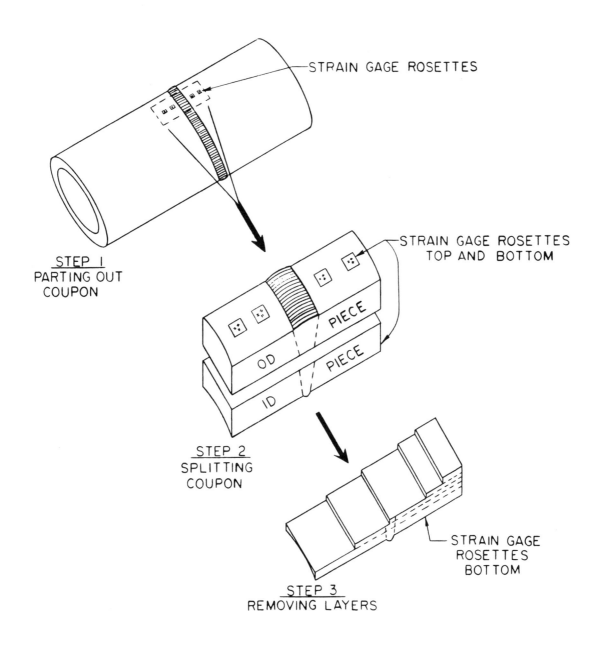

STRAIN GAGE ROSETTES

STEP I
PARTING OUT
COUPON

STRAIN GAGE ROSETTES
TOP AND BOTTOM

OD PIECE

ID PIECE

STEP 2
SPLITTING
COUPON

STRAIN GAGE
ROSETTES
BOTTOM

STEP 3
REMOVING LAYERS

Figure 4. Three-Step Laboratory
Process for Determining Through-
Wall Residual Stresses

361

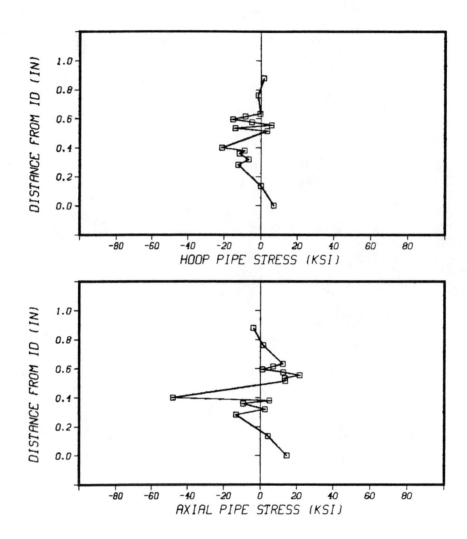

Figure 5. Plot of Through-thickness
Stress Measured on a Stress Relieved
Calibration Plate of 304 Stainless
Steel. Fluctuations are due Primarily
to Induced Machining Stresses During
Splitting.

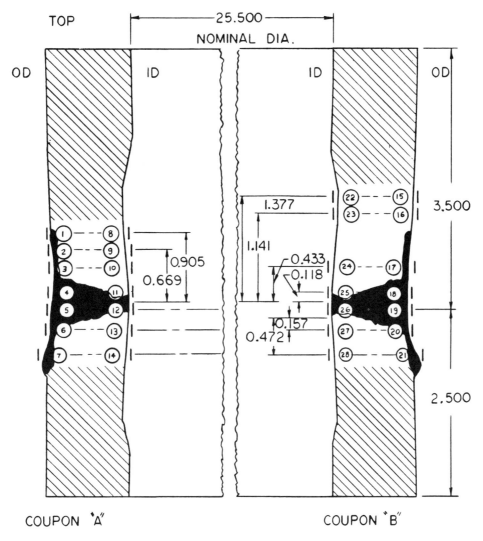

COUPON 'A"   COUPON "B"

①—ROSETTE IDENTIFICATION NO.

ALL DIMENSIONS ARE IN INCHES

Figure 6. Typical Gage Layout
28-Inch Through-Wall

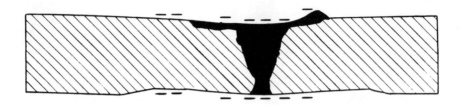

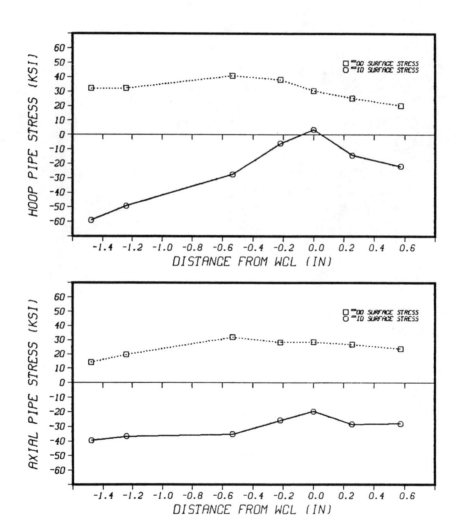

Figure 7. Summary Plot of Surface
Residual Stress Determined From the
Through-thickness Analysis of a
28-inch Pipe Butt Weld, with a Special
HSW Process.

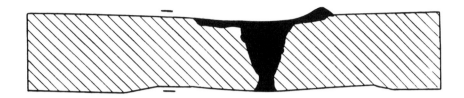

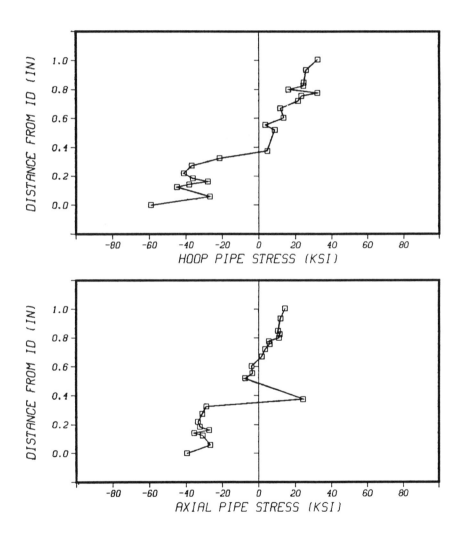

Figure 8. Through-thickness Residual
Stress Measured 1.377 inches from ID
WFL for a 28-inch Pipe Butt Weld, with
a Special HSW Process.

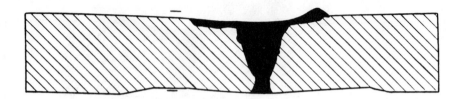

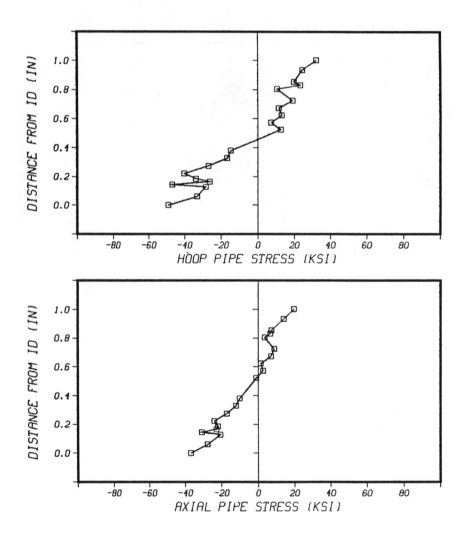

Figure 9. Through-thickness Residual Stress Measured 1.141 inches form ID WFL on a 28-inch Pipe Butt Weld, with a Special HSW Process.

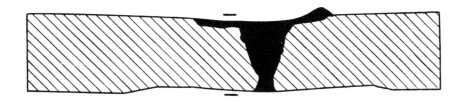

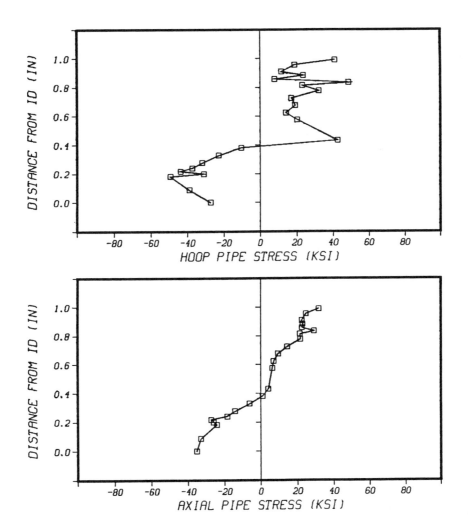

Figure 10. Through-thickness Residual Stress Measured 0.433 inches from ID WFL for a 28-inch Pipe Butt Weld, with a Special HSW Process.

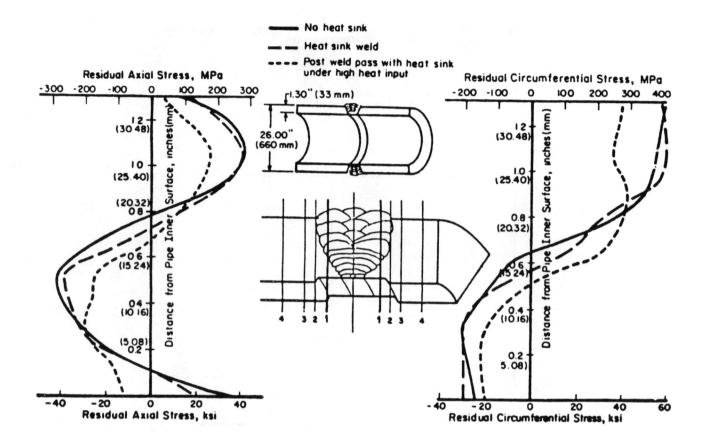

Figure 11. (From Reference 14) Graphs showing axial and circumferential stress distribution predictions. This information was calculated for the through-wall thickness of a 26-inch pipe after a post-weld heat-sink pass at location 1, which is 8mm from the WCL.

RADIAL BORE CRACKS
IN ROTATING DISKS

Abdurahman A. Sukere

Mechanical and Aerospace Engineering
University of Missouri-Columbia
Independence, Missouri, USA

ABSTRACT

An electro-optical technique, based on the method of caustics, for determining the stress intensity factor of radial cracks in rotating disks is presented. The measurement strategy, test set-up, and experimental procedure are discussed. The technique is employed in the determination of the opening mode stress intensity factor of radial bore cracks in rotating disks and the results compared with existing analytical estimates.

INTRODUCTION

Machinery such as turbines, turbocompressors, generators and flywheels operate at very high speeds. The high centrifugal force that results can cause catastrophic failures. These failures are known to be caused by flaws or crack-like defects which are considered to be initially present in the components or initiated by repeated stressing and grow finally to critical size [1].

In order to assess the structural reliability of cracked structures it is necessary to know both the strength of the cracked component and the rate at which the cracks grow under in-service fatigue loads. Both the strength and crack growth depend upon the stress intensity factor K. Therefore, an accurate determination of this parameter is of utmost importance in the attempt to optimize the dimensions and insure the inservice safety of rotating structures.

Many attempts have been made to apply the methods of fracture mechanics, which have proved successful in solving other problems of strength and safety. Rooke and Tweed [2,3] considered a finite elastic disk with an internal radial crack subjected to a constant loading, and a rotating elastic disk with an edge crack, and determined the stress intensity factor and crack energy in terms of the solution of a Fredholm integral equation. Williams and Isherwood [4] developed a method for an approximate calculation of strain-energy release rates for a finite rotating disk with a radial crack emanating from a central hole. Winne and Wundt [5] adapted Bowie's [6] (infinite plate solution) strain energy release rate formula to study the bursting behavior of disks. Bueckner and Giaever [7] considered a ring shaped rotor which has a notch or crack in its inner periphery and solved the problem using Muskhelishvili's method. Owen and Griffiths [8] and Riccardella and Bamford [9] used two-dimensional finite elements analysis to determine the mode I stress intensity factor for cracks radiating from a central hole in a finite disk. Bluel et al. [10] used photoelastic methods to determine the mode I stress intensity factors for the simple case of rotating solid disk containing radial internal cracks. Marloff et. al, [11] also used photoelastic methods to determing stress-intensity factors of cracks in rotating disks.

This paper describes an electro-optical technique, based on the method of caustics, for determining the stress intensity factor of radial cracks in rotating disks. The technique is employed in the determination of the mode I stress-intensity factor of radial bore cracks in rotating disks and the results compared with existing analytical estimates.

BASICS OF SHADOW OPTICS

The method of caustics was introduced by Manogg [12], then later developed further by Theocaris [13]. The basis of the method is that a light ray passing through a stressed plate is deviated from its straight path partly due to thickness variation and partly due to the change in refractive index caused by the stress optic effect. If the plate contains a crack, due to the strong thickness and refractive index variation at the region close to the crack tip, the rays are scattered and concentrated along a strongly illuminated curve on the reference plane placed some distance from the specimen. This singular curve is variously called caustic, stress corona, or shadow spot.

Considering the optical configuration in Figure 1, the equation of the caustic for

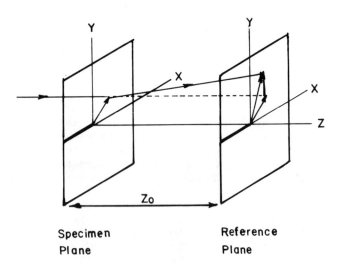

Fig. 1 - Optical configuration for the method of caustics

the case of a straight edge crack in a tensile field is given by [13]

$$x = \lambda r_o (\cos\theta + \frac{2}{3} \cos \frac{3\theta}{2})$$

$$(1)$$

$$y = \lambda r_o (\sin\theta + \frac{2}{3} \sin \frac{3\theta}{2})$$

where

$$r_o = (\frac{3}{2} \frac{C}{\lambda})^{2/5}$$

$$(2)$$

$$C = \frac{1}{\sqrt{2\pi}} z_o t \frac{\nu}{E} K_I$$

Here r_o is the radius of the initial curve in the specimen plane, λ is the image magnification factor, z_o is the distance between the object and the reference plane, K_I is the opening mode stress intensity factor and t, ν and E are respectively the thickness, Poisson's ratio and modulus of elasticity of the specimen.

From equation 1 it can be shown that

the transverse diameter, D_t, of the caustic is related to the radius, r_o, through the following relation

$$r_o = \frac{D_t}{3.17\lambda}$$

$$(3)$$

Substituting this value of r_o in relation 2 one obtains an expression for the stress intensity factor in terms of the transverse caustic diameter

$$K_I = \frac{1.671E}{z_o t \nu} (\frac{1}{\lambda})^{3/2} (\frac{D_t}{3.17})^{5/2}$$

$$(4)$$

Thus, the determination of K_I is a simple matter of measuring the transverse caustic diameter.

EXPERIMENTAL CONSIDERATIONS

The test setup of Figure 2 was used to implement the measurement strategy. The principle elements of the system are: a disk specimen, a driving plate attached to the shaft of a 25,000 rpm tool grinder motor, a 5 mw HeNe laser, a photomultiplier tube, a trigger circuit and an oscilloscope. A

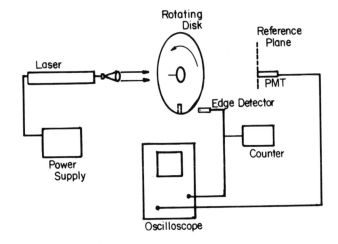

Fig. 2 - Schematic of the experimental setup.

Dayton model 4X797B speed control unit was used to infinitely vary the motor speed. The angular speed of the disk was monitored using an HP5300 universal counter.

The configurations of the disk specimens and drive plate are shown in Figure 3. The specimens in the present investigation were 305 mm diameter, 3 mm thick disks, with a bore diameter of 76 mm, machined of PMMA cast acrylic sheet with mass density

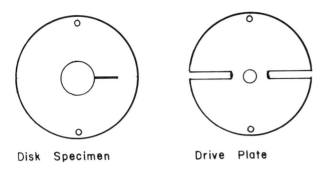

Disk Specimen   Drive Plate

Fig. 3 - Configurations of disk specimen and drive plate.

ρ = 1.19 x 10³ kg/m³. The radial cracks emanating from the bore hole were simulated by 0.3 mm wide radial saw cuts of various lengths. The disk specimens were rotated by the drive plate via two steel pins which transmitted rotation through the drive holes in the specimen.

As mentioned earlier when a light beam impinges on the loaded specimen in the close vicinity of the crack a shadow spot with a bright halo is formed on the reference plane as shown in Figure 4. In determining the stress intensity factor one need only measure the transverse diameter of the shadow and the distance between the specimen plane and the reference plane. The transverse diameter of the caustic is monitored by a photomultiplier tube whose window is covered except for a thin slit with a width far less than the width of the crack image on the reference plane. As the shadow spot

passes over this slit, the intensity seen by the photomultiplier tube varies and the result is an intensity profile of the shadow spot as shown in Figure 5. The photomultiplier tube used in this experiment is an RCA type 4840 tube at an operating voltage of 1200 v.

The oscilloscope was used to display and manipulate the intensity profile of the shadow spot. The oscilloscope is triggered using a TRW type OPB703A reflective object sensor which sensed the edge formed by a contrasting aluminum foil and black tape radially glued on the back of the disk on a line perpendicular to the plane of the crack. Since the measurement of the transverse diameter of the shadow spot is critical one needs to expand the time scale of the trace for better resolution, and the improvement would be maximized if the intensity profile of the shadow spot is at the beginning of the trace. This function was accomplished by placing the reflective edge sensor on a travelling stage. Thus by spatially moving the edge sensor the position of the shadow spot profile on the trace can be varied.

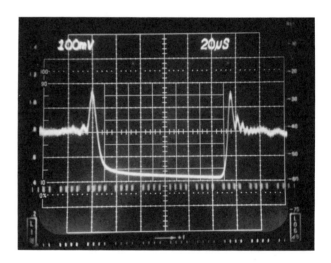

Fig. 5 - Intensity profile of a shadow spot

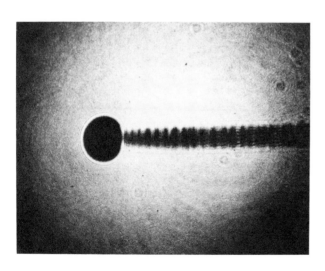

Fig. 4 - Shadow spot surrounding a mode I loaded crack tip.

THEORETICAL CONSIDERATIONS

Considering the geometry and coordinates shown in Figure 6, the stresses in a spinning disk with bore radius a, and rim radius b are [14]

$$\sigma_r = \sigma_o \left[1 + \frac{a^2}{b^2} - \frac{a^2}{r^2} - \frac{r^2}{b^2}\right] \quad (5)$$

$$\sigma_\theta = \sigma_o \left[1 + \frac{a^2}{b^2} + \frac{a^2}{r^2} - \left(\frac{1+3\nu}{3+\nu}\right)\frac{r^2}{b^2}\right] \quad (6)$$

$$\tau_{r\theta} = 0 \quad (7)$$

where,

$$\sigma_0 = (\frac{3+\nu}{8}) \rho\omega^2 b^2 \qquad (8)$$

where ν is Poisson's ratio, ρ is the mass density of the disk material, r the radial distance, ω the angular velocity and σ_0 is the stress which would exist at the center of a solid disk. For radially oriented through the thickness cracks the loading by the hoops stresses $\sigma_\theta(r)$ is, of the simple mode I type but varies along the crack length.

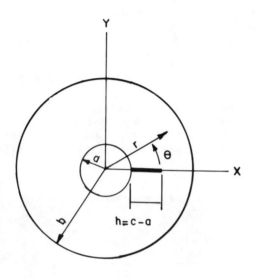

Fig. 6 – Rotating disk containing a radial bore crack.

An approximate method for determining K_I has been suggested by Williams and Isherwood [4]. The basis of their method is to define an effective stress acting on the crack as

$$\sigma^1 = \frac{3\sigma_m + \sigma_n}{4} \qquad (10)$$

where

$$\sigma_n = \frac{1}{b-c} \int_a^b \sigma \, dr \qquad (11)$$

$$\sigma_m = \frac{1}{c-a} \int_a^c \sigma \, dr \quad \text{(no keyway)} \qquad (12)$$

$$\sigma_m = \frac{1}{c-a^1} \int_{a^1}^c \sigma \, dr \quad \text{(keyway of depth } a^1-a) \qquad (13)$$

The stress intensity factor is then

given as

$$K_I = \frac{\sigma^1 [\pi (c-a)]^{\frac{1}{2}}}{\sigma^1 [\pi (c-a^1)]^{\frac{1}{2}}} \qquad (14)$$

The values for σ_n and σ_m are obtained by integrating equation 6. It should be noted that this approximation predicts the same K_I value for one crack as for two opposite cracks.

Bueckner and Giaever [7] obtained a numerical solution for the stresses in an infinite disk containing a single notch at the bore from which they derived the following expression for the stress-intensity factor

$$\frac{K_I}{\sigma_0} = \pi \sqrt{\pi(c-1)/(c+1)^3} \; f(h/a) \qquad (15)$$

where c = (h+a) and f(h/a) is a geometric function. Bowie [6] analyzed the case of an infinite rectangular plate under a uniform stress field and containing both single and double radial cracks emanating from the periphery of a bore hole. Winne and Wundt [5] adapted Bowie's analysis to the notched disk problem and showed an equivalence of stress fields provided the overall crack length 2c is less than 0.25 of the outer disk diameter. Bowie's expression for the stress-intensity factor has the form

$$\frac{K_I}{\sigma_0} = \sqrt{\pi a} \, / \, F(h/a) \qquad (16)$$

where the function F(h/a) is a different geometric function for the single and double notched configurations.

RESULTS AND DISCUSSION

The procedure used to obtain the transverse caustic diameter is schematically shown in Figure 7. A circular tab of known diameter is radially affixed at the same radial distance as the crack tip. As the image of the shadow spot and the calibration tab pass the slit of the photomultiplier tube they each create a pulse. Thus,

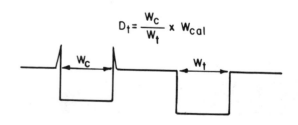

Fig. 7 – Scheme used for determining the transverse diameter of the shadow spot.

measuring the width of the two pulses is sufficient in determining the caustic diameter. An alternative scheme would be to determine the velocity at the crack tip position and multiply by the pulse width of the shadow spot. Our experiment shows the difference in the two approaches to be within ± 1%.

The experimental data was obtained by sequentially incrementing the angular speed a small amount, waiting for the speed to get stable, recording the pulse width of the shadow spot and the calibrator and the corresponding frequency.

Once the transverse diameter D_t of the caustic formed at the crack tip has been determined, the values of the stress intensity factors K_I can be computed using relation 4. For our experiments the constants entered in this relation had the following values: $z_o = 1.735$ m, $\lambda = 1.0$, $E = 2944$ MPa, $\nu = 0.35$.

Figure 8 shows a comparison of the experimental results with results obtained using the Williams and Isherwood technique. This figure shows the variation of K_I/ω^2 (MPa√m) as a function of h/a. The experimental data points represent average values. For a disk without a keyway the approach considerably overestimates the values of K_I at the higher values of h/a but which are considerably better for shorter

cracks. Note that the finite element solution of Reference 8 suggests the approximation to be good for short crack lengths while that of Reference 9 suggests the approximation to be good for longer cracks. The results of the comparison of the stress-intensity factor of a 44 mm long crack emanating from a 13 mm long 5 mm wide square keyway is shown in Figure 9 where K_I(MPa√m) is plotted as a function of frequency.

The results of the comparison of the experimental results with the results of Bueckner and Giaever's solution and with Bowie's solution is shown in Figure 10 where $K^* = K_I/\sigma \sqrt{h}$ is plotted as a function of $\sqrt{h}/a$. The results show that the rigorous infinite disk solution of Bueckner and Giaever compares favourably with the experimental results for a wider range of crack lengths, while the comparison with the results based on the adaptation of Bowie's solutions is poor. It should be noted the present disk geometry violates the restriction of c < 0.25b for equation 16.

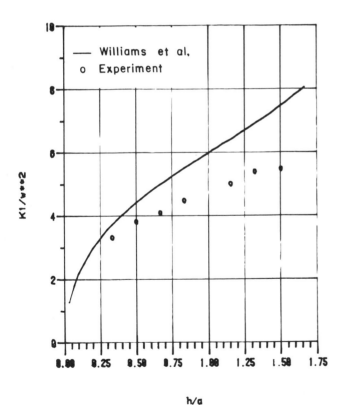

Fig. 8 – Stress-intensity factors for radial bore cracks in rotating disks.

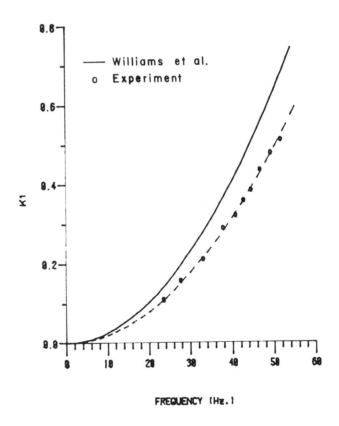

Fig. 9 – Results of a bore crack emanating from a keyway.

373

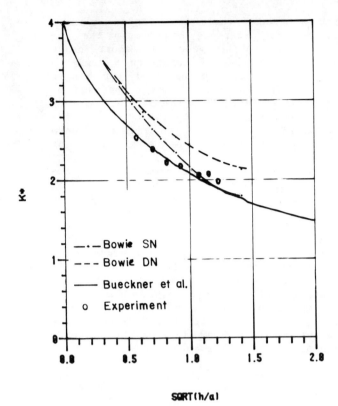

Fig. 10 - Variation of dimensionless stress –intensity factor with relative crack length. (SN single notch; DN double notch)

CONCLUSIONS

An electro-optical technique based on the method of caustics, for determining the stress-intensity factor of radial cracks in rotating disks is presented. The technique is employed in the determination of the opening mode stress-intensity factor for radial bore cracks in rotating disks. The experimental results are compared to available analytical approximations. The results show Bueckner and Giaver's infinite disk solution gives good estimates of K_I for a wider range of h/a than the other analytical estimates.

REFERENCES

1. Yukawa, S., D. P. Timo and A. Rubio, "Fracture Design Practices for Rotating Equipment," Fracture, H. Liebowitz, ed., Academic Press, Vol. 5, p. 65, (1969).

2. Rooke, D. P. and J. Tweed, Intnl. J. Engrg. Sci., 10, 323-335, (1972).

3. Rooke, D. P. and J. Tweed, Intnl. J. Engrg. Sci., 11, 279-283, (1973).

4. Williams, J. G. and D. P. Isherwood, J. Strain Analysis, 3, 17-22, (1968).

5. Winne, D. H. and B. M. Wundt, Trans. A.S.M.E., 80, 1643-1656, (1958).

6. Bowie, O. L, J. Mathematics and Physics, 25, 60-71, (1956).

7. Bueckner, H. F. and I. Giaever, Zeitschrift für Augewandt Mathematik, Vol. 46, p. 265, (1966).

8. Owen, D. R. J. and J. R. Griffiths, Intnl. J. of Fract., 9, 471-476, (1973).

9. Riccardella, P. C. and W. H. Bamford, Journal of Pressure Vessel Technology, Trans. ASME, p. 279, (1974).

10. Blauel, J. B., J. Beinert and M. Wenk, Experimental Mechanics, 17, 3, (1977).

11. Marloff, R. H., M. M. Leven, T. N. Ringler and R. L. Johnson, Experimental Mechanics, 11, 529-539, (1970).

12. Manogg. P., "Schattenoptische Messung der Specifischen Bruchenergie Während des Bruchvorgans bei Plexiglas," in Proceedings, International Conference on the Physics of Non-Crystaline Solids, Delft, p. 481, (1964).

13. Theocaris, P. S., J. Appl. Mech. 37, Trans. ASME, Series E92, p. 409, (1970).

14. S. Timoshenko, and J. N. Goodier, "Theory of Elasticity," 3rd edition, McGraw-Hill, New York, (1970).